Advanced Structured Materials

Volume 34

Series editors

For further volumes:
http://www.springer.com/series/8611

Germán Ferreira
Editor

Alternative Energies

Updates on Progress

Springer

Editor
Germán Ferreira
CIRCE—Centre of Research for Energy
Resources and Consumption
Zaragoza
Spain

ISSN 1869-8433 ISSN 1869-8441 (electronic)
ISBN 978-3-642-40679-9 ISBN 978-3-642-40680-5 (eBook)
DOI 10.1007/978-3-642-40680-5
Springer Heidelberg New York Dordrecht London

Library of Congress Control Number: 2013954561

Printed on acid-free paper

Springer is part of Springer Science+Business Media (www.springer.com)

Contents

Municipal Solid Waste

Energy Recovery from the Organic Fraction Based on Anaerobic Digestion

S. Astals, M. Romero-Güiza and J. Mata-Alvarez

Abstract Municipal solid waste production has increased over the past years. The European Union Waste Framework Directive establishes different hierarchy levels for its management, where prevention and recycling appear as the most convenient management strategies for the organic fraction of MSW. This fraction can be either recycled by composting or by anaerobic digestion followed or not by composting. Anaerobic digestion has the advantage of producing energy instead of consuming it and it is widely described in this chapter. First, a revision of the state of the art of its implementation at full scale is carried out. Then, a detailed description of the environmental and operational factors affecting the process performance is described. Finally, methods for improving the yields of the anaerobic digestion process are considered, which includes biological and physical pre-treatment technologies together with the basis and possibilities of anaerobic co-digestion.

Abbreviations

AD	Anaerobic digestion
AcoD	Anaerobic co-digestion
CHP	Combined heat and power unit
EU	European Union
HRT	Hydraulic retention time
LCFA	Long chain fatty acid
MAD	Mesophilic anaerobic digestion
MSW	Municipal solid waste
MBT	Mechanical–biological treatment
OLR	Organic loading rate
RDF	Refuse derived fuel
SOF	Stabilized organic fraction

S. Astals · M. Romero-Güiza · J. Mata-Alvarez (✉)
Department of Chemical Engineering, University of Barcelona, C/Martí i Franquès, no. 1, 6th floor 08028 Barcelona, Spain
e-mail: jmata@ub.edu

G. Ferreira (ed.), *Alternative Energies*, Advanced Structured Materials 34,
DOI: 10.1007/978-3-642-40680-5_1, © Springer-Verlag Berlin Heidelberg 2013

SRT	Solid retention time
TAD	Thermophilic anaerobic digestion
UASB	Upflow anaerobic sludge blanket
VFA	Volatile fatty acid
VS	Volatile solids
WWTP	Wastewater treatment plant

1 Introduction

A wide range of organic materials, also named biomass, can be used to produce energy, among others: wood, wood wastes, forest wastes, livestock wastes, agricultural wastes (straws, stalks, prunings, etc.), energy crops, wastes from food and paper industries, municipal solid wastes (MSW) and sewage sludge. Altogether, these biomass resources represent the 67 % of primary production of renewable energy. As can be seen in Fig. 1, which shows the evolution of all kinds of renewable energy between 2000 and 2010, wood and wood wastes occupies the first place (48 %), followed by hydropower (19 %).

The volume of renewable energy produced within the European Union (EU) increased by 60 % between 1999 and 2009, equivalent to an average increase of about 5 % per year [1]. In this matter, the biomass coming from MSW constitutes a significant source of alternative energy since it is available in every municipality. This energy can basically be recovered using thermal or biological processes as depicted in Table 1.

On the one hand, thermal technologies include incineration, as the most developed thermic technology, as well as gasification and pyrolysis. On the other hand, biological technologies include methods mainly focussed on the production of hydrogen or methane. Of course, the application of one or other technology depends on the waste characteristics and, obviously, biological methods are only applicable to the biodegradable organic fraction of MSW. As a matter of fact, thermal technologies are useful for wastes containing high percentages of non-biodegradable matter, whereas biological methods are more appropriate for highly biodegradable rich-moisture wastes. The present chapter is devoted to organic matter methanisation, most known as anaerobic digestion or biomethanisation, a technology that has grown in the last 30 years and that now can be considered completely mature.

1.1 Municipal Solid Waste Production and Management Policies

Our present society consumption patterns lead to the production of huge and constant amounts of waste. On average, each of the 500 million inhabitants of the

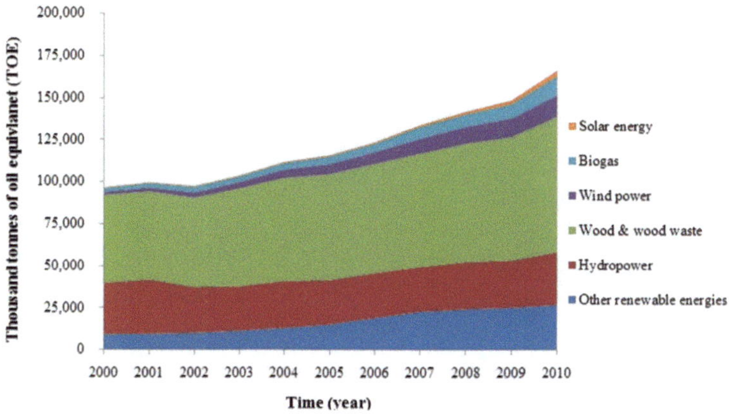

Fig. 1 Evolution of the different sources of renewable energy consumption in Europe (adapted from [1])

Table 1 Main technologies to produce energy from the municipal solid wastes

Technologies	Thermal conversion	Incineration (the most widespread)
		Gasification/Plasma (in development)
		Pyrolysis and liquefaction (in development)
	Biological conversion	Anaerobic digestion (quite widespread)
		Hydrogen fermentation (in development)

EU produces around 500 kg of MSW per year. Specifically, MSW comes from households, small businesses, offices, schools, hospitals, government buildings, and so on. But it also includes waste from parks and street cleaning [2]. On the one hand, Fig. 2a shows the evolution of the MSW production during the period 1995–2010. As can be seen, it has been growing continuously, although in the last years a reduction has been registered, mainly due to the economic crisis affecting the EU. On the other hand, Fig. 2b shows the evolution of the main treatment technologies during the same period of time. As can be seen, landfilling was and is the main option; however, the percentage of waste disposed has been reduced from 68 to 38 %. In contrast, biological treatments (composting and anaerobic digestion) have more than doubled its treatment capacity during this period.

It is important to highlight that the waste management policy in the EU is aiming, at long term, to turn Europe into a recycling society. In this vein, the Waste Framework Directive, revised in 2008 [3], constitutes a modernised approach to waste management, where wastes are considered as a valued resource instead of an unwanted burden. The present directive brings five hierarchy levels, where waste management shall follow the following priority: prevention, reuse, recycling, other forms of recovery and, as last option, disposal of waste in landfills.

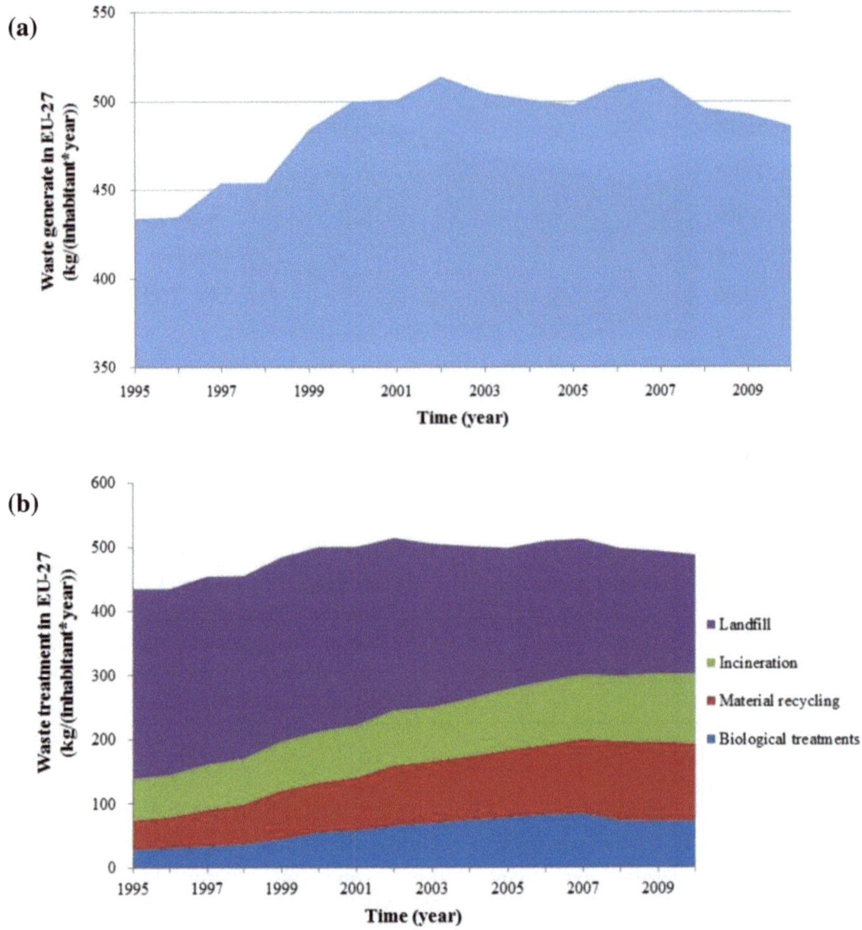

Fig. 2 Evolution of the municipal solid waste. **a** Production. **b** Treatment

Prevention involves minimizing material and/or energy consumption and can be achieved by developing more efficient technologies for production and by optimizing the use of resources.

Product reuse involves the multiple use of a product or component. Its reuse can be for the original purpose they were conceived for or for an alternative one, with or without reconditioning. Actually, reuse can also be encouraged by improving the durability of products, discouraging the purchase of disposable items provides a feasible alternative for reuse exists.

The third step in the waste management hierarchy is recycling, which means to feed the waste into a process to give a new use for that material. The recovery of organic material present in the MSW should be included within the category of recycling, since both composting and anaerobic digestion, main organic matter processing technologies, result in a stabilized product that can be used as fertilizer or soil conditioner. Anaerobic digestion is the main topic of this chapter and will be developed in the following sections.

The Waste Framework Directive also includes an option entitled "other recovery". Incineration, with energy recovery, constitutes the most widespread technology to recover energy from waste. Other technologies under development for energy recovery include gasification/plasma and pyrolysis.

Last option in this hierarchy is landfilling. If the waste cannot be further recovered, it is placed in landfills, areas specially located, designed and controlled, to store partially inert materials and to ensure compliance with regulations.

1.2 Collection, Sorting and Transport

A very important issue for all the technologies treating the organic fraction of MSW is the sorting procedure used to select the different materials to be recycled. Obviously, the best results are obtained when sorting is carried out at source by the households. Concerning the organic fraction, to obtain good output through an aerobic treatment, and also to be qualified as compost, the European legislation requires MSW source separation. In contrast, if the aerobic treatment is applied with undifferentiated waste (non source-sorted) the result material is named as stabilized organic fraction (SOF). As a result, compost can be used for agricultural purposes as a fertilizer, whereas SOF can only be used for other less exigent purposes, such as daily cover for landfills [4]. As a consequence, if good outputs of the composting plant are desired, including or not a previous AD process, it is necessary to promote source sorting for the organic fraction of MSW. At the present time, several systems are available and applied to collect sorted municipal solid wastes. A basic one consists in separating the humid parts (kitchen and garden wastes) from the dry ones (plastic, paper, glass, etc.). Therefore, the humid part, rich in organic matter, can be transformed through an aerobic treatment in good quality compost. However, in most systems, more than two containers are normally present, within reasonable distance from each dwelling, in order to simplify the use of the sorting system. In this matter, with the purpose of facilitate citizens identification and location of waste correct receptacles, the containers are different in colour, size and/or marked for a given type of material. In the present system, each waste collection point consists, depending on the local characteristics, of 3–6 containers allocated for the sorting of: organic waste, glass, plastic and packaging, cardboard and paper and/or garden waste. Citizens use bags or baskets for carrying waste to the containers or to place recyclables in the adequate container. It should be pointed out that the waste collection scheme is very important as MSW collection procedures have a great impact on the efficiency of the recycling process, and even more on the biological process.

On the one hand, recycling of clean materials offer always the best results and reduces the quantity of the refused waste, i.e. the remaining waste fraction after recycling materials, including organic matter. In opposition, if the MSW is collected in an undifferentiated way, and then is sorted in a mechanical-biological treatment (MBT) plant, the quality of the recycled materials is always worse than the one obtained by a sorting performed by households. However, more complex systems require to educate and inform the households. Education is essential to obtain a high degree of purity through source waste segregation, whereas periodic information campaigns must also be devised to remind the importance of source sorting, especially in towns in which multi-story building predominate. On the other hand, transport is an important cost to be considered when the collecting strategy is designed since, in developed countries, it can represent 60–70 % of the total management costs. Nowadays, in large modern towns, collection is carried out in special vehicles adapted to circulate on town streets and facilitate waste collection from the specially designed containers. Moreover, depending on the scenario these vehicles can be very simple or complex, such as the rear-loading compactor trucks used in many industrialized cities. Once finished the primary collection, these vehicles transport the waste to a transfer station, where more large-powerful vehicles transfer the collected waste to the treatment plant. In some cases, a short storage period may follow the primary collection as well as a pre-processed in order to remove certain undesired materials and/or to perform some basic operations, such as shredding, compacting or screening.

1.3 Technologies to Stabilise the Organic Fraction of the Municipal Solid Waste: Anaerobic Digestion

The organic fraction of the MSW can be either treated by composting or by anaerobic digestion followed or not by composting. Anaerobic digestion (AD) has the advantage of producing energy instead of consuming it, but the investment required as well as the process complexity is higher. The decision to adopt one or another solution depends, among other factors, on the quality and quantity of the organic stream and the availability and practice of the land where the digestate or compost is to be spread. However, if a good output is to be obtained it is very important to operate the plants with source sorted organic fraction, since stabilized outputs from MBT plants, where the organic fraction is selected from undifferentiated MSW cannot be used in agriculture. Therefore, they have to be used in other less exigent destinations or even dumped in landfills. Moreover, plants using source sorted organic fraction are less prone to have failures due to mechanical problems. However, whether treating source sorted organic fraction or mechanically selected organic fraction, it is indispensable to set a sorting pre-treatment in the plant. So, inappropriate materials present in the organic stream are removed and the input organic fraction enriched.

AD is the decomposition of organic matter carried out by microorganisms in strict anaerobic conditions to produce biogas, a mixture of methane and carbon dioxide. The production of biogas is a very important issue as it is a source of renewable energy. AD technology applied to the organic fraction of MSW is a rather new application as it was developed in the 1980s [5].

1.4 Evolution of Anaerobic Digestion Technology for Municipal Solid Waste

The possibility of digesting the organic fraction of MSW has been studied for more than 80 years, but it was not until the 1970s that the first's demonstration plants were built. However, today, it is possible to find MSW anaerobic digestion plants around the world, even though most of them have been developed and built in the EU. The large implementation of the AD in Europe is a result of the financial support for projects in the field of alternative energy sources and energy savings. In this context, the production and utilisation of biogas from urban, industrial or livestock wastes began to be important in the 1980s and, consequently, the commercial use and with itself the development and evolution of AD technology [5].

The first AD plants were constructed for the treatment of mixed MSW, as no source sorting collection existed at the time. In the 1990s, the rise of source sorting collection promoted the implementation of a large number of AD plants. The fact that the feedstock derived from source sorting collection was much cleaner stimulated the development and adaptation of more conventional digestion technologies, sometimes with mixed success. To be specific, the AD plant capacity rate increased at a rate of around 30 tonne per year during the period between 1900 and 1995, whereas the rate of increase averaged 150 tonne per year during the period 1996–2000. More than one thousand tonne, available or under construction, could be treated in 2000 by 53 plants across Europe [6]. Moreover, in 1999, with the implementation of the landfill directive [7] a favourable scenario for the growth and development of anaerobic digestion of MSW was generated. As a result, the MSW anaerobic digestion capacity in EU tripled between 2000 and 2010. Finally, in recent years, a reduction of the number of projects, due the difficult economic situation, has been observed. Nevertheless, that does not change the future prospects of AD due the major advantage over others treatment and the great diversity of biogas applications. At the present time, more than 7 million of tonne of MSW per year are treated by 212 plants in the EU. Countries having the largest annual capacity installed are Germany and Spain with 2 and 1.6 million tonnes of capacity, respectively. However, if the inhabitants are taken into account, Switzerland becomes the country with the highest installed annual capacity, with 49,000 tonnes per million of inhabitants (Fig. 3).

During the development of the AD for MSW more than thirty types of processes have been commercialized. Nevertheless, most of the plants are concentrated in five companies: BTA, Dranco, Kompogas, Ros Roca and Valorga. Moreover, nowadays, the development of new processes, like percolation processes, occupies a significant part of the market. The majority of the MSW digestion plants are constituted by the same steps: reception, selection, pre-treatment of the organic matter, digestion, post-treatment (Table 2).

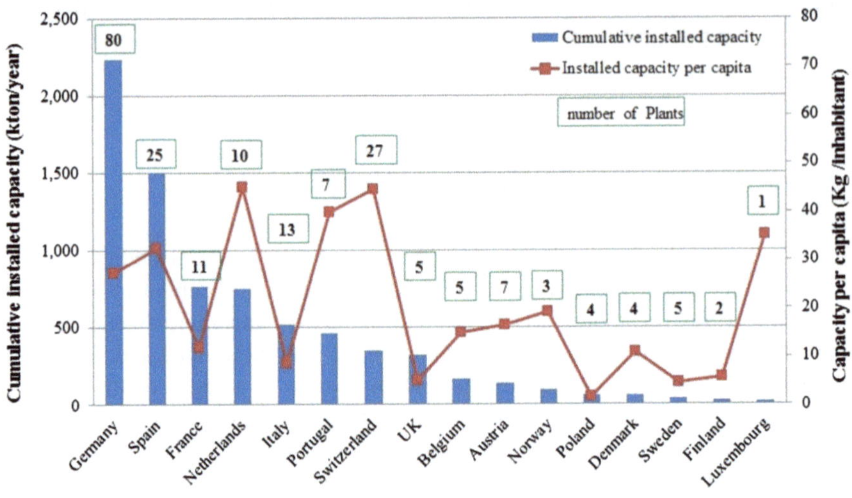

Fig. 3 Installed capacity and numbers of plants in EU (adapted from [8])

Table 2 Processes that can be found in a MSW anaerobic digestion plant

Step	Processes
Reception	Reception point and storage of the MSW
	Elimination of bulky and harmful objects
	Homogenisation and feeding of the preparation process
Selection	Classification and separation of the materials
	Recovery of recyclable materials
Pre-treatment	Elimination of precipitable inerts (sand, clay, glass, etc.)
	Elimination of flotating materials (plastics, expanded polystyrene, etc.)
	Particle size reduction
	Adjustment of the water content
Digestion	Conversion of the organic matter into biogas and digestate
Post-treatment	Biogas refining
	Digestate solid–liquid separation
	Water treatment
	Solid composting

However, the technology applied for each company is mainly distinguished by the conditioning of the organic matter and design and operation of the digestion step, since the other steps have standard equipment widespread in most plants.

The design and operating conditions of the anaerobic digesters are the most important process parameters. Therefore, the available technologies can be classified as function of the: temperature, moisture content, number of stages, feedstock origin and co-digestion (mixing wastes). However, the number of stages and the co-digestion has become less relevant since over the 90 % of the installed capacity is done in single stage digester and only few, 27 out of 212, plants include other wastes in the digester influent.

Hence, temperature, moisture level and MSW quality are more spread among the digestion plants. MSW digesters are usually operated in two temperature ranges, mesophilic (35–40 °C) and thermophilic (50–55 %) (see Sect. 2.2.4). Normally, digester heating is done by energetic recuperation from the biogas combined heat and power (CHP) unit; however, some pants can also use boilers. Thermophilic systems are associated to faster and larger biodegradation, but the higher energy requirements as well as the greatest need of process control makes mesophilic conditions the most widely used, with a 60 % of the installed capacity.

The solid concentration, moisture level, classifies the processes into dry (>15 % solids) and wet (<15 % solids). The wet processes, Ros Roca and BTA, have some of the pre-treatment steps in wet and, therefore, large amount of waste and water have to be moved. In contrast, the dry processes, Dranco, Kompogas and Valorga, have less complex pre-treatments systems (water is added just prior the feeding) but, the high solid concentration and high viscosity of the waste requires heavy work equipment.

Taking into account the collecting system used to obtain the organic material of the MSW the feedstock can be classified as:

- Biowaste: organic matter from source sorting collection.
- Mixed waste: organic matter from non source-sorted collection
- Residual waste: organic matter remaining after the collection of biowaste

Moreover, the type of waste to be treated has a great influence on the pre-treatment and the final destination of the digestate. At the present time, the plants are equally distributed between biowaste and mixed waste. However, this parameter is mainly influenced by the regional political conditions.

2 Basic Principles of Municipal Solid Waste Anaerobic Digestion

Anaerobic digestion is a biochemical process which, in the absence of oxygen, decomposes the organic matter into biogas, mainly methane and carbon dioxide, and a digestate, a mixture of partially degraded organic matter, anaerobic biomass and inorganic matter. The conversion of the organic matter into biogas is a process which involves several serie-parallel reactions and different groups of microorganisms (bacteria and archaea). In this vein, the variety and complexity of the organic matter present in the MSW implies a complex metabolic pathway before the organic matter is transformed into biogas. Additionally, the performance of the anaerobic digester is highly related with the structure of its microbial community [9].

2.1 Steps of the Anaerobic Digestion Process

The degradation of the organic matter has been typical divided in four steps, namely: (1) disintegration and hydrolysis, (2) acidogenesis, (3) acetogenesis and (4) methanogenesis; where the starting point and degradation pathway depends on the nature of the organic matter (Fig. 4).

2.1.1 Disintegration and Hydrolysis

Disintegration and hydrolysis step include non-biological and extra-cellular biological processes mediating the breakdown and solubilisation of complex organic matter to soluble compounds [10]. In this step, the organic matter clusters are disintegrated into macromolecules (i.e. carbohydrates, proteins and lipids) and then, those macromolecules are hydrolysed to soluble compounds. Specifically, the extra-cellular enzymes (cellulases, proteases and lipases) excreted by the fermentative bacteria solubilise carbohydrates, proteins and lipids to mono- and disaccharides (sugars), alcohols, amino acids and long chain fatty acids (LCFA) among others.

Many studies have concluded, due to the large fraction of organic matter that must be solubilised before its methanisation, that the disintegration and hydrolysis step is the rate-limiting step of the MSW digestion process. However, the

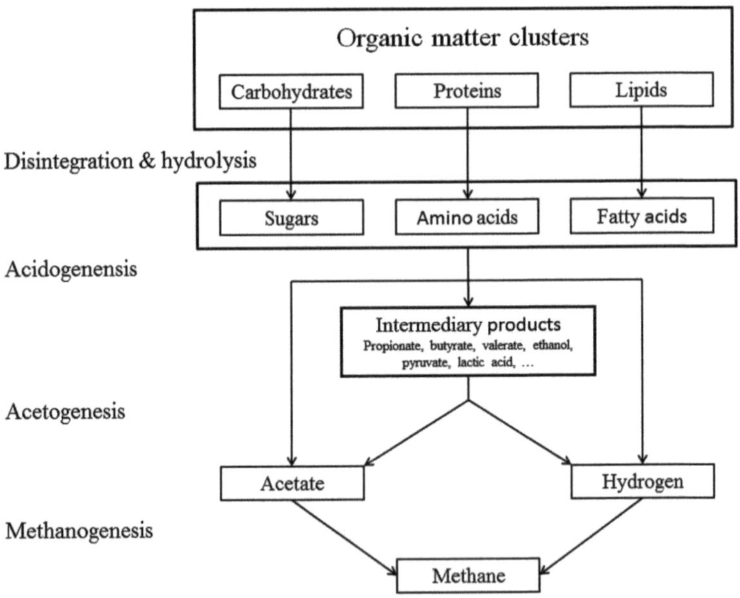

Fig. 4 Simplified scheme of the MSW anaerobic biodegradation (adapted from [10])

solubilisation rate is affected by several parameters such as particle size, pH, temperature, biomass concentration or the intrinsic substrate characteristics [11].

2.1.2 Acidogenesis

Acidogenesis, also known as fermentation, is carried out by a large group of facultative fermentative bacteria. In this stage, the fastest of the AD process, the soluble compounds obtained from the disintegration and hydrolysis are able to be transported inside the bacteria and converted mainly to volatile fatty acids (VFA; i.e. acetate, propionate, butyrate, valerate), ethanol, pyruvate, lactic acid, ammonia, hydrogen sulphide, hydrogen and carbon dioxide. It should be noted that the acidogenesis of sugars and amino acids is carried out without an electron acceptor or donor, whereas LCFA are oxidised using hydrogen ions as electron acceptors. The main product of all acidogenesis reactions is acetate, however, the accumulation of hydrogen and/or acetate in the digester medium can promote the formation and accumulation of more reduced compounds such as propionate and butyrate. Therefore, VFA concentration and evolution can be used as indicator of process performance (see Sects. 2.2.3 and 2.3.4).

2.1.3 Acetogenesis

The volatile fatty acids, excluding acetate, and other products from the acidogenesis stage are converted by obligate hydrogen-producing acetogens to acetate, hydrogen and carbon dioxide which are appropriate substrate for the methanogenic biomass. It is well known that acetogenesis reactions are only thermodynamically possible when the hydrogen concentration in the digester medium is low. Consequently, the syntrophic relationship between acetongens and hydrogenotrophic methanogens (hydrogen degraders) is of utmost important to regulate the hydrogen concentration and, therefore, the whole digestion process.

2.1.4 Methanogenesis

The last stage of the anaerobic process is carried out by methanogenic archaea, which convert the end products of the previous reactions into biogas. In a MSW digester, the major methane production ($\sim$70 %) is generated by aceticlastic methanogens, which split the two carbons of the acetate; one is reduced to methane and the other is oxidised to carbon dioxide ($CH_3COOH \rightarrow CH_4 + CO_2$). Two different types of aceticlastic methanogens, mutually exclusive, dominate as function of the ammonia and VFA concentration in the digester medium. *Methanosaeta*, characterised by its filaments, dominate when the VFA and the ammonia concentration are low whereas *Methanosarcina*, characterised by its clumps, dominate when the volatile fatty acids and the ammonia concentration are high

Fig. 5 Fluorescence in situ hybridisation picture of Methanosaeta (*left*) and Methanosarcina (*right*). Photographer: Miriam Peces

[12] (Fig. 5). Minor methane production ($\sim$30 %) is produced by hydrogenotrophic bacteria, which used hydrogen as electron donor and carbon dioxide as electron acceptor to produce methane ($4H_2 + CO_2 \rightarrow CH_4 + 2H_2O$). Finally, even been negligible, methyl groups can also be converted to methane ($CH_3OH + H_2 \rightarrow CH_4 + H_2O$).

2.2 Environmental and Operational Factors Affecting the Anaerobic Digestion Performance

As a biological process, anaerobic digestion behaviour and efficiency is highly influenced by the environmental and operational conditions of the system.

2.2.1 Nutrients

There are many substances, organic and inorganic, which are indispensable for the anaerobic biomass growth and metabolism. Not considering the obvious presence of organic carbon, there is the requirement of nitrogen, phosphorous and sulphur (macronutrients) and several metals (micronutrients) like iron, nikel, cobalt, magnesium, calcium, sodium, selenium, copper, etc. It is important to highlight that the presence of micronutrients in small quantities can stimulate the activity of the anaerobic biomass. However, if a certain limit concentration is surpassed their presence can slow down the growth or even cause severe inhibition [13]. Moreover, the combination of some metal ions or the combination of micronutrients with other compounds present in the digester medium, like ammonia, can create antagonistic or synergistic phenomena in the digestion process. As the MSW present high diversity and amounts of macro- and micronutrients, the later phenomena are more probable to take place in a MSW digester.

2.2.2 pH

Each of group of microorganisms, responsible of the different anaerobic steps, has a different optimum pH range. Fermentative bacteria can survive in a wide range of pH between 4 and 9; even though the optimum pH is reported to be around 5–6. In contrast, the methanogenic bacteria present a narrow survival pH range which is reported to be between 6.0 and 8.5, with and optimum around the neutrality [14]. Since the methanogens are the most sensitive and the key microorganisms of the process, the digester are design and operated to achieve a pH between 7 and 8. It should be taken into account that a drop in the pH or the presence of toxic compounds mainly affects the methanogens activity, whereas the acid-forming bacteria, much more resistant, can still degrade more organic matter. As a result, the acid concentration increase and the methanogens become more inhibited: phenomena that can lead to digester failure. However, the pH is not a good parameter to monitor the performance of an anaerobic digester, because in a MSW digester the pH remains stable, due to its high alkalinity, until the digester is inhibited by the accumulation of hydrogen and VFA.

2.2.3 Alkalinity and Volatile Fatty Acids

The alkalinity, or buffer capacity, is the capacity of the digester medium to neutralise the acids formed during the process and, therefore, to mitigate pH changes. The alkalinity of a digester is mainly given by few acid-base pairs, mainly carbon dioxide–bicarbonate, ammonium–ammonia, dihydrogen phosphate–hydrogen phosphate and unionised–ionised volatile fatty acids. The ratio between the total and partial alkalinity, parameters determined by a titration method, are normally used to monitor digester stability. However, for systems with high alkalinities, like MSW digesters, the ratio between the intermediate (the difference between total and partial alkalinity) and partial alkalinity is more adequate, since it is more sensitive to digester changes [15].

Volatile fatty acids, which typically include acetate, propionate, butyrate and valerate, are the main intermediates of the AD process. Therefore, its concentration and evolution is very used to monitor digester performance and stability. Under stable conditions, mesophilic anaerobic digesters (MAD; 35–40 °C) normally present lower acid concentration, been acetate the main acid, than thermophilic anaerobic digesters (TAD; 52–57 °C), where propionate is the main VFA. Moreover, the ratio between the VFA concentration and the total alkalinity is also used to monitor digester stability. All the aforementioned ratios are based on the same concept: if the acid concentration exceeds the buffer capacity the methanogens activity in the digester is likely to be inhibited.

2.2.4 Temperature

Temperature is one of the most important physical parameters in the AD process, since it directly affects the kinetics and the thermodynamics of the reactions and the growth and metabolism of the biomass. As a result, temperature determines the degradation pathway and biomass dynamics in the digester. Even though, AD can take place between 10 and 65 °C, most digesters are operated under the two optimal ranges: mesophilic (around 35 °C) or thermophilic (around 55 °C) conditions. Due to the higher process stability and the lower energy requirements about the 60 % of the MSW digesters are, at the present time, operated at mesophilic conditions (see Sect. 1.4). However, the need to improve the process feasibility, by means of increasing the biogas yield and reducing the digestate costs has increased the interest in the thermophilic conditions. Specifically, TAD offers some potential advantages over the conventional MAD: an increase of the biological and chemical reaction rates, an increase of the organic matter removal, a higher solubilisation of the particulate organic matter and a better hygienisation. Nevertheless, some drawbacks are unavoidable: an elevated energy requirement for heating the digester, a higher risk of process destabilisation, a poor digestate dewaterability and a higher odour potential.

2.2.5 Mixing

In MSW stirred tank digesters, mixing is important to provide contact between the organic matter and the biomass, uniform temperature, prevent the formation of a scum layers and avoid particles depositions on the bottom of the reactor. However, the required mixing level (mechanical, pump recirculation or gas recirculation) may vary as function of the solids concentration and digester design [16]. In general, rapid mixing is known to break the biomass aggregates and so destroy syntrophic relationships between microorganisms. In contrast, low mixing levels distributes the feed adequately, do not disrupt biomass flocs and minimise the energy requirements.

2.2.6 Hydraulic and Solid Retention Time

The hydraulic retention time (HRT) and the solid retention time (SRT) are usually used as digesters design parameters. Specifically, HRT represents the average period of time during which the waste has remained in the digester, whereas SRT represent the average time that the anaerobic biomass has stayed in the system. In digesters without recirculation or supernatant withdrawal the HRT and SRT are equal. However, a recirculation of the digestate may increase the SRT and avoid the risk of biomass washout. Typically, MAD are operated with a SRT between 15 and 20 days, whereas TAD are operated with SRT between 10 and 15 days. Also used as design parameter, the organic loading rate (OLR) is the amount of organic matter introduced in the digester (kilograms of volatile solids) per day and cubic meter of digester. The OLR can be increased/decreased by reducing/increasing the HRT or increasing/reducing the organic matter concentration of the digester feed, respectively.

2.3 *Inhibitors of the Anaerobic Digestion Process*

There are many substances that at a given concentration inhibit the anaerobic biomass activity, especially methanogens. However, the reported threshold values vary significantly from one study to another. The differences can be attributed to the characteristics of the substrates and anaerobic biomass, the environmental conditions and the adaptation periods.

2.3.1 Oxygen

Oxygen is an inhibitory compound for acetogens and methanogens, which are strictly anaerobic microorganisms but not for the fermentative bacteria, which are facultative bacteria (i.e. can live either in aerobic or anaerobic conditions). Unintentionally, all MSW digester are exposed to low oxygen doses. However, the fermentative bacteria protect the strictly anaerobes from oxygen exposure since they become aerobic when oxygen is present (oxygen reactions are more energetically favourable) and switch back fermentative when the oxygen level is negligible.

2.3.2 Substrate Competition

Sulphate and nitrate in the digester medium are used as electron acceptor by sulphate-reducing and the nitrate-reducing bacteria, respectively. Both groups of bacteria compete with the methanogens and the acetogenic biomass for the substrate, which is converted to carbon dioxide instead of methane. Moreover, sulphate-reducing bacteria convert sulphate to sulphide, which is inhibitory for all the microorganisms involved in the anaerobic process. In order to avoid corrosive problems in the process but mainly in the CHP unit some MSW plants add iron(III) chloride in the digester medium to precipitate iron sulphide and, therefore, reduce the percentage of hydrogen sulphide in the biogas.

2.3.3 Ammonia

Nitrogen ammonia is produced, by biological degradation of the nitrogenous organic matter, in the acidogenesis step and remains in the digester medium in two forms, acid (NH_4^+) and basic (NH_3), which are in equilibrium depending mainly on temperature and pH. Although both forms have been reported as inhibitors of methanogenic activity, the capacity to diffuse into the cell, causing proton imbalance and/or potassium deficiency, made NH_3 the most harmful form [17]. However, this inhibition did not lead to a process instability, since the interact ion between NH_3, VFA and pH led the AD to an "inhibited steady state", which is a condition where the process is running stable but with lower methane yields [18]. Due to the higher protein hydrolysis and temperature, ammonia inhibition is especially distinct in thermophilic digestion.

2.3.4 Volatile and Long Chain Fatty Acids

High VFA concentrations can cause inhibition to anaerobic microorganisms. Normally, the VFA inhibition is coupled with low pH inhibition, since the undissociated species, which are able to diffuse into the cell, are the more toxic. Therefore, VFA inhibition is linked with the pH and the alkalinity of the system. Moreover, high acetate concentration inhibits propionate and butyrate acetogenesis as well as aceticlastic methanogenesis, whereas propionate is known to inhibit methanogenesis.

Saturated and unsaturated LCFA (such as palmitic, stearic and oleic acids) are inhibitor of the methanogenesis step, mainly affecting acetoclastic archaea. LCFA inhibition is likely to occur when treating fatty wastes. Specifically, LCFA accumulates on the surface of the cell, leading to the cell membrane no longer being able to perform important functions, such as protecting the cell and transportation of materials in and out of the cell [19]. However, several studies have shown that LCFA is reversible and therefore, after a lag phase, anaerobic biomass is able to convert them to methane.

2.3.5 Cations and Heavy Metals

Cations (Na, K, Mg, Ca) are indispensable micronutrients for the anaerobic biomass, however, they are inhibitory at high concentrations. Under high cation concentration biomass cell trends to dehydrate as they try to dilute ambient salt concentration. The cations concentration in the digester depends on: the cations concentration in the influent, the cations released during organic matter degradation and the cations supplied by chemicals (e.g. to regulate digester pH).

Heavy metals are present at relatively high concentrations in MSW, where cadmium, chromium, cobalt, copper, iron, lead, mercury, nickel and zinc are reported to be the most toxic ones. As aforementioned, some of them are indispensable as micronutrients (see Sect. 2.2.1); however, as they are non-biodegradable compounds, they tend to be accumulated in the digester medium. The chemical binding of heavy metals to the enzymes and subsequent disruption of the enzyme structure and function are the main cause of inhibition [14]. Moreover, the presence of several heavy metals can lead to synergic and antagonistic effects. However, at the present time, little is known about those phenomena.

2.3.6 Xenobiotics

Xenobiotics (complex organic compounds) are released in large quantities due to human activities; consequently, they are found in MSW. Many xenobiotics substances have been reported as inhibitory of the AD process [13]. Surfactants (detergents: AES and LAS), solvents (alcohols, ketones, benzene and toluene), phenols, pesticides (halogentedphenols and nitrophenols), phthalates esters (added to plastics) and medicines (antibiotics) are the most frequents in the MSW.

2.4 Final Destination of the Digestate

The European legislation, like the landfill, waste and the waste framework directives [3, 7, 19], is trying to promote the recycling of the organic matter through biological process and diminish the amount incinerated or dumped in a landfill. Regarding the MSW, nowadays, the combination of AD and composting seems the best option to recover energy and material from the MSW. The present scenario is also a result of the AD plants operation, where the prevalence of efficiency criteria for biogas production instead of digestate stability, lead to low HRT of the material in the digester and consequently a digestate that is not completely exhausted in terms of easily biodegradable organic matter. The quality of the digestate and, therefore, its recycling options are a result of three main factors: feedstock origin, digestion process and digestate post-treatment [20]. Moreover, the addition of a co-substrate in the feedstock is an important factor to take into account, since it represents a decrease of the HRT and, therefore, it is likely to obtain a less stabilised digestate [15]. Additionally, "clean" industrial wastes (like food processing or pharmaceutical) can dilute the heavy metals concentration in the digestate, whereas the addition of sewage sludge or animal manure can raised the need for effective sanitation procedures during the operation of AD or composting plant. At the present time, there is not an EU specific legislation for MSW digestates; even though the 2nd draft of the biological treatment of biowaste [21] can be used as reference. Some of the values suggested are in the same order of magnitude than the ones present in the legislation for sewage sludge and animal manure digestates [22–24]. In any case, the use of digestate/compost derived from the MSW anaerobic digestion into the soil may depend on: (1) Chemical properties: even though the only restriction limit is the one stabilised by the nitrate directive [25], from an agricultural point of view is very interesting to consider other parameters like pH, conductivity, density, nutrient content, etc. (2) Stability and maturity: the use unstable compost/digestate can cause N-immobilisation and/or oxygen exhaustion because of an excessive increase in soil microbial activity. However, there is little agreement about which methodology (dynamic or static respirometric index, dissolved organic carbon or VFA concentration, residual biogas potential) should be used to determine digestate stability. Additionally, compost maturity is related with the presence of phytotoxic compounds, which can affect the growth and/or yield of a plant or crop. (3) Hygienisation: because MSW are known to contain pathogens (Salmonella, *Escherichia coli*, etc.), the digestate must be safe for people and animals in order to be recycled. Otherwise, new ways of transmission of pathogens between people and animals could be established [26]. Thermophilic digestates are known to fulfil the EU higienisation requirements. In contrast, mesophilic digestates have to be pasteurised or composted priot its use in land. (4) Heavy metals and inerts: heavy metals (Cd, Cr, Cu, Hg, Ni, Pb, Zn) concentration can limit or even prohibit the use of a MSW digestate on land, since they accumulate on plants, animals and soil. Moreover, the presence of inerts (sand, glass, plastics, etc.) must be avoided.

Taking into account all the aforementioned factors, it is clear that only the anaerobic digestion of biowaste (organic matter from source sorting collection) can be used in some land applications whereas the digestate from mixed waste and residual waste (organic matter selected through a MTB process) can only be used in restricted applications, for instance land reclamation or landfill daily cover.

3 Methods for Improving Anaerobic Digestion Yields

As shown in Sect. 2, many parameters influence the performance of a MSW anaerobic digester. However, the yields of the process are mainly related with the biomass characteristics, the enzymatic activity and the physical characteristics of the substrate, which manifest themselves through three measurable parameters: methanogenic activity, biodegradability and solubilisation rate [27]. Therefore, enhancing the AD process performance can be done through the analysis and optimization of one or more of those three parameters. To be specific, two options are currently available: (1) improve the AD operating conditions and (2) pretreatments that favour waste biodegradability and solubilisation rate. Moreover, a completely different approach to improve the yield of the AD is the introduction of a co-sustrate, an option that has gained momentum in the last years.

3.1 Pre-treatments

The digester operating conditions, which affect the rate and the yield of the process, are mainly decided on the design stage, although some changes and improvements should always be considered. However, when digester is in operation at its optimal conditions, there is still an option to improve, which is the addition of a pre-treatment. The majority of the pre-treatments (physical, chemical and/or biological) are focused on the breakdown of the organic matter into smaller particles and/or soluble compounds.

3.1.1 Biological Pre-treatments

Rotary drum reactor (RDR), design to emulate the initial stages of composting (thermophilic phase), is one of the most widespread MBT, being Bedminster, Biomixer and Dano the most important companies [28]. Through the combination of microorganisms, aeration and agitation, RDR force the breakdown of the organic matter into small particles, which not only will improve the AD performance but also the separation of recyclable and undesired materials. The RDR consist of a long inclined reactor, where the MSW remains between one and 3 days and the temperatures can raise up, due to the biological activity, to 55–70 °C. The reactor is followed by a trommel for waste selection. The RDR design depends on the desired retention time, the amount and origin of the waste, air supply methodology and the drum's slope and rotational speed. Zhu et al. [29] evaluated the influence of the residence time and the type of waste over the anaerobic biodegradability and biogas potential. The results showed that the RDR can be used as an effective pre-treatment of the MSW. However, the optimum treatment time will be affected by the waste composition, since 1-day treatment time showed the maximum biogas yield for MSW but the lower for the mixture between MSW and paper, which optimum was for 2–3 days treatment time.

Another option to enhance the AD performance is the addition of hydrolytic enzymes to the digester medium [30]. However, although methane yields improvements around 10 % have been reported, the purchase cost makes it generally unfeasible. In this field and with the aim of increasing the enzymatic activity in the digester medium, Fernandez-Güelfo et al. [31] mixed mature compost with the MSW, with and without prior incubation, since it typically presents a consortium of aerobic and anaerobic microorganisms capable to solubilize partially-biodegradable organic matter. Specifically, mixing a 2.5 % (in volume) of compost with MSW, without or with compost pre-incubation, led to an increase of the methane production of 60 and 73 %, respectively.

3.1.2 Physical Pre-treatments

The physical pre-treatment includes a large variety of options; nevertheless, the most common treatments are: mechanical particle size reduction, low and high temperature and ultrasonication.

Mechanical size reduction is a widely use pre-treatment in AD plants, which can be carried out by shredding or grinding in wet or dry conditions. The main advantage of reducing the particle size is an increase of the kinetics and stability of the AD, whereas, typically, not significant increase in the specific methane production is obtained [32]. The reduction of the particle size also improves the digester medium (e.g. better heat transfer and mixing) and reduces the risk of clogging. In some plants, a light crusher is used in the first steps of the mechanical selection in order to improve material selection and opening bags. However, in most plants size reduction is done at the end of the pre-treatment process, just before AD. A high speed rotating water mill, typically called hidropulper, is the most applied in wet AD. In the hidropulper, the MSW is shred, homogenised and diluted to obtain a waste a suspension; likewise, some systems are able to remove a light and heavy refuse. Furthermore, in wet and dry AD crushing machines are also frequently used. Usually, in wet systems the water content is adjusted in the grinding, whereas in dry systems the water is mixed after the trituration. Several companies offer crushing machines, but most of these technologies are adaptations of mining, agricultural and food industry applications. As an example, the company Biogas-weser-ems GmbH & Co. offers a new biowaste AD scheme based on energy crops AD technology [33].

Thermal pre-treatments, which can be carried our between 65 and 200 °C, are considered a feasible option to improve organic matter solubilisation and digestate higienization. The low temperature pre-treatment is based on a two-stage AD, where the first digester is operated between 65 and 80 °C and at low-HRT (between 1 and 4 days). Under these conditions the effect of the biological (high enzymatic activity) and the temperature pre-treatments are combined. However, as shown by the large investigation and implementation, high temperature pre-treatments are preferred, maybe because of lower time required. In that sense, Liu et al. [34], who treated biowaste at 175 °C during 1 h, were able to solubilise the 60 % of the organic matter. However, an about 10 % reduction of the methane potential was reported, due to organic matter mineralisation and/or the formation of refractory compounds. It is important to highlight that the temperature and time of this system must be optimised since higher temperatures and times will lead to higher organic matter solubilisation, but also lower methane potentials. The company Cambi™ uses a scheme, called thermal hydrolysis process, for the thermal pre-treatment of MSW. The system consist in three reactors in series prior to AD: (1) a pulper at 97 °C, (2) the main reactor which operates at about 150 °C and (3) a reactor operated at 107 °C and low pressure. As a result of the temperature and pressure fluctuations, the water particles into the organic material change from liquid to gas, favouring the solubilisation of the organic matter. Finally, the AD is carried out at mesophilic and wet conditions [35].

Ultrasonication causes a localized pressure drop (below the evaporating pressure) in the aqueous phase, which results in the formation of cavitation bubbles. Then, several bubbles join together and grow. Finally, the bubbles implode realising a shock wave that propagates through the medium and contributes to the organic matter solubilisation. A study comparing five different systems, combinations of sonicated and unsonicated biowaste coupled with mesophilic AD and wit/without hydrogen production, showed an increase of the particulate matter removal from 36 to 44 % and an increase of the methane production from 1.6 to 2.1 m^3 $CH_4/(L_{digester}$*day) [36].

3.1.3 Chemical Pre-treatments

Several chemicals, with different characteristics, have been investigated to improve the MSW biodegradability such as oxidants (hydrogen peroxide, ozone), acids (sulphuric and hydrochloride acid) and alkalis (sodium and calcium hydroxide). However, it should be considered the addition of chemicals is likely to enhance the ion concentration in the digester medium, which can inhibit the process (see Sect. 2.3). In this vein, Shahriari et al. [37] evaluated the combination of two pre-treatments over MSW: microwave (physical) and hydrogen peroxide (chemical) at different temperature range, from 115 to 175 °C. The combination of pre-treatments, although showed a high organic matter solubilisation, did not lead to a significant biogas production improvement, probably due to the formation of refractory compounds and organic matter mineralisation. Additionally, Torres and Espinosa [38] studied the influence of the calcium hydroxide concentration and treatment time over MSW methanisation. The authors obtained an optimum value of 62 mEq $Ca(OH)_2$/L for 6 h, which supposed a 170 % increase of the specific methane production, when compared with the untreated one.

3.2 Anaerobic Co-Digestion

Anaerobic co-digestion (AcoD) consists of the anaerobic digestion of a mixture of two or more waste with complementary characteristics, so that the biogas production and the organic matter removal are enhanced through their joint treatment; and thus, it is not the digestion of the different types of waste present in the MSW [39]. Even though, the improvement of the biogas yield is mainly consequence of the increase in the organic loading rate (ORL), when possible it is important to choose the best co-substrate and blend ration in order to: (1) favour positive interactions, i.e. positive synergisms, macro- and micronutrient equilibrium and moisture balance; (2) dilute inhibitory and/or toxic compounds (ammonia, lipids, heavy metals), (3) optimize methane production and (4) enhance digestate stability. As can be seen in Fig. 6, minor modifications need to be done in the AD plant (reception tank, piping and dosing bomb) to include the majority of co-substrate in the digester feed supply, whereas many economic advantages, apart from the higher electricity sell, are obtained from the fact of sharing equipment and general infrastructures cost. However, some drawbacks also need to be pointed out: (1) the cost of waste transfer from the co-substrate generation point to the AD plant, (2) the risk of spreading poisonous substances and (3) the harmonisation of different policies of the waste generators [40].

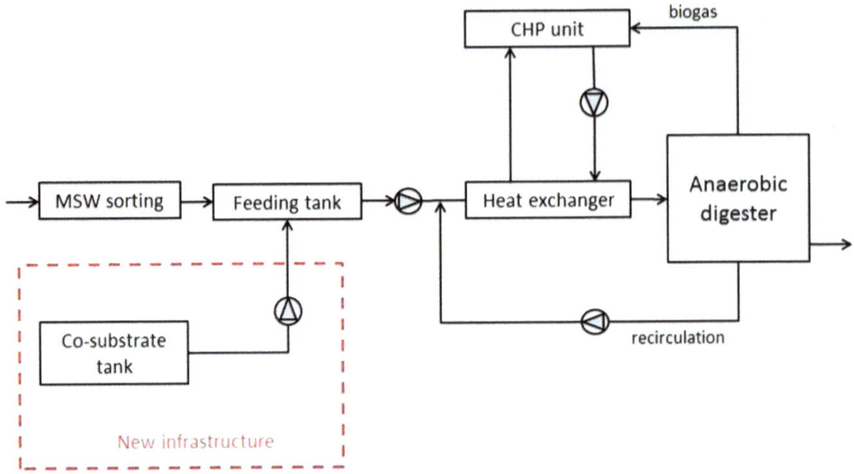

Fig. 6 Modifications needed to implement a co-substrate in an existing AD plant

It is well known that one of the main issues for the AcoD lies in balancing the carbon-to-nitrogen (C/N) ratio of the feed, where the optimum value is between 20 and 70. MSW presents a wide variety of C/N ratio which can vary between 10 and 60, depending on MSW origin, season, sorting among others. When a MSW digester influent has a low C/N ratio a carbon rich co-substrate can help to enhance the ratio, and in most cases, reduce the ammonia concentration in the digester medium, inhibitory of the methanogenesis step (see Sect. 2.3.3). In contrast, MSW that present high C/N ratios are more appropriated to be co-digested with sewage sludge and/or animal manure, substrates that normally present lower C/N ratio. However, the right combinations of other factors in the mixture like macro- and micro-nutrients, pH/alkalinity, inhibitors/toxic compounds, humidity and biodegradable organic matter are also relevant. In this topic, MSW digesters are robust and stable, with pH values around 7.5 and high alkalinities, 5–10 g $CaCO_3$ L^{-1}. Consequently, they are able to accept industrial highly biodegradable wastes which produce large amounts of VFA, such as pharmaceutical by-product, glycerol or food processing wastes. However, when dealing with quickly biodegradables substrates it is indispensable to investigate the maximum percentage of co-substrate, since random or heuristic decisions on the ratio between waste streams could have negative effects on the digester medium and, therefore, a significant reduction of the methane production [40].

Finally, it should be noted that MSW present all the macro- and micronutrients necessary for the AD process. In fact, the addition of a co-substrate can be useful to dilute the concentration of some micronutrients, which concentration is closer or above the reported threshold values. In this matter, it is of utmost importance to avoid the introduction of inhibitory/toxic compounds for the anaerobic biomass and/or for the digestate final destination, since it directly affect the efficiency and economic feasibility of the whole process.

3.2.1 Co-Digestion Between Municipal Solid Waste and Sewage Sludge

Co-digestion between MSW and sewage sludge (SS) is, by large, the most popular research in the AcoD field [39]. In the MSW-SS system, both basic components of the sewage sludge play an important role in co-digestion. The higher nitrogen content of the waste activated sludge can supplement a possible deficit of nutrients in the MSW, whereas the higher biodegradability of the primary sludge provides an additional contribution to the increase in biogas production. However, typically, MSW have been used as co-substrate for SS since many wastewater treatment plants (WWTP) anaerobic digesters are oversized and therefore are operated at a low OLR [39]. The majority of the studies have shown the goodness, at mesophilic and thermophilic conditions, of the mixtures between MSW and SS. The enhancement of the process performance has been generally related with the optimisation of the C/N ratio and the high biodegradability of the MSW. However, some drawbacks need to be pointed out when the MSW are introduced to a SS digester. Pahl et al. [41] reported an accumulation of zinc, lead and nickel in the digester when MSW was introduced as co-substrate (70, 250 and 220 %, respectively). The same authors also claimed that MSW material need to be converted into slurry with water and/or sewage sludge, to allow the material to be conveyed in the existing pipework and pumps. Moreover, the danger that sediment at the bottom represent for the digester operation should be considered [41].

3.2.2 Co-Digestion Between Municipal Solid Waste and Industrial Wastes

MSW has also been co-digested with several industrial wastes, being animal manure the most reported one. The increase of the biogas production, between 50 and 100 %, when co-digesting MSW and animal manure has been attributed to several factors. Li et al. [42] and Macias-Corral et al. [43] concluded that the synergistic effect was due to the complementary characteristics of both substrates (alkalinity and nutrient balance) which improved waste biodegradability. Similarly, Zhang et al. [44, 45] suggested that the trace elements (Co, Mo, Ni, Fe, etc.) supplemented by manure was the key factors to enhance MSW digestion performance. These trace elements, which were present in negligible quantities in the MSW under study, are necessary for activating and maintaining the enzymatic activity of the anaerobic biomass. Moreover, Macias-Corral et al. [43] also suggested that the biodegradability of the MSW could be improved by the presence of cellulose degrading bacteria in the manure. Regarding the mixture between MSW and manure, Zhang et al. [45] introduced card packing to reduce the nitrogen ammonia and to increase the OLR of the digester, as a result a more stable AD process was obtained.

The mixtures between MSW and an agro-industrial waste have also been widely investigated. For example, Wang et al. (2012) investigated the effect of the mixture proportion between MSW and destiller's grain. Fernandez et al. [46], who investigated the possibility of co-digesting MSW and animal and vegetable fat from a food industry, concluded that once the anaerobic biomass is adapted to fats the LCFA inhibition can be avoided, therefore good biogas yields and organic matter removals are obtained. Finally, Zhang and Banks [47] studied the feasibility of co-digesting MSW, at different OLR, with either blood or a mixture of pig intestines with flotation fat. The results showed that the high nitrogen ammonia present in the blood inhibited, even at low OLR, the AD process, whereas the mixture between MSW and mixture of pig intestines with flotation fat showed high VFA concentration when the OLR reached 3 kg VS/(m^3digester*day).

References

1. Sturc, M.: Renewable energy: analysis of the latest data on energy from renewable sources. Statistics in focus Report 44/2012. European Union (2012)
2. Eurostat Statitstics.: Renewable energy statistics. European Union. http://epp.eurostat.ec.europa.eu/statistics_explained/index.php/Renewable_energy_statistics (2010). Accessed 7 Nov 2012
3. Directive 08/98/EC.: Waste Framework Directive. European Union, 19 Nov 2008
4. Baird, J., Hutton, M.K., Savage, A.L., Hipkin, A., Cruz, P., MacLeod, I.: Preparing new ground: how to put stabilized biowaste to proper use in landfill restoration. Waste Manag. World. http://www.waste-management-world.com/index/display/article-display/_printArticle/articles/waste-management-world/volume-7/issue-7/features/preparing-new-ground-how-to-put-stabilized-biowaste-to-proper-use-in-landfill-restoration.html(2012). Accessed 10 Nov 2012
5. Cecchi, F., Traverso, P.G., Mata-Alvarez, J., Clancy, J., Zaror, C.: State of the art of R & D in the anaerobic digestion of municipal solid waste in Europe. Biomass **16**, 257–284 (1988)
6. De Baere, L.: Anaerobic digestión of solid waste: state-of-the-art. Water Sci. Technol. **41**, 263–273 (2000)
7. Directive 99/31/EC.: The landfill of waste. European Union, 26 Apr 1999
8. De Baere, L., Mattheeuws, B.: Anaerobic Digestion of the organic fraction of municipal solid waste in Europe. OWS. http://www.ows.be/downloads/Anaerobic%20Digestion.pdf. Accessed 24 Sept 2012
9. Demirel, B., Scherer, P.: The roles of acetotrophic and hydrogenotrophic methanogens during anaerobic conversion of biomass to methane: a review. Rev. Environ. Sci. Biotechnol. **7**, 173–190 (2008)
10. Batstone, D.J., Keller, J., Angelidaki, I., Kalyuzhnyi, S.V., Pavlostathis, S.G., Rozzi, A., Sanders, W.T., Siegrist, H., Vavilin, V.A.: The IWA anaerobic digestion model no 1 (ADM1). Water Sci. Technol. **45**, 65–73 (2002)
11. Veeken, A., Hamelers, B.: Effect of temperature on hydrolysis rates of selected biowaste components. Bioresour. Technol. **69**, 249–254 (1999)
12. Karakashev, D., Batstone, D.J., Trably, E., Angelidaki, I.: Acetate oxidation is the dominant methanogenic pathway from acetate in the absence of methanosaetaceae. Appl. Environ. Microbiol. **72**, 5138–5141 (2006)
13. Chen, J., Cheng, J., Creamer, K.S.: Inhibition of anaerobic digestion process: a review. Bioresour. Technol. **99**, 4044–4064 (2008)
14. Apples, L., Baeyens, J., Degreve, J., Dewil, R.: Principles and potential of the anaerobic digestion of waste-activated sludge. Prog Energ Combust. **34**, 755–781 (2008)
15. Astals, S., Nolla-Ardèvol, V., Mata-Alvarez, J.: Anaerobic co-digestion of pig manure and crude glycerol at mesophilic conditions: Biogas and digestate. Bioresour. Technol. **110**, 63–70 (2012)

16. Stroot, P.G., McMahon, K.D., Mackie, R.I., Raskin, L.: Anaerobic codigestion of municipal solid waste and biosolids under various mixing conditions-I. Water Res. **35**, 1804–1816 (2001)
17. Kayhanian, M.: Ammonia inhibition in high-solids biogasification: an overview and practical solutions. Environ. Technol. **20**, 355–365 (1999)
18. Angelidaki, I., Ahring, B.K.: Anaerobic thermophilic digestion of manure at different ammonia loads: effect of temperature. Water Res. **28**, 727–731 (1994)
19. Directive 06/12/EC.: Waste Directive. European Union, 5 Apr 2006
20. Holm-Nielsen, J.B., Al Seadi, T., Oleskowicz-Popiel, P.: The future of anaerobic digestion and biogas utilization. Bioresour. Technol. **100**, 5478–5484 (2009)
21. Environment DG, EU.: Working document on the biological treatment of biowaste, 2nd draft. European Union (2001)
22. Directive 86/278/EEC.: The protection of the environment, and in particular of the soil, when sewage sludge is used in agriculture. European Union, 12 June 1986
23. Regulation EC 1774/2002.: Guidelines for applications for new alternative methods of disposal or use of animal by-products under regulation. European Union, 3 Oct 2002
24. Environment DG, EU.: Working document on sludge 3rd draft. European Union (2000)
25. Directive 91/676/EEC.: The protection of waters against pollution caused by nitrates from agricultural sources. European Union, 12 Dec 1991
26. Sahlstrom, L.: A review of survival of pathogenic bacteria in organic waste used in biogas plants. Bioresour. Technol. **87**, 161–166 (2003)
27. Sanders, W.T., Veeken, A.H., Zeeman, G., Van-Lier, J.B.: Analysis and optimization of the AD process of OFMSW. In: Mata-Alvarez, J. (ed.) Biomethanization of the Organic Fraction of Municipal Solid Waste. IWA Publishing, London (2003)
28. Zhang, R., Gikas, P., Zhu, B., Lord, J., Choate, C., Rapport, J., El-Mashad, H., Jenkins, B.: Integration of rotatory drum reactor and anaerobic digestion technologies for treatment of municipal solid waste. CalRecycle. http://www.calrecycle.ca.gov/Publications/Default.aspx (2010). Accessed 24 Sept 2012
29. Zhu, B., Gikas, P., Zhang, R., Lord, J., Jenkins, B., Li, X.: Characteristics and biogas production potential of municipal solid waste pretreated with rotary drum reator. Bioresour. Technol. **100**, 1122–1129 (2009)
30. Bochmann, G., Herfellner, T., Susanto, F., Kreuter, F., Pesta, G.: Application of enzymes in anaerobic digestion. Water Sci. Technol. **56**, 29–35 (2007)
31. Fernandez-Güelfo, L.A., Álvarez-Gallego, C., Sales-Márquez, D., Romero-García, L.I.: The effect of different pretreatments on biomethanation kinetics of industrial Organic Fraction of Municipal Solid Waste. Chem. Eng. J. **171**, 411–417 (2011)
32. Zhang, Y., Banks, C.J.: Impact of different particle size distributions on anaerobic digestion of the organic fraction of municipal solid waste. Waste Manag. (2012)
33. Biogas weser-ems.: Biogas-weser-ems GmbH & Co. http://www.biogas-weser-ems.de/en/technologie.php (2012). Accessed 25 Sept 2012
34. Liu, X., Wang, W., Gao, X., Zhou, Y., Shen, R.: Effect on thermal pretreatment on the physical and chemical properties of municipal biomass waste. Waste Manag. **32**, 249–255 (2012)
35. Cambi.: http://www.cambi.no/wip4/plant.epl?cat=10645&id=195088 (2012). Accessed 25 Sept 2012
36. Elbeshbishy, E., Nakhla, G.: Comparative study of the effect of ultrasonification on the anaerobic biodegradability of food waste in single and two-stage systems. Bioresour. Technol. **102**, 6449–6457 (2011)
37. Shahriari, H., Warith, M., Hamoda, M., Kennedy, K.: Anaerobic digestion of organic fraction of municipal solid waste combining two pretreatment modalities, high temperature microwave and hydrogen peroxide. Waste Manag. **32**, 41–52 (2012)
38. Torres, M.L., Espinosa-Lleréns, M.C.: Effect of alkaline pretreatment on anaerobic digestion of solid waste. Waste Manag. **28**, 2229–2234 (2008)

39. Mata-Alvarez, J., Dosta, J., Macé, S., Astals, S.: Codigestion of solid wastes: a review of its uses and perspectives including modelling. Crit. Rev. Biotechnol. **31**, 99–111 (2011)
40. Astals, S., Ariso, M., Galí, A., Mata-Alvarez, J.: Co-digestion of pig manure and glycerine: experimental and modelling study. J. Environ. Manage. **92**, 1091–1096 (2011)
41. Pahl, O., Firth, A., MacLeod, I., Baird, J.: Anaerobic co-digestion of mechanically biologically treated municipal waste with primary sewage sludge—a feasibility study. Bioresour. Technol. **99**, 3354–3364 (2008)
42. Li, X., Li, L., Zheng, M., Fu, G., Lar, J.S.: Anaerobic co-digestion of cattle manure with corn stover pretreated by sodium hydroxide for efficient biogas production. Energy Fuels **23**, 4635–4639 (2009)
43. Macias-Corral, M., Samani, Z., Hanson, A., Smith, G., Funk, P., Yu, H., Longworth, J.: Anaerobic digestion of municipal solid waste and agricultural waste and the effect of co-digestion with dairy cow manure. Bioresour. Technol. **99**, 8288–8293 (2008)
44. Zhang, L., Lee, Y.W., Jahng, D.: Anaerobic co-digestion of food waste and piggery wastewater: focusing on the role of trace elements. Bioresour. Technol. **102**, 5048–5059 (2011)
45. Zhang, Y., Banks, C.J., Heaven, S.: Co-digestion of source segregated domestic food waste to improve process stability. Bioresour. Technol. **114**, 168–178 (2012)
46. Fernández, A., Sánchez, A., Font, X.: Anaerobic co-digestion of a simulated organic fraction of municipal solid wastes and fats of animal and vegetable origin. Biochem. Eng. J. **26**, 22–28 (2005)
47. Zhang, Y., Banks, C.J.: Co-digestion of the mechanically recovered organic fraction of municipal solid waste with slaughterhouse wastes. Biochem. Eng. J. **68**, 129–137 (2012)

Combustion Behavior of Novel Energy Crops in Domestic Boilers: Poplar and Brassica Experiences

Maryori Díaz-Ramírez, Fernando Sebastián Nogués, Javier Royo and Adeline Rezeau

Abstract In the Mediterranean countries, several reasons promote cultivation of energy crops with respect to other kind of biomass. Specifically, they are expected to enhance the biomass heating opportunities in the energy market. Focused on that goal, Spanish research efforts have involved the evaluation of all the bioenergy chain steps, from fuel production to its transformation into energy in the conversion system. In this work, special emphasis have been placed on the assessment of two crops (*Brassica carinata* and *Populus sp.*) combustion requirements and adaptability level of a novel 250 kW_{th} Spanish grate fired technology to cope with their conversion according acceptable level of gaseous emissions and efficiency. Finally, further research steps, needed to improve the novel fuel conversion system performance, have been presented seeking for a more definite introduction of this heating system in the Spanish bioenergy market.

1 Perspectives of the Energy Crops Use as Biofuels for the Heating Sector in Europe: The Spanish Case

Biomass fuels are one of the most important energy resources. They constitute approximately 14 % of the global primary energy, the fourth largest following coal, oil, and natural gas [1]. In addition, bioenergy is considered to be the

M. Díaz-Ramírez (✉) · F. S. Nogués · A. Rezeau
Natural Resources Division-BERA Group, Centre of Research for Energy Resources and Consumption (CIRCE), Mariano Esquillor 15—Edificio CIRCE, España, Spain
e-mail: mdiaz@unizar.es

M. Díaz-Ramírez · J. Royo
Campus Río Ebro, University of Zaragoza, E-50018 Zaragoza, Spain

M. Díaz-Ramírez · J. Royo
Department of Mechanical Engineering, University of Zaragoza, María de Luna 3 E-50018 Zaragoza, Spain

G. Ferreira (ed.), *Alternative Energies*, Advanced Structured Materials 34,
DOI: 10.1007/978-3-642-40680-5_2,

renewable energy source with the largest unexplored potential in the European Union (EU) [2, 3].

In the last decades, several types of biofuels have been incorporated to the biomass market. Main driving forces to the recent participation of these sources have been related to growing concerns about environmental pollution, especially due to greenhouse gases (GHG), and also those related to fossil fuel supply uncertainties. In addition, limitations concerning biomass availability in specific regions and increasing energy demands have enforced to consider a broad variety of novel sources. It currently involves sources from agro-industrial activities, such as olive oil and almond production, as well as dedicated energy crops, which includes short rotation coppice, perennial grasses and herbaceous species.

With respect to dedicated energy crops, it can be stated that the bioenergy potential contribution of this biomass type is enormous. Taking land availability and projected biomass yield into account, the largest bioenergy potential from the dedicated energy crops will be found in seven of the European Union Member States (Spain, France, Germany, Italy, the United Kingdom, Lithuania and Poland) [4–7]. Specifically, in the Mediterranean countries, energy crops are seen as an attractive alternative to satisfy local energy demands because the possibilities they offer to enhance the local biomass heating opportunities. Moreover, their production is expected to provide collateral benefits to this region, for instance, social sustainability by creating new employment opportunities.

Particularly, in Spain, research efforts for energy production based on dedicated energy crops have mainly been focused on the assessment of inputs and outputs of each chain step involved in the energy crop production processes, from fuel production to its transformation into energy [8]. Emphasis has been placed not only to identify the most suitable fuels adapted to local Mediterranean edapho-climatic conditions but also to learn how to handle and use these materials for heating supply purposes.

Local research in these issues started as part of the "PSE On Cultivos" R&D&I national project carried out between 2005 and 2012. It was defined as singular and strategic project for development, demonstration and viability evaluation of the commercial production of energy from dedicated crops in Spain, in terms of their use for heating, electricity and transport.

With regard to the energy crop application for the domestic heating, this project entailed the assessment of the energy crop properties as well as the development of suitable thermal conversion technology adapted to characteristics of the most promising energy crops produced in Spain.

Concerning the energy crop properties assessment, a complete fuel characterization was done considering both physical and chemical aspects. Energy crops cultivated today can generally be described as "problematic" fuels mainly due to both their high ash content and their composition of ash forming matter in comparison to biofuels used traditionally, such as stemwood based assortments [9–11]. Consequently, a higher tendency to cause fouling, slagging and/or corrosion problems is expected from their thermochemical conversion.

Experiences carried out as part of the "PSE On Cultivos" framework have involved one short rotation coppice, *Populus sp.* (poplar) and one herbaceous *Brassica carinata* (brassica) energy crops, among other raw materials proposed as potential biofuels in the Spanish "PSE On Cultivos" energy schemes.

On the other hand, in order to evaluate the Mediterranean energy crops applicability for heating purposes, a medium biomass grate-fired unit has been developed. The technological features existing in this conversion unit are been based on both previous knowledge from stemwood combustion and specific experimental research carried out with the energy crops assessed as part of the "PSE On Cultivos" domestic heating framework. In spite of grate firing combustion units has been suggested as the most suitable ones to satisfactory thermal conversion of the novel biofuels, significant impacts attributed to fuel quality factors, such as ash content and ash chemical composition, might affect the performance of grate technologies. The intrinsic fuel characteristics might limit the reliability level of the combustion unit, basically affected by type of fuel as well as operation and maintenance of the conversion equipment, among others attributes [12, 13]. Consequently, knowledge of fuel quality influence on the conversion system performance and adaptability level of the novel technology to manage the fuel characteristics was considered of importance as part of the heating generation "PSE On Cultivos" framework.

Based on knowledge acquired by the Spanish experience, the main objective of this chapter is to highlight some important considerations that should be taking into account to satisfactorily convert the more troublesome energy crops for heating purposes. Main characteristics of both the thermal conversion system developed and tested fuels in the "PSE On Cultivos" framework are primarily described. Secondly, an evaluation of the thermal and emission performance during brassica and poplar conversion is presented. Special emphasis is given on analyzing the ash effects on the Spanish fixed-grate system performance, and how to manage them particularly, in terms of practical aspects to take into account for providing high combustion quality. Therefore, research from this work may contribute to identify key factors to consider during grate conversion of the novel crops besides offering useful information to define emission levels or develop reliable conversion units able to cope with the energy crop characteristics.

2 Thermal Conversion of Two Energy Crops for Heating Purposes: Brassica and Poplar Experiences

Knowledge about the dedicated energy crops properties and their conversion requirements to attain a high acceptability level of the final users is limited mainly because of the recent introduction of the novel biofuels into the energy sector. Most of the available information in literature regarding the thermal conversion of dedicated energy crops has been acquired under quite well controlled laboratory conditions, and published full-scale experiences are scarce.

Therefore, gaining knowledge related to the understanding of the combustion behavior of the novel crops is foreseen of relevance for the sustainable development of an energy crop market based on high efficiency, low costs and low environmental impact levels. Considering these aspects, a detailed assessment of combustion experiences carried out for the two energy crops, brassica and poplar, as part of the "PSE On Cultivos" framework, is presented here. Main tested conditions and results are summarized in the following.

2.1 Fuel Properties

A detailed fuel characterization has been carried out to the novel crops. A briefly description of the most significant fuel properties (physical and chemical aspects) will be presented in this section.

The herbaceous energy crop, brassica and the short rotation coppice, poplar, produced as part of the Spanish research project PSE "On Cultivos", were both cultivated and pelletized in Spain. The pelletization plant is located at the Renewable Energy Development Center (CEDER-CIEMAT), in Spain (Lubia, Soria). Biomass raw materials used in the pelletizing process consisted of brassica straw (grains included) and poplar stem (branches, leaves and bark were not included). Pellets had a diameter of 6 mm and a length of 15–20 mm. In Fig. 1, pictures of brassica and poplar raw materials and densified as pellets are showed.

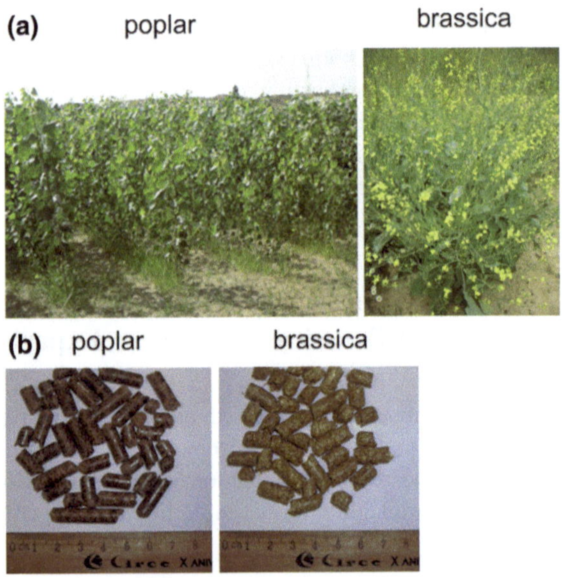

Fig. 1 Brassica and poplar raw materials and densified as pellets

Table 1 Fuel characteristics. Values are given in weight percent dry basis (wt % d.b.) except for heating value, moisture and ultimate analysis, which are given in MJ/kg d.b., weight percent wet basis (wt % w.b.), and dry basis ash free (wt % d.b.a.f), respectively

Parameter		Poplar	Brassica
Calorific value			
Higher heating value, HHV	MJ/kg d.b.	19.49	17.46
Lower heating value, LHV	MJ/kg d.b.	18.13	16.31
Proximate analysis			
Ash content in pellets	wt % d.b.	2.8	10.7
Fixed carbon	wt % d.b.	15.2	18.8
Volatile matter	wt % d.b.	82.0	70.4
Moisture	wt % w.b.	5.0	4.9
Ultimate analysis			
C	wt % d.b.a.f.	50.8	49.7
H	wt % d.b.a.f.	6.1	6.2
N	wt % d.b.a.f.	0.1	1.3
O	wt % d.b.a.f.	42.9	42.2
S	wt % d.b.a.f.	0.16	0.36
Cl	wt % d.b.a.f.	0.02	0.26
Percentage of dry substance in ashes			
SiO_2	wt % d.b.	15	22
Al_2O_3	wt % d.b.	1.7	3.0
CaO	wt % d.b.	37	26
Fe_2O_3	wt % d.b.	0.6	1.2
K_2O	wt % d.b.	8.3	15
MgO	wt % d.b.	4.8	1.7
MnO	wt % d.b.	0.06	0.05
Na_2O	wt % d.b.	0.9	1.6
P_2O_5	wt % d.b.	4.5	6.1
SO_3	wt % d.b.	11	7.3
ZnO	wt % d.b.	0.01	0.08

The main fuel properties of the tested pellets are listed in Table 1. In addition, the ash chemical composition of both fuels, given on a mole of ash-forming element per GJ of dry solid fuel, is provided in Fig. 2.

As shown in Table 1, these two biomass types exhibit a broad variation range of their properties and, consequently, a different combustion behavior is expected for them. According to Table 1, moisture content is rather low and similar between fuels. Based on ash content required for standardized woody pellets defined by the European norm (EN) 303-5 [14] as <0.5 wt % d.b. and by the DINplus standard set at ≤ 0.5 wt % d.b., the energy crops samples are characterized by a larger ash content, and can be generally grouped as ash-rich fuels. The ash condition in these two fuels also affects the V/CF ratio being lower for the herbaceous, and consequently, fuel reactivity differences are expected between them during their thermal conversion.

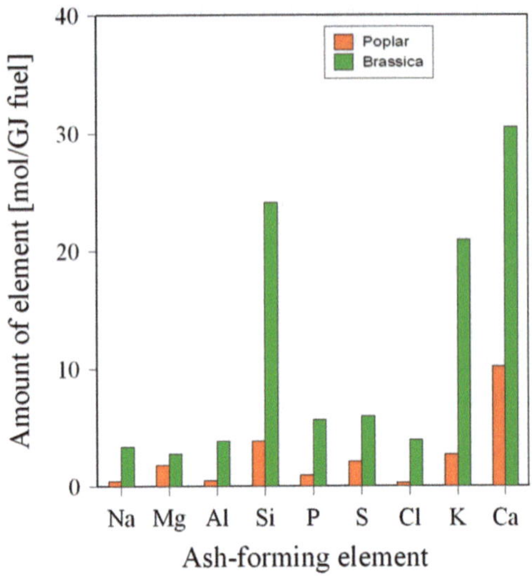

Fig. 2 Brassica and poplar inorganic constituents. Amount of element is calculated based on the LHV of the dry solid fuel. Amount of Cl in ash at 550 °C was assumed as 100 % of the content in fuel obtained by the ultimate analysis

Concerning ultimate analysis, compared on a dry ash free basis, N content for poplar was according to limiting values for the DINplus quality standard set at $\leq$0.3 wt % d.b. On the contrary, brassica yielded a larger N content, and hence, more significant impacts on the NO_X emission level are expected to this fuel.

In addition, results from the ultimate analysis indicated that brassica depicted the worst condition in terms of other troublesome elements, such as Cl and S. They play an important role on particulate matter and acidic gaseous emissions as well as corrosion phenomenon. With regard to the ash composition, although concentrations vary over quite a broad range, major ash components among fuels are Si, Ca and K. A second group of the ash-forming elements is suggested to be formed by Mg, P and S.

2.2 Conversion System Characteristics

As mentioned previously, a 250 kW_{th} (referred to its useful output) grate-fired thermal conversion system (currently, still under development and improvement stages) has been used to perform the combustion tests for the two energy crops, as part of the "PSE On Cultivos" domestic heating framework. The current state of technological advance available in this conversion unit is based on design modifications and improvements aimed to manage the specific combustion properties and ash chemical characteristics of the novel dedicated crops. The work presented here is a part of the development process of this Spanish technology.

The equipment was initially composed of a biomass burner, consisting of six horizontally positioned and separated cast iron grates connected to a horizontal and modular cast iron heat exchanger, which was designed and normally used for gaseous and liquid fuels. The main characteristics of this earlier design, as well as the further improvements carried out on the system to achieve the current state of development of the 250 kW_{th} grate-fired thermal conversion unit can be found elsewhere [15–18].

Combustion experiences performed with the pelletized ash-rich energy crops in the previous designs were considered as preliminary steps for all the improvements defined to the conversion unit presented here. As shown in Fig. 3, the 250 kW_{th} grate-fired thermal conversion system is mainly divided into four zones: burner (I), combustion chamber (II) and two heat exchanger sections (III and IV).

The burner zone is constructed according to the main principles for the combustion of pellets. In this system, fuel feeding (see Fig. 3, 1) is carried out by using a screw with a rotation speed control to provide the optimum adjustment of the fuel flow to the required load condition. The burner consists of two stationary main grates (see Fig. 3, 5). The first, which is located at the upper part of the burner, is divided into four horizontally positioning small steps, resembling a ladder (with nozzles in the riser part), whereas the second is also flat but is longer to ensure complete char burnout. Primary air (see Fig. 3, 2) is provided as under-feed air through each grate.

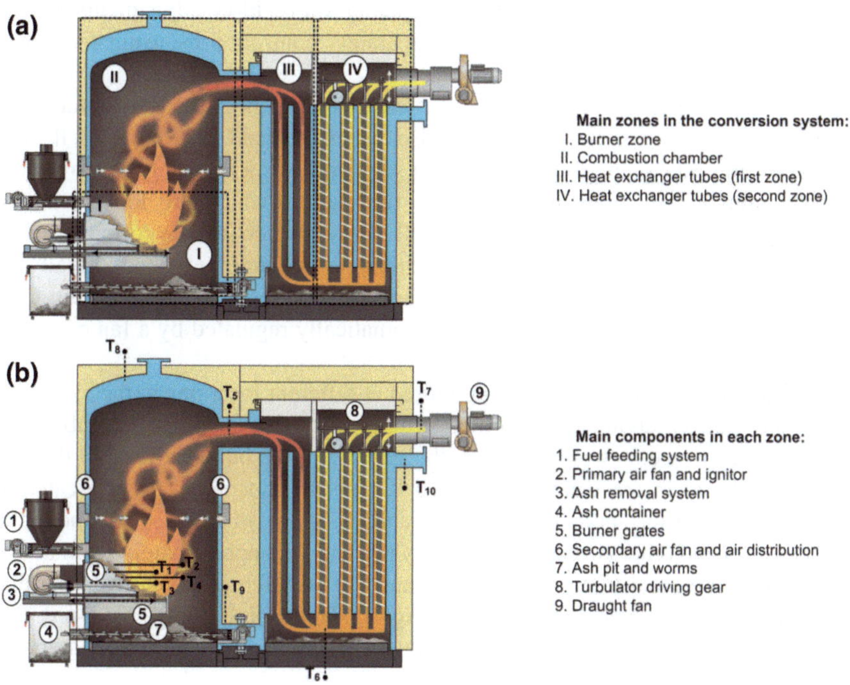

Fig. 3 Schematic figure of the experimental setup. Scheme presents the main zones in the conversion system and the main components in each zone

The air injection nozzle arrangement and their number differ in each grate step in order to supply as much air as possible according to requirements for each thermal transformation process that fuel particles undergoes during combustion over the grate (i.e., heating up, drying, devolatilization and char oxidation phases).

This burner is also designed with an automatic ash removal system (see Fig. 3, 3), which scrapes ashes and other combustion residues away from the second grate and moves them into the ash pit located inside the combustion chamber (see Fig. 3, 7). The ash pusher (on/off) working period (i.e., frequency and duration of the cleaning periods) can be adjusted according to the fuel properties' requirements during continuous operation of the conversion unit. It serves to both guarantee the required time for char combustion and avoid ash accumulation and severe slag and sintering occurrence over the grate.

Bed movement conditions on the grate are also linked to the primary air adjustments. In this manner, an adequate control of the ash pusher working periods and air condition contributes to enhance air–fuel contact over the grate, according to the fuel ash properties under a continuous operation mode of the conversion unit.

In order to facilitate complete mixing, as far as possible, between the air and volatilized matter released from the fuel bed on the grate, secondary air is distributed into the combustion chamber through several nozzles located in a surrounding channel above the grate (see Fig. 3, 6), where secondary air is partially preheated by heat transferred to the boiler walls. Flow adjustments carried out on primary and secondary airs are separately controlled by using frequency meters.

Besides the boiler walls in the combustion chamber, the integrated heat exchanger (see Fig. 3, III and IV) is also water jacketed. Shedding of fly ash deposited on heat exchanging surfaces is also possible as a result of the movement of the automatically controlled up/down turbulators, which are located in the second heat exchanger zone (see Fig. 3, 8). They also contribute to increase the flue gas turbulence and, therefore, the heat transfer rate into the system. Finally, in order to control the amount of air supply and flue gas residence time during their path in the system, the force draught is automatically regulated by a fan equipped with a frequency meter (see Fig. 3, 9).

Temperature was continuously monitored at different locations in the system to gain information about the thermal profile of the unit during combustion. Measuring instruments were placed in the vicinity of the burner grates, in the flue gas channel and in the water steps. Two K-type thermocouples were located at the upper zone of the secondary grate (T_1 and T_2, approximately 24 cm above the longer grate) and other two at the proximity (T_3 and T_4, approximately 6 cm above the longer grate), and three more (T_5, T_6 and T_7) at different positions in the flue gas path through the heat exchanger (see Fig. 3). Three PT 100 were provided, one positioned in the water-in step (T_8) and at two positions in the water-out step (T_9 and T_{10}) (see Fig. 3).

2.3 Combustion Tests in the 250 kW_{th} Spanish Fixed-Grate Technology

As part of the experimental test campaigns, two different type of combustion tests were performed to each fuel, i.e., preliminary and reproducibility tests.

Firstly, preliminary combustion tests were performed to adjust operation conditions to requirements of each fuel. Basically, under the fixed design characteristics of the novel grate technology, adjustments were addressed on four specific combustion parameters: load input, lambda factor, air distribution and pusher working period. The screw feeding frequency used to regulate load input, the working period of the ash pusher for adjusting residence time and air–fuel contact on the grate, the excess air flow as means of lambda factor and the primary/secondary air distribution as means of lambda distribution were considered as the main parameters to be controlled.

Tested conditions were established seeking an output as closer as possible to the nominal value, and for the best performance of the system with regard to unburnt matter (i.e., loss of ignition matter in solid residue and gaseous unburnt quantified as CO concentration), and thermal efficiency. In these tests, combustion quality was defined in terms of CO concentration in flue gases and thermal efficiency. European limits established to standardized high quality woody biofuels by the European standard EN 303-5:1999, and the more restrictive Austrian limits defined by the agreement according to Art.15a B-VG were considered to assess combustion quality obtained during the tests.

In addition, the four combustion parameters aforementioned were also adjusted to avoid the total saturation of the burner section in short test periods by severe sintering and slagging formation over the grate working at as far as possible load input. Nevertheless, main portion of the residues was clearly fused ash to large blocks (see Fig. 4). A more detailed explanation concerning criteria considered to define the operation adjustments carried out to the system can be found elsewhere [19, 20].

Secondly, reproducibility tests were carried out at the selected operating conditions by the preliminary tests. Data was collected for further calculation of combustion performance parameters after a steady-state regime was reached at the fixed operating conditions determined by the preliminary experiments. The concentrations of O_2, CO and NO within the exhaust gas were continuously measured with electrochemical sensors (Flue gas analyzer Testo 350XL), just after the second heat exchanger exit, during all the tests. CO and NO_X concentrations estimated as mg/Nm^3 were normalized at 10 % oxygen in the dry flue gas (d.g.) and their specific emission values were calculated in relation to the energy input in fuels. NO_X emissions were quantified as NO_2 equivalents to be compared to standards limits proposed by the Austrian restriction to other standardized fuels than woody pellets. The European standard EN 303-5:1999 suggests measurement of NO_X emissions although no limits are defined for them yet [14].

After each experiment, solid residues, which remained on the burner grates, and the ones scraped by the pusher into the ash pit were collected. Sintering degree in

Fig. 4 Bottom Ash accumulated on the grates after brassica combustion experiments

the collected samples was assessed both through visual inspection and by a simple strength test. Unburnt content in solid combustion residues was determined as the difference between 100 % and the ash content obtained from a representative sample of the gathered bottom ash heated at 550 °C according to the standardized method CEN/TS 14775:2004 [21].

2.4 Sampling of Residual Bottom Ash and Sintering Degree Assessment

After each experiment, the solid combustion residues, which remained on the burner grates, and the ones scraped by the pusher into the ash pit were gathered. Sintering degree in the collected samples was assessed both through visual inspection and by a simple strength test.

2.5 Experimental Results During Brassica and Poplar Conversion

Generally, combustion conditions were controlled and relatively stable for each fuel. A summary of results obtained for the four assessed combustion parameters (i.e., load input, lambda factor, air distribution and pusher working period), during the best condition identified to each fuel is presented in Table 2.

Table 2 Operation parameters attained to each fuel maximum load during best condition at stationary combustion phase

Parameter		Poplar	Brassica
Load input	kW_{th}	257	174
Useful output	kW_{th}	230	157
Pusher on (duration)/off (frequency) working period	Adim.	4/100	5/50
Primary lambda	Adim.	0.95 ± 0.01	1.47 ± 0.03
Total lambda, λT	Adim.	1.93 ± 0.09	2.07 ± 0.13

As seen, significant differences with respect to the four selected operating parameters were identified between the crops to manage the fuel properties under the fixed design conditions of the 250 kW_{th} technology tested in this work.

Load input was eventually lowered to the herbaceous compared to the woody crop, up to around 37 % less than the nominal output. This condition was selected as a first measure to control saturation of the burner grates by bottom ashes. Nevertheless, a higher ash removal was also needed to cope with the brassica ash characteristics (total amount of residues and sintering tendency) despite the lower load input used. In general, around 30 % for poplar and up to 50 % for brassica of in-going ash was estimated to be retained as bottom ash inside the burner section and was characterised by a severe slagging degree.

Both frequency and duration of the pusher movements were significantly increased for the herbaceous, up to around 50 % the values applied to poplar. Ash removal adjustments were limited by the carbon burnout control for high combustion quality (i.e., loss of ignition matter was less than 0.01 kg/kg solid residue, d.b.). Because of this condition, partial saturation of the grates could not be avoided. As a result, primary air requirements increased for the herbaceous to minimize the reduction of an effective oxidizer-fuel contact by the ash condition on the grates (quality and quantity).

Concerning combustion quality (CO emissions and thermal efficiency), it was generally according to the Austrian requirements (limits defined by the European norm EN 303-5 are less restrictive). Evolution curves for CO specific emissions and for thermal efficiency at T_7 (temperature in the flue gas at the exit of the conversion system) during all the tested period at maximum load attainable for each fuel are showed in Fig. 5. Comparison of results during the best condition reached to each fuel is presented in Table 3.

Considering Fig. 5, stability of CO emissions and opening degree of the curves were different for the two energy crops. This condition might indicate the higher sensibility of the system during conversion of brassica. As it has been suggested in literature, fuel combustibility is affected by fuel properties, such as ash content and V/CF ration [23–27]. Based on the fuel composition comparison (see Table 1), brassica exhibited the worst fuel quality with regards to these two parameters.

As previously mentioned, ash accumulation on the grates minimizes an effective air to fuel contact. Furthermore, a lower V/CF ratio negatively impacts fuel combustibility. To increase conversion rates of unburnt species, additional air is

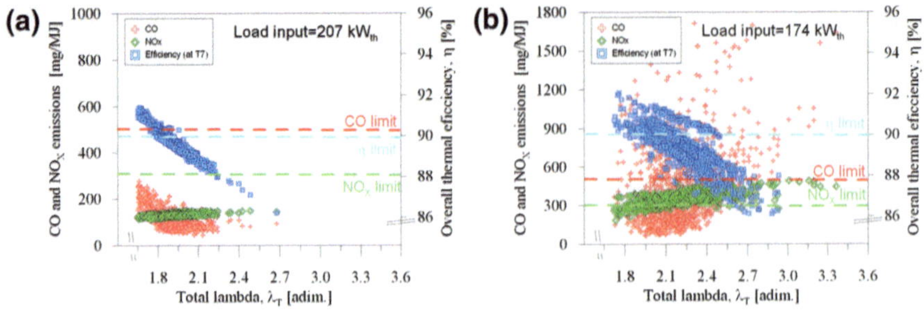

Fig. 5 Evolution curves for CO and NO_X specific emissions and for thermal efficiency at T_7 (temperature in the flue gas at the exit of the conversion system) with regards total lambda factor during all test period at maximum load attainable to each fuel. Limits corresponds to requirements defined by the Austrian law art. 15a B-VG 1994 to CO emissions for high quality stemwood assortments and to NO_X emissions for standardized fuels other than high quality woody biomass [22]

Table 3 Thermal efficiency, gaseous emissions and temperature in the exhaust gases attained to each fuel maximum load during best condition at stationary combustion phase

Parameter		Poplar	Brassica
Oxygen content	O_2 vol % d.g.	10 ± 1	11 ± 1
CO concentration	mg/Nm^3 at 10 % O_2	186 ± 32	285 ± 96
CO concentration	mg/MJ	91 ± 15	138 ± 47
NO_X concentration	mg/MJ	130 ± 3	326 ± 18
Thermal efficiency at T_7	%	89.7 ± 0.4	89.5 ± 0.6
Thermal efficiency at 120 °C	%	92.3 ± 0.4	92.1 ± 0.4
Temperature in the exhaust gases, T_7	°C	155 ± 1	164 ± 1

needed. Consequently, for this fuel, higher amounts of excess air distributed as primary air were demanded (see Table 1). This suggests that the residual ash negatively influenced the combustion performance of the system. The higher the bottom ash amount, the lower the fuel combustibility because of the worsening of air to fuel contact on the grate being necessary to add more primary air to increase combustion rates. Similar effect has been also suggested in literature for several types of ash-rich fuels [23–27]. Nevertheless, an increase of combustion rates has the knock-on effect of increasing slagging occurrence during combustion of the fuels with remarkable ash sintering tendency as brassica (temperature profile in the system was between 850 and 1100 °C). If the fuel ashes exhibit a high sintering tendency, considerably residence time on the high temperature zone subsequently results into their melting. This effect was partially controlled by the ash pusher because of burnout control needs as mentioned previously.

As showed in Table 3, gaseous emissions and thermal efficiency were under the Austrian requirements for the best condition defined by the combustion parameters presented in Table 2 (see also Fig. 5).

Based on data for the best condition attainable to each fuel (see Table 3), CO emission level fulfilled the European norm EN 303-5 requirements set at 1,200 mg/Nm^3 at 10 % O_2 d.g. to the strictest class in terms of thermal efficiency (class 3) for combustion of standardized high quality woody pellets in boilers of nominal output <300 kW_{th}. Results also fulfilled the Austrian limits set at 500 mg/MJ for other standardized non-woody biofuels based on bark, straw, cereals and mixtures (agripellets) fired in automatically boilers up to 400 kW_{th} (CO limit values increase up to 750 mg/MJ at partial load, 30 % nominal heat output) [22].

Thermal efficiency was also according to the EN 303-5 requirements, set at 81 %, and the Austrian limit, set at 90 %.

Under these combustion conditions, NO_X emissions were acceptable and according to the N content in fuels. It has to be considered that the Austrian limit is proposed to other standardized fuels than woody pellets with a nitrogen content ≤0.7 wt % d.b [22], which is 60 % lower than nitrogen content found to the herbaceous crop tested in this work. Main differences between the fuels are related to the NO_X formation mechanism. During biomass combustion for heating application, NO_X emissions are mainly formed by oxidation of nitrogen chemically bound in the fuel matrix. Additional NO_X formation routes, such as thermal and prompt NO_X mechanisms, are suggested to have a lesser significant contribution because of temperature profile typically found to biomass heating units [28]. Based on these considerations, NO_X formation during commercial scale combustion has been directly linked to nitrogen content in fuel, although, implemented air conditions, i.e., amount of air supply and distribution can also influence conversion of fuel nitrogen.

3 Future Research Needs Focused on Commercial Heating Application of Energy Crops in the Spanish Bioenergy Market

Taking into account the current conversion unit design and also results from the combustion experiences with the pelletized energy crops presented here, further research lines are identified towards the commercial heating application of this novel fuel type into the Spanish bioenergy market.

In contrast to stemwood based assortments, energy crops are generally more troublesome fuels owing to their higher ash composition and concentration of ash forming and trace elements. A lower thermal efficiency, undesired emissions and unscheduled shut down to required maintenance are counted as the main ash effects. Consequently, combustion technologies should be adapted to manage the energy crop properties not only to provide working periods at high thermal efficiency to the users but also to fulfill European restrictions that have been established to control air pollutants emitted by the heating sector.

Because of significant differences may exist among fuel properties of standardized sources and novel energy crops, established level of emission pollutants might imply one drawback for the crops but also a driving force for research and development. As a first step, specific knowledge of the novel energy crop properties and influence of both design features and operation conditions on their environmental behavior is needed. Consequently, experiences with other dedicated crops are required to gaining a more general knowledge about characteristics of this biofuel type and their specific conversion needs.

Secondly, research efforts have to be addressed on strategies to control or minimize main drawbacks of the novel fuels in terms of possible impacts on design and operation of the existing conversion technologies in the heating market. Based on results obtained to the two crops assessed in this work, brassica and poplar, limitations of the current system design features to manage the ash properties might prevent to burnt fuels with high sintering tendency and high N content.

In addition, control of pollutants resulting from their combustion may adversely impact on the overall conversion system performance. For instance, control of the CO and NO_X followed opposite trends under tested conditions for poplar and brassica being required a trade-off between their emission levels to attain combustion according to the European requirements. Although Fuel-N converted to NO_X greatly depends on N content in fuels, nitrogen oxidation level also increased because of lambda factors used to favor the complete burnout. Based on these conditions, a more detailed study on the air distribution system, regarding the total number of air injection nozzles, their dimensions and their arrangement, can be necessary.

Computational fluid dynamics (CFD) studies may help to provide knowledge about each thermal transformation process which fuel particle undergoes during its path in the combustion unit and, therefore, to improve understanding of the process going on the system. In addition, accordingly to the current state of the technology developed as part of the "PSE On Cultivos" R&D&I framework and results presented in this work, CFD research might have useful contributions to technologic developments. Improvements of the burner and combustion chamber design features (geometries and the air supply system) as well as application of measures, such as flue gas recirculation, might serve as a good option to reduce emissions and to control bed temperature, and hence, sintering occurrence on the grates. It seems that there are two spaces of the Spanish conversion technology to the complexity of the novel fuels are still possible.

On the other hand, an increasing interest on the utilization of biomass fuels in the heating sector may contribute to hugely increase aerosols emissions to air affecting ambient air quality. Exposure to particulate matter, especially fine particulate mode (PM $<$ 2.5 μm), is associated with a wide range of diseases, including respiratory infection, lung cancer and bronchus [28–30]. In addition, particulate matter also impacts global climate change. Therefore, sources of these kind of emissions and factors that contribute to their formation have to be controlled. Minimization of these pollutants can be reached by optimization of operation conditions (primary measures), and by incorporation of special equipments (secondary measures), such as filtering systems adapted to the conversion unit, may be an option to fulfil environmental requirements. This kind of study has not been considered in the present work mainly due to this type of research was out of scope in the "PSE On Cultivos" R&D&I framework. However, it is a clear research line that is going to be studied in the framework of projects that are currently being developed in Spain in order to guarantee a sustainable energy crop thermal conversion.

On the other hand, standards are also required to minimize uncertainties related to comparison of raw materials properties from different laboratory data or different producers. There is a greater need for standardized engineering practice in the sampling and analysis of biofuels, as well as in the interpretation of analytical data for biomass fuels. Some protocols have been defined to provide a common basis for analysis. However, due to the huge diversity of biomass sources they are still incomplete with respect to the definition of property evaluations that can or should be applied for each biomass type.

Limiting values presented in standards included here have been developed based on biofuels used traditionally such as stemwood based assortments. Therefore, special attention should be paid to emissions and the risk of deposit effects (for instance, slagging and corrosion inside the boiler) when more troublesome fuels, such as the dedicated energy crops assessed here are combusted.

In many European countries there are neither national quality standards nor good practice guidelines. The reasons of this situation are several, for instance, few raw material available, competitive price of gas natural or inexistent pellets market. In Spain, the National Technical Standardisation Committee on Solid Biofuels (AEN/CTN 164) is working on developing standards for novel opportunity fuels. Considering all aforementioned issues presented here, a sustainable utilization of the novel biofuels may be carried out with high efficiency, low costs and low environmental impacts.

4 Summary and Conclusions

Although still are some questions that have to be solved related to energy crops global sustainability or to their impacts to other markets, energy crops are expected to increase bioenergy contribution in the energy market, especially in areas where other type of less problematic and even less controversial fuels as residual biofuels are not available.

One of sectors in which energy crops production is more interesting is the heating sector since the availability of the usual biofuels employed in this sector (generally stemwood based assortments) seems that is not going to be enough to satisfy the increasing demand and also because the profitability that this market can offer to the whole crop chain is better than the one that other alternatives do. Nevertheless, the thermochemical conversion of these novel fuels shows some drawbacks that have to be overcome before achieving their final introduction in the market. Fuel quality is foreseen as a critical parameter.

Compared to the traditionally used stemwood assortments, conversion of these troublesome fuels might undergo some limitations, in terms of specific design features and operation conditions requirements of the system to attain high quality combustion. A higher tendency to cause ash-related effects, for instance, slagging, fouling and/or corrosion effects as well as undesired particulate emissions is expected from their thermochemical conversion. In this sense, combustion units have to be designed taking into account existing differences that exist among traditional fuels and energy crops, mainly related to their composition and impurities that can accompany them in order to guarantee both efficient thermal performance and low emissions.

Within the framework of the "PSE On Cultivos" R&D&I Spanish project, efforts have been addressed to gain knowledge related to the aforementioned issues. Two energy crops one herbaceous, Brassica carinata, and one short rotation coppice, Populus sp., were selected and their combustion performance assessed in a novel medium scale fixed-grate combustion technology (250 kW_{th}) adapted to the two representative energy crops. In this research, special emphasis has been placed on the assessment of adaptability level of a Spanish combustion technology to cope with the fuel complexity.

Main differences between the two crops properties are not only attributed to results by the proximate and ultimate analyses but also due to the characteristics they exhibit in terms of the ash composition. A higher ash content and sintering tendency among fuels lead to lowering the load input. Nevertheless, a reasonably high load was achieved to the ash-rich energy crops. It corresponded to a load input between 20 and 30 % lower than maximum mean value obtained to the system.

During the fuel conversion, it seems that fuel properties such as ash content and V/CF ration influenced the combustibility of the fuels. Consequently, besides changes on load and pusher working period, air adjustments were also required to achieve good combustion quality during the energy crop conversion.

Concerning air supply, higher amounts of primary air were needed not only to improve combustion quality when ash content and sintering degree tendency increased among the tested fuels because of air passing were blocked, but also to satisfactory convert char proportion in these fuels limiting NO_X emission reduction based on air staging strategies, i.e., non-substoichiometric conditions were kept in the grate area. Nevertheless, NO_X emission level was under the Austrian requirements.

Deviation for brassica was greatly influenced by the Fuel-N content. Although Fuel-N converted to NO_X greatly depends on N content in fuels, nitrogen oxidation level also increased because of lambda factors used to favor the complete gaseous burnout. Therefore, the control of the CO and NO_X opposite trends under tested conditions required a trade-off between their emission levels.

Based on an adequate regulation of these combustion parameters an acceptable level of emissions and thermal efficiency was attainable to each fuel according to requirements established by the European norm EN-303-5 and the Austrian agreement Art.15a B-VG. In general, differences among emissions level from fuels were basically dependent on fuel quality and limitation of the system to cope with the properties of the most troublesome fuel, the herbaceous one.

Considering these results, further improvements steps have been suggested involving design and operational issues of the 250 kW_{th} grate conversion unit. Due to the high variability level expected to the energy crops properties, experiences with other dedicated crops are required to gaining a more general knowledge about characteristics of this biofuel type and their specific conversion requirements.

The results that have been achieved allow to stating that when design and operation considerations are adapted to specific properties of energy crops, both high thermal efficiencies and low emissions (under regulations) will be reached. Although further improvements should be considered in the future, these results are another step on the road to sustainable energy crops pellets final penetration in the market. By attaining this goal, technological requirements to guarantee the energy crop thermochemical conversion in an efficient and cost-effective way can be also identified, and hence, ash-related problems faced by the novel biofuels combustion might be controlled or minimized.

Acknowledgments The authors would also like to express their gratitude to the Spanish Education and Science Ministry for the financing the PSE "On Cultivos" project, reference PS-120000-2005-6. Within the framework of this project, the development of the grate fired thermal conversion system has been economically supported and Mediterranean pellet fuels have been provided.

Acknowledgement is also given for the financing of the 'Small-Scale Trigeneration based on Mediterranean Energy Crops and Residual Biomass Combustion' project, reference ENE2008-03194/ALT. Within the framework of this project some of the work presented in this paper has been developed. The authors would like to extend special thanks to the LASIAN Tecnología del Calor S.L., especially to all the employees involved in this work.

References

1. Hall, D.O., Scrase, J.I.: Will biomass be the environmentally friendly fuel of the future? Biomass Bioenergy **15**(4–5), 357–367 (1998)
2. Kavalov B., Peteves S.D.: BIOHEAT Applications in the European Union: An Analysis and Perspective for 2010. The Netherlands (2004)
3. Hoogwijk, M., Faaij, A., van den Broek, R., Berndes, G., Gielen, D., Turkenburg, W.: Exploration of the ranges of the global potential of biomass for energy. Biomass Bioenergy **25**(2), 119–133 (2003)
4. Berndes, G., Hansson, J.: Bioenergy expansion in the EU: Cost-effective climate change mitigation, employment creation and reduced dependency on imported fuels. Energy Policy **35**(12), 5965–5979 (2007)
5. de Wit, M., Faaij, A.: European biomass resource potential and costs. Biomass Bioenergy **34**(2), 188–202 (2010)
6. van Dam, J., Faaij, A.P.C., Lewandowski, I., Fischer, G.: Biomass production potentials in Central and Eastern Europe under different scenarios. Biomass Bioenergy **31**(6), 345–366 (2007)
7. Ericsson, K., Nilsson, L.J.: Assessment of the potential biomass supply in Europe using a resource-focused approach. Biomass Bioenergy **30**(1), 1–15 (2006)

8. Gómez, A., Zubizarreta, J., Dopazo, César, Fueyo, N.: Spanish energy roadmap to 2020: Socioeconomic implications of renewable targets. Energy **36**(4), 1973–1985 (2011)
9. Monti, A., Di Virgilio, N., Venturi, G.: Mineral composition and ash content of six major energy crops. Biomass Bioenergy **32**(3), 216–223 (2008)
10. Vassilev, S.V., Baxter, D., Andersen, L.K., Vassileva, C.G.: An overview of the chemical composition of biomass. Fuel **89**(5), 913–933 (2010)
11. Jenkins, B.M., Baxter, L.L., Miles, T.R.: Combustion properties of biomass. Fuel Process. Technol. **54**(1–3), 17–46 (1998)
12. Sultana, A., Kumar, A.: Ranking of biomass pellets by integration of economic, environmental and technical factors. Biomass Bioenergy **39**, 344–355 (2012)
13. Wang, J.-J., Jing, Y.-Y., Zhang, C.-F., Zhao, J.-H.: Review on multi-criteria decision analysis aid in sustainable energy decision-making. Renew. Sustain. Energy Rev. **13**(9), 2263–2278 (2009)
14. European standard EN 303-5: Heating boilers for solid fuels, hand and automatically stoked, nominal heat output of up to 300 kW. European Committee for Standardization (CEN) (1999)
15. Díaz, M., Boström, D., Boman, C., Royo, J.: Effect of fuel composition on ash transformation and emissions during combustion of pellets from herbaceous and woody energy crops. In: Proceedings of the 17th European Biomass Conference and Exhibition, Hamburg, Germany, pp. 1219–1224 (2009)
16. Rezeau, A., Díaz, M., Arqué, H., Quintero, A., Sebastián, F., Royo, J.: Improvement of a biomass burner and its connection to a modular diesel fuel boiler and analysis of the new biomass boiler concept efficiency and emissions. In: Proceedings of 15th European Biomass Conference and Exhibition, Berlin, Germany, pp. 1678–1683 (2007)
17. Díaz, M., Rezeau, A., Sebastián, F., Royo, J.: Environmental performance of a novel small-scale biomass boiler fuelled with native Spanish re-sources. In: Proceedings of 16th European Biomass Conference and Exhibition, Valencia, Spain, pp. 1447–1450 (2008)
18. Rezeau, A., Díaz, M., Sebastián, F., Royo, J.: Operation and efficiencies of a new biomass burner when using pellets from herbaceous energy crops. In: Proceedings of 16th European Biomass Conference and Exhibition, Valencia, Spain, pp. 1447–1450 (2008)
19. Díaz-Ramírez, M., Sebastián, F., Royo, J., Rezeau, A.: Combustion requirements for conversion of ash-rich novel energy crops in a 250 kW_{th} multifuel grate fired system. Energy **46**, 636–643 (2012)
20. Díaz-Ramírez, M., Nogués, F.S., Royo, J., Rezeau, A.: Influence of bottom ash-rich energy crops characteristics on fixed grate combustion performance in: Impacts of fuel quality on power production and the environment. Puchberg, Austria (2012)
21. (CEN) E.C.f.S. CEN/TS 14775:2004 Solid biofuels. Method for the determination of ash content (2004)
22. Zeng, T.: National conditions-Austria. IEE/09/758/SI2.558286—MixBioPells. WP4/D 4.3 (2011)
23. Porteiro, J., Patiño, D., Collazo, J., Granada, E., Moran, J., Miguez, J.L.: Experimental analysis of the ignition front propagation of several biomass fuels in a fixed-bed combustor. Fuel **89**(1), 26–35 (2010)
24. Sippula, O., Hytönen, K., Tissari, J., Raunemaa, T., Jokiniemi, J.: Effect of wood fuel on the emissions from a top-feed pellet stove. Energy Fuels **21**(2), 1151–1160 (2007)
25. Houshfar, E., Løvås, T., Skreiberg, Ø.: Experimental investigation on NOx reduction by primary measures in biomass combustion: Straw, peat, sewage sludge, forest residues and wood pellets. Energies **5**(2), 270–290 (2012)
26. Tissari, J., Sippula, O., Kouki, J., Vuorio, K., Jokiniemi, J.: Fine particle and gas emissions from the combustion of agricultural fuels fired in a 20 kW burner. Energy Fuels **22**(3), 2033–2042 (2008)
27. Porteiro, J., Patiño, D., Moran, J., Granada, E.: Study of a fixed-bed biomass combustor: influential parameters on ignition front propagation using parametric analysis. Energy Fuels **24**(7), 3890–3897 (2010)

28. Nussbaumer, T.: Combustion and co-combustion of biomass: fundamentals, technologies, and primary measures for emission reduction. Energy Fuels **17**(6), 1510–1521 (2003)
29. Wiinikka, H., Gebart, R.: Critical parameters for particle emissions in small-scale fixed-bed combustion of wood pellets. Energy Fuels **18**(4), 897–907 (2004)
30. Shen, G.F., Yang, Y.F., Wang, W., Tao, S., Zhu, C., Min, Y.J., Xue, M.A., Ding, J.N., Wang, B., Wang, R., Shen, H.Z., Li, W., Wang, X.L., Russell, A.G.: Emission factors of particulate matter and elemental carbon for crop residues and coals burned in typical household stoves in China. Environ. Sci. Technol. **44**(18), 7157–7162 (2010)

Dual-Fuel (Natural Gas/Biodiesel) Engines: Fundamentals, Performance and Environmental Impact

Fernando José da Silva, Antonio Gilson Barbosa de Lima, Yoge Jerônimo Ramos da Costa, Celso Rosendo Bezerra Filho and Marcelo Bezerra Grilo

Abstract This chapter briefly focuses on the use of biodiesel and natural gas in dual-fuel diesel engines, providing information about natural gas and biodiesel, and their characteristics, engine performance and exhaust gas emissions. The theoretical and experimental study provides analysis of a commercial compression ignition engine operating in dual mode with three fuels: natural gas, diesel and biodiesel at different load and biodiesel blends. On the basis of the results, the effects of air conditions, and the type and quantity of fuel used over the engine performance and environmental impact are reported, and viability of natural gas and biodiesel as alternative fuels for diesel engines is verified.

F. J. da Silva
Department of Mechanical, Federal Institute of Alagoas (IFAL), Maceió, AL 57020-510, Brazil
e-mail: fernando.jose@ifal.edu.br

A. G. B. de Lima (✉) · Y. J. R. da Costa · C. R. B. Filho · M. B. Grilo
Department of Mechanical Engineering, Federal University of Campina Grande, Av. Aprígio Veloso, 882, Bodocongó, Campina Grande, PB 58429-900, Brazil
e-mail: gilson@dem.ufcg.edu.br

Y. J. R. da Costa
e-mail: yoge@dem.ufcg.edu.br

C. R. B. Filho
e-mail: celso@dem.ufcg.edu.br

M. B. Grilo
e-mail: grilo@dem.ufcg.edu.br

G. Ferreira (ed.), *Alternative Energies*, Advanced Structured Materials 34,
DOI: 10.1007/978-3-642-40680-5_3,

1 Introduction

Energy is defined as the ability to produce work and it can be found in different forms such as electrical, radiant, thermal, motion, sound, chemical, mechanical, nuclear and gravitational. Energy can be stored, converted and/or amplified depending on the application [1].

Energy sources can be fossil (petroleum, coal, betumes, natural gas, shale oil, etc.), renewable (alternative) (biomass, hydro, wind, solar, geothermal, marine, hydrogen, etc.) and fissile (uranium, thorium, etc.). Depending on the source different fuels can be obtained, for example, diesel and gasoline are derived from petroleum and biofuel (ethanol, methanol, biodiesel, hydrogen, methane, etc.) is produced from biomass. Biodiesel (vegetable oil) can be produced, from different agricultural products, for example, soybean, palm, coconut, sunflower seed, rapeseed, peanut kernel and cottonseed.

Globally, in view of different factors such as fast depletion of fuel resources, increasing consumption of fossil fuel, uncertainties with the future availability of fossil fuel, increasing world energy demand, energy and cost saving, increase in pollutants emissions (CO_2, HC, NO_x, SO_x, etc.) which affects respiration system and the nervous system of people, and producing skin diseases and climatic changes (greenhouse, heating Earth), many studies have been directed for investigations on using different sources of energy instead of oil and its derivatives [2–9]. Among them we can cite natural gas and biofuel (biodiesel).

Diesel fuel has been used for decades in engines for mechanical and/or electrical energy generation. In the scenarios of political considerations, energy and cost saving and reduction of pollutant emissions, the natural gas and biodiesel appears both as the best alternative fuels to be used in diesel engines because diesel fuel can be partially or totally replaced by them. Thus, appear the concept of dual-fuel engines.

2 Fuels

2.1 Natural Gas

2.1.1 What's Natural Gas?

Natural gas, in itself, might be considered a very uninteresting gas—it is colorless, shapeless, and odorless in its pure form. However, natural gas is combustible, and when burned it gives off a great deal of energy, but, unlike other fossil fuels, natural gas is clean and emits lower levels of pollutants into the air.

From the chemical viewpoint, natural gas consists of a mixture of inorganic gases and saturated hydrocarbons that occurs in gaseous state at environmental temperature and pressure. The composition of natural gas depends on the factors involved in the gas fields, production process, collects, conditioning and transport. Methane is the main constituent but it may also contain other hydrocarbons such as ethane, propane, butane, pentane, hexane, heptane and octane as well as non-hydrocarbon compounds such as traces of gaseous "impurities" and non combustibles [10, 11]. Typical analysis of natural gas in Brazil is given by: Methane (CH_4) 70–90 %; Ethane (C_2H_6), Propane (C_3H_8) and Butane (C_4H_{10}) 0–20 %; Carbon Dioxide (CO_2) 0–8 %; Oxygen (O_2) 0–0.2 %; Nitrogen (N_2) 0–5 %; Hydrogen sulphide (H_2S) 0–5 %; Rare gases (Ar, He, Ne, Xe) trace. Nitrogen, helium, carbon dioxide and trace amounts of hydrogen sulfide, water and odorants can also be present. Natural gas also contains and is the primary market source of helium. Mercury is also present in small amounts in natural gas extracted from some fields. Organosulfur compounds and hydrogen sulfide are common contaminants which must be removed prior to most uses. Gas with a significant amount of sulfur impurities, such as hydrogen sulfide, is termed sour gas; gas with sulfur or carbon dioxide impurities is acid gas.

Processed natural gas that is available to end-users is tasteless and odorless, however, before gas is distributed to end-users, it is odorized by adding small amounts of odorants (mixtures of t-butyl mercaptan, isopropyl mercaptanthiol, tetrahydrothiophene, dimethyl sulfide and other sulfur compounds), to assist in leak detection. Processed natural gas is, in itself, harmless to the human body; however, natural gas is a simple asphyxiant and can kill if it displaces air to the point where the oxygen content will not support life.

2.1.2 Why Natural Gas?

Natural gas is a vital component of the world's energy supply. It is one of the cleanest, safest, and most useful of all energy sources. According to BP Statistical Review of World Energy [12], world proved reserves of natural gas in 2011 were of the order of 208.4 trillion cubic meters. The major sources are located in Russia (21.4 %) and the Middle East (Iran, Qatar, Saudi Arabia, United Arab Emirates, and Iraq: 36.4 %). Brazilian proved reserves are about 500 billion cubic meters. In 2011, world production reached 3.27 trillion cubic meters per year, 3.1 % higher than in 2010. Russia (18.5 %) and the United States of America (20.0 %) were the largest producers.

The Brazilian agency ANP (National Agency of Petroleum, Natural Gas and Biofuels) reported that, in 2011, Brazil produced 24.1 billion cubic meters of natural gas [13]. In 2012, this production was 25.6 billion cubic meters. Brazil's largest natural gas production occurs in the Campos Basin, in the State of Rio de Janeiro, from offshore fields. Most onshore production occurs in the States of Amazonas and Bahia, although the natural gas produced is mostly for local consumption due to the lack of transportation infrastructure. However, it is hoped that several new transport infrastructure projects will reduce the problems associated with the increased production in these regions. Petrobras is the largest producer of natural gas in Brazil. The company reportedly controls over 90 % of Brazil's natural gas reserves, and it is also the largest wholesale supplier of natural gas.

Brazilian regulations allow each state to maintain a monopoly on natural gas distribution in their corresponding territory, but many states have begun to partially privatize the distribution companies. Natural gas has many residential, commercial and industrial uses. The industrial sector is the largest consumer of natural gas in Brazil, representing about 80 % of total domestic consumption. However, the two fastest growing sectors are thermal electricity generation and vehicular compressed natural gas (CNG).

Natural gas appear as an attractive energy source to be used as a fuel in internal combustion engines due to favorable physical and chemical properties, to be available in great quantities in many location of the world, and to have efficient and clean combustion (substantial reduction of emissions). Thus, it is evident the economical and environmental potential benefits of using natural gas fuel in diesel engine.

2.2 *Biodiesel*

2.2.1 What's Biodiesel?

Basically, from a chemical viewpoint, biodiesel (from Greek bio, life plus diesel from Rudolf Diesel) is a very safe, eco-friendly, alternative energy fuel made from renewable biological sources compound of a mixture of mono alkyl esters of fatty acids and derived from oleaginous plants oils (edible or non edible vegetable oils) or animal fats [1, 14]. An essential feature of a biodiesel fuel is that its fatty acid profile corresponds to that of its parent oil or fat. All vegetable oils and animal fats consist primarily of triglycerides (also known as triacylglycerols). Triglycerides have a three-carbon backbone with a long hydrocarbon chain attached to each of the carbons. Thus, biodiesel fuels derived from different feedstocks may have significantly different compositions and properties [6, 15]. It is made through a chemical reaction called transesterification that breaks the oil molecules (with or without catalyst) and transforms new or used oil into biodiesel plus glycerine [1, 14].

Biodiesel has been in use in many countries such as The United States of America, Malaysia, Indonesia, Brazil, Germany, France, Italy and other European countries. Globally, there are more than 350 oil-bearing crops identified as potential sources for biodiesel production. Currently, more than 95 % of the world biodiesel is produced from edible oils such as rapeseed (84 %), sunflower oil (13 %), palm oil (1 %), soybean oil and others (2 %).

Globally, there are many efforts to develop and improve vegetable oil properties in order to approximate the properties of diesel fuels. It has been remarked that high viscosity, low volatility and polyunsaturated characters are the mostly associated problems with crude vegetable oils [2].

Biodiesel is does not present a unique but a range of compound. The chemical formulae range is C_{14}–C_{24} methyl esters or $C_{15-25}H_{28-48}O_2$. One typical molecular formula of the biodiesel is $C_{18.74}H_{34.43}O_2$ [1, 16].

For identifying biodiesel we use the international nomenclature Bxx, where xx represent the volumetric concentration of biodiesel in the diesel/biodiesel blend. Then, B100 correspond to pure biodiesel and B20 correspond to blend with 20 % of biodiesel and 80 % of pure diesel (no biodiesel). In Brazil, B5 correspond to commercial diesel [1, 6].

2.2.2 Why Biodiesel?

Biodiesel is one of the most promising alternative fuels to be used in diesel engines because it is renewable and sustainable, highly biodegradable, minimal toxicity, it has similar chemical and physical properties to that of mineral diesel, high availability and relatively cost equivalent to diesel fuel. Further, biodiesel produces less pollutant to the environment when burned, it is less nocive to human health, and it provides better lubricity as compared to that of diesel fuel (reduced engine wear and friction). It does not contain carcinogens, such as poly-aromatic hydrocarbons and nitrons poly-aromatic hydrocarbons, it causes significant improvement of rural economic potential and promotes sustainable rural development, it can recycle the CO_2 from their combustion through photosynthetic ways, it has lower sulfur than the diesel fuel, the cetane number is similar or close to that of diesel fuel. Heating values of various vegetable oils are nearly 90 % to those of diesel fuel [1, 4, 6, 7, 9, 17–22]. On the other hand, as disadvantages biodiesel has high viscosity, high molecular weight, and low volatility. Thus, in some cases can lead to problems such as severe engine deposits (carbon), injection cooking, and piston ring sticking, clogged filters, especially in low temperature, and thus it affects engine durability [5, 7, 19, 21].

The majority of studies have found sharp reductions in exhaust emissions with biodiesel as compared to diesel fuel. The more accepted reasons in reduction of emissions particularly CO, CO_2, hydrocarbons, SO_2, particulates, and smoke can be attributed to the presence of sufficient oxygen in biodiesel. Biodiesel contains about 10 % oxygen while diesel has no oxygen content [17, 23]. Thus, biodiesel appears as an appropriated alternative fuel to be used in diesel engine.

3 Dual-Fuel Diesel Engine

The energy crises which the world faces, due to the petroleum price, has stimulated the use of alternative fuels to minimize the emission of toxic gases which cause great harm to human health and contributes to the increase of the greenhouse effect. Several researches have been done along the last few years with the objective of finding alternative fuels to reduce the level of pollution on Earth. Since then the search for renewable energy sources has been intense.

The researches which have been going on since 1980 have presented the biodiesel and natural gas as renewable sources of energy which may contribute to meet the energetic demand of countries all over the world. Studies also show that the use of biodiesel is very effective in reducing the exhaust gases. As far as the Brazilian energetic matrix is concerned both the biodiesel and the natural gas have significant importance, especially after the crisis caused by the electric energy shortage which occurred in Brazil in 2001.

Thermal engines are generally divided in two categories: compression-ignition and spark-ignition engines. In compression-ignition engines (diesel engines), air is compressed at pressures and temperatures at which the injected liquid fuel fires easily and burns progressively after ignition. When simultaneously more than one fuel is used in diesel engines, they are called as dual-fuel engines.

Dual fuel engines present different good attributes: (a) operate with more than one fuel source, (b) it has gained popularity because they reduce the amount of diesel fuel used, (c) they reduce pollutant emissions, thus providing improvement of air quality, (d) does not require modification in diesel engines (or require appropriate relatively simple conversion) to operate in dual mode, (e) full original power capacity (in dual or pure diesel modes), (f) Diesel pilot fuel provides lubrication to valves and rings, when combined with natural gas. Because of these advantages, dual-fuel diesel engines are becoming popular in many parts of the world.

Because of the important characteristics of natural gas and biodiesel reported in the literature, they are being used as alternative fuels in dual-fuel engine.

However, Natural gas combustion is characterized by a long ignition time delay and cannot be used directly as a fuel for an internal combustion diesel engine. Therefore, some type of ignition aid is required [24]. Then, today, dual-fuel engine can be operated with a mixture of air, diesel (or yet biodiesel) and natural gas that is compressed and then fired by ignition at the end of the compression phase. According to Mansour et al. [25] the advantage of this type of engine resides in the fact that it uses the difference of flammability of two fuels. The presence of the gaseous fuel influences both pre-ignition and post-ignition processes in a complex manner, depending mainly on the fuel used, its concentration and operating conditions [26].

The performance of dual-fuel engine has been investigated by many researches with promising results. These studies have been carried out to compare engine performance, combustion characteristics, and emissions when gas natural and pure diesel or blended biodiesel fuels are used instead of conventional petroleum diesel fuels.

Exhaust emissions and engine performance depend largely on the combustion, air turbulence, air–fuel mixture quality, injector pressure, actual start of combustion and many other factors that make test results vary from one engine to another. In addition, it can vary depending on the quality and origin of fuel as well as engine operating parameters like speed, load, engine design characteristics, etc. [17].

Sahoo et al. [8] report a review related to research carried out by various scientists about the effect of engine operation and design parameters (load, speed, compression ratio, pilot fuel injection timing, pilot fuel mass inducted, intake manifold conditions) and the type of gaseous fuel on the performance of dual fuel (gas-diesel) engines. Comment about performance, combustion and emission characteristics of different dual-fuel engines which use natural gas, biogas, produced gas, methane, liquefied petroleum gas, propane, etc., as gaseous fuel are presented. The authors reveal that thermal efficiency of dual-fuel engines improve either with increased engine speed, or with advanced injection timings, or with increased amount of pilot fuel.

Papagiannakis et al. [27] and [28] present studies about the effects of the engine parameters (total air-fuel ratio and inlet air temperature) on performance and emissions of dual diesel engines (natural gas and diesel). According to the authors, the increase of intake air temperature could be a promising solution for improving engine efficiency and reducing CO emissions. The use of natural gas as a supplement for liquid diesel fuel permits to control both NO and soot emissions on existing direct injection diesel engine, requiring only slight modifications of the engine structure.

Carlucci et al. [29] reports an experimental investigation and combustion analysis of a direct injections dual-fuel diesel-natural gas engine. In this research the effect of compressed natural gas (methane), and diesel fuel injection pressure and quantity of fuel injected during the pilot injection on the combustion development and engine performance (emissions and fuel consumption) were analyzed. It was verified that an analysis of the rate of heat release is not sufficient to explain the effect of each of the injection parameters on the pollutant emissions.

Nabi et al. [22] presents a theoretical and experimental investigation about pollutant emissions by using neat diesel and diesel-biodiesel blends in a four stroke naturally aspirated direct injection diesel engine. According to these authors, compared with conventional diesel fuel, when diesel-biodiesel blends are used lower carbon monoxide including smoke and higher NO_x emissions are found. With application of the technique EGR NO_x emissions decreased for both fuels.

Shahid and Jamal [9] reports a literature review of the use of biodiesel in compression ignition engines. According to the authors, experiments with raw biodiesel as fuel did not show the satisfactory results. They authors recommend the use of biodiesel mixed with diesel, in order, to reduce problem like injection coking and piston ring sticking. On the other hand, they authors reports that the lubricant properties of the biodiesel are better than diesel and it can help to increase the engine life. Besides, this fuel is environment friendly and produces much less pollutant emissions (NO_x, HC and zero SO_x) when compared to diesel. They recommend the use of B100 in urban area.

Research that discusses the influence of biodiesel (rapeseed oil) on the injection, spray, and diesel engine characteristics is presented by Kegl [21]. The focus is to reduce harmful emissions when compared with the use of mineral diesel as fuel in diesel engine. Thermal and physical properties such as viscosity, density, and surface tension, and sound velocity are determined and compared with those of mineral diesel. The results indicate that, by using biodiesel, harmful emissions (NO_x, CO, HC and smoke) can be reduced, ~25, ~25, ~30, and ~50 %, respectively, under optimized injection pump timing condition.

Murugesan et al. [7] presents a review study about the prospects and opportunities of the use of biodiesel as fuel in diesel engines. The authors have presented the advantages and disadvantages of the use of this fuel in diesel engine. According to these authors the use of biodiesel in a conventional diesel engine results in substantial reduction in unburned hydrocarbon, carbon monoxide, particulate matters pollutants emissions and oxide of nitrogen. Further, the sustainability of injection timing for diesel engine operation with biodiesel and blends are presented and discussed. The authors report that B20 blend is the best alternative fuel for diesel engine.

Misra and Murthy [30] conducted a review about the use of straight vegetable oils in compression ignition engine. According to the authors, many researchers state that straight vegetable oils in small percentage blend with diesel when used in lower capacity diesel engines have shown great promise with regards to the thermal performance as well as exhaust emissions. The authors has reported that lube oil dilution, high carbon deposits, ring sticking, scuffing of the engine liner, and injection nozzle failure are the major problems associated with direct use of straight vegetable oils in compression ignition engine.

Sidibé et al. [5] present a literature review about the use of crude filtered vegetable oil as fuel in diesel engines. Information about the type and quality, physical and chemical characteristic, and production parameters of these fuels are detailed and discussed. According to these authors the main differences in performance between straight vegetable oil and diesel oil are: approximate 10 % drop in power and deposits in the combustion chamber in direct-injection engines. The research has revealed that for the correct use of straight vegetable oil in diesel engine it is better to perform upstream adaptation or modification to the engine or inside the engine combustion chamber (piston modification).

From scientific literature, Fazal et al. [17] present a review related to biodiesel feasibility automobiles. Information about material compatibility (wear and corrosion) engine performance, pollutant emissions, and engine durability are presented. The authors have reported that the use of biodiesel as fuel reduced emissions and increased moving parts sticking, injector coking and filter plugging. Because of auto-oxidation, hygroscopic nature, higher electrical conductivity, polarity and solvency properties, the use of biodiesel in diesel engine can cause corrosion of metal and degradation of elastomers. On the other hand, higher concentration of oxygen in biodiesel improves lubricity and combustion, and reduces emissions while it slightly increases NO_x emissions.

Xue et al. [19] report a review about the use of biodiesel on engine performances and emissions. Factors such as engine power, engine economy, durability and regulated emissions (NO_x, CO, HC and CO_2) and non-regulated emissions (formaldehyde, acetaldehyde, benzene, toluene, xylene, etc.) are presented and analyzed. According to these authors the use of biodiesel will lead to the reduced engine power, but which can be accepted commonly, and biodiesel engine economy is affected by engine type and its operating conditions (load, speed, injection timing, injection pressure, etc.). They also reported that fuel consumption and NO_x emissions increase when biodiesel is used in diesel engine instead of mineral diesel.

No [31] presents a review study about inedible vegetable oils (Jatropha, karanja, mahua and linseed oils) and their derivatives (pure, blends, biodiesel blends, etc.) to be used as liquid alternative fuels in combustion ignition engine. According to this author biodiesel and its blends generally cause an increase in NO_x emissions and a decrease in HC, CO and PM compared to diesel. It was reported that a diesel engine without any modification would run successfully on a blend of 20 % vegetable oil and 80 % diesel fuel without damage of engine parts.

Misra and Murthy [32] relate in a review study different additives used to improve the cold flow properties of biodiesels and the performance of a diesel engine and its emissions while using additive blended biodiesels. According to the authors: (a) the low-temperature flow properties of biodiesel fuels are less favorable than diesel fuel, but when blended with additives like ethanol, kerosene, methanol, and orange oil, the cold flow performance is improved, and (b) ethanol seems to be a good additive as the power produced is comparable to the diesel engine operation and it has a lot of advantage over the biodiesel operation in respect of CO, HC, and NO_x pollutant emissions.

Atabani et al. [2] report an extended and comprehensive review of the use of biodiesel as an alternative energy resource and its characteristics. Information about the extraction methods, production technologies, physical and chemical properties, advantages and disadvantages and economical viability of biodiesel fuel is presented and discussed. The authors conclude that biodiesel can be more effective if used as a complement to other energy sources, and thus they recommend more research and technological development related to biodiesel.

Lacour et al. [3] conducted a study related to biogas production system. Topics about pollutant emissions, energy demands and production are presented and discussed, in order to produce methane for use in dual fuel engines. According to the authors for high loads, dual fuel efficiency is generally higher than diesel efficiency while for low loads, the dual fuel efficiency is lower due to difficulties in controlling gas combustion. The dual fuel engine efficiency ranged 17.5–28.5 % while diesel engine efficiency ranged 19.0–26.5 %. The energy substitution rate is around 83 % of diesel replaced by methane in an engine tractor with nominal power 110 kW.

Gupta et al. [20] report a review about biofuels for the use in gas turbine. Biofuel such as biodiesel, bio-ethanol, bio-methanol, pyrolysis oil, biogas, synthetic gas (dimethyl ether) and hydrogen are discussed. According to the authors based on properties and availability, the different biofuels can be used in gas turbine for power generation. Based on the pollutant emissions, the authors recommend the use of pure hydrogen in gas turbine because of its zero carbon emissions and low polluting characters.

From the reported works, we can see that different researches have been developed and recent studies have presented results which show the viability of the use of a mixture of diesel (or yet biodiesel) and natural gas, as the fuel for diesel engines. These facts have contributed for studies to be done using the three kinds of fuels: diesel, biodiesel and natural gas simultaneously, aiming at evaluating its potentiality in generating electric energy as well as to reduce the emissions of pollutants in the atmosphere. Thus, following, this chapter reports information about the emissions of exhaust gases in diesel engines operating with a mixture of diesel, biodiesel and natural gas.

4 Natural Gas and Biodiesel as Alternative Fuels to CI Engine

4.1 Experimental Procedure

In order to evaluate the performance and environmental impact of the dual-fuel engine, several experiments with natural gas and diesel, natural gas and biodiesel, and natural gas and biodiesel blends as fuels were performed. The experiments were realized using the following equipments: Electric-mechanical system (engine/electrical generator), different sensors to measure temperature, pressure and mass flow rate of liquid and gas fuels, data acquisition system and gas analyzer system, as illustrated in Figs. 1, 2, 3. More detail about these pieces of equipment can be found in the literature [33–35].

Fig. 1 Electric thermo-generation system

Fig. 2 Pressure and temperature sensors for air

Fig. 3 Pressure sensor for lubricating oil

The electromechanical system is composed of a 8.3L CUMMINS 6CTA turbo charger diesel-engine with mechanical power of 188 kW at 1,800 rpm, coupled with an Onan Genset electrical load bank of 150 kW.

In the operation of the diesel engine in the dual form it has been used the compressed natural gas in its primary form, and the mixture of diesel and biodiesel that is the complementary element which promotes the ignition and burn of the fuel. In the dual mode, the compressed natural gas is introduced in the combustion chamber, together with the air. Modification in the diesel engine was unnecessary.

The experiments were made using different engine load, with the diesel engine operating in rotation of 1,800 rpm and a replacement rate of 85 %. During the experiments we used cottonseed biodiesel supported by Centro de Tecnologias Estratégicas do Nordeste (CETENE), located in Caetes, Pernambuco State, Brazil. The cottonseed oil is abundantly produced in Brazil. The standard diesel and the natural gas were obtained in the fuel stations of Campina Grande, Paraiba State, Brazil. The natural gas used in the experiments, was coming from the Brazilian system of gas distribution, through the Companhia Paraibana de Gás—PBGás (Paraiba state, Brazil).

The experiments were made with a composition of fuels containing 85 % of natural gas and 15 % of a mixture of diesel and biodiesel. In the composition of the liquid mixture, the biodiesel had a participation, which varied from 10 % (B10) to 30 % (B30) in volume. It was also carried out experiments with standard diesel (0 % gas and 5 % biodiesel), under the same conditions of experimentation done when a mixture of diesel, biodiesel and natural gas, was used as fuel so that the results of the emissions could be compared. During the pilot test the engine was initialized and, after 15 min, the emissions (CO, CO_2, NO, NO_x, and SO_2) and other parameters were measured. This initial period of time is necessary to stabilize the engine.

According to the ANP (National Agency of Petroleum, Natural Gas and Biofuels, Brazil) [13] Standing no. 07/2008, since January 2010, biodiesel has been added to the diesel in the proportion of 5 % in volume. Then, in order to calculate the amount of biodiesel to be added to the diesel and determine the mixtures B10, B20 and B30, the following equation was used:

$$\Delta V_b = \frac{\frac{V_b(\%)}{100}\left(V_b^{\circ} + V_d^{\circ}\right) - V_b^{\circ}}{1 - \frac{V_b(\%)}{100}} \tag{1}$$

where:

ΔV_b is the volume of biodiesel to be added in the present mixture, to obtain the new mixture,

$V_b(\%)$ is the volume of biodiesel desired in the new mixture,

V_b^{o} is the volume of biodiesel of the present mixture,

V_d^{o} is the volume of diesel of the present mixture

Table 1 Basic composition of diesel and natural gas used

Fuel	Chemical composition (in volume)							Source
Diesel	$C_{12}H_{26}$ 98.53 %	S 1.47 %						[36]
Natural gas	CH_4 89.42 %	C_2H_6 7.24 %	C_3H_8 0.16 %	C_4H_{10} 0.18 %	N_2 1.27 %	CO_2 1.66 %	O_2 0.08 %	[37]

The chemical composition of the natural gas and diesel fuel used in the experiment are presented in Table 1. These values are representatives of typical commercial fuels supplied in Campina Grande City, Paraiba State, Brazil.

4.2 Results and Analysis

Emissions of pollutant depend on the quality of the fuel, the fuel consumption and the type of engine in study. Fuel consumption varies under different load conditions. Besides, with the same load and under different operation conditions (pure diesel and in dual mode), the specific consumption varies. Therefore, to compare pollutant emissions and engine performance is a difficult task. However, some results obtained may be compared with results supplied in the literature. All results have been obtained in stable operation (non-knocking) conditions. In the dual mode (Biodiesel blends plus natural gas), the exhaust gas temperature have changed from 350 to 478 °C.

Table 2 shows a sample of the collected data in the experiments.

The raw material significantly interferes in the final product, the biodiesel, as well as in the emissions of NO_x ($NO + NO_2$) originated when it is burned. So, one of the factors which contribute to increasing the nitrogen oxide emission rate is the degree of unsaturation of the biodiesel. The higher the level of unsaturation, there is reduction on the viscosity and on the amount of cetane, whose consequence is low quality fuel [15].

The nitrogen oxide, one of the main pollutants of the atmosphere, is released as the result of the combustion of any substance which contains nitrogen. These emissions are released into the atmosphere mainly by the industry and by internal combustion engines. In a research made with biodiesel, NO is the predominant element in the formation of NO_x and NO_2, it complements the composition with much smaller amounts [38]. The work done with pure diesel, B100 from palm oil and B20 [39], concluded that on average, biodiesel fuels tend to emit higher levels of NO emissions.

The formation of oxides of nitrogen is generally favored by the increased oxygen concentration at higher temperatures. Therefore, at lean fuel air ratios, when oxygen is available in abundance, the effect of temperature is expected to predominate. A slight decrease in the combustion temperature is thus expected with lean dual fuel operation, since the delay period is extended [26].

Table 2 Experimental data

Sample	Power (kW)	Inlet water temperature (°C)	Output water temperature (°C)	Oil lubricating temperature (°C)	Air pressure in the intake manifold (bar)	Ambient pressure (bar)	Motor speed (rpm)	Mass flow rate of diesel/biodiesel (kg/h)	Mass flow rate of air (kg/h)	Mass flow rate of gas (kg/h)	Ambient temperature (°C)	Oil fuel temperature (°C)
B5	40	65	82	n/a	1.41	0.95	1804	12.54	801.37	0.00	29.70	n/a
	60	68	85	n/a	1.60	0.95	1804	17.20	910.36	0.00	30.40	n/a
	80	70	86	n/a	1.82	0.95	1805	21.08	1031.79	0.00	30.60	n/a
	100	68	90	n/a	1.96	0.95	1792	25.07	749.09	0.00	30.90	n/a
B10 + Natural gas	40	62	82	62	1.45	0.95	1795	1.75	826.57	13.81	31.80	46.70
	60	64	84	66	1.59	0.95	1795	2.51	916.85	15.76	31.80	53.10
	80	65	86	70	1.73	0.95	1793	2.90	977.28	18.42	32.00	56.40
	100	66	86	76	1.89	0.95	1792	3.45	1079.16	19.85	32.10	59.80
B20 + Natural gas	40	62	81	59	1.47	0.95	1792	1.90	826.57	12.33	36.00	52.00
	60	62	83	62	1.61	0.95	1793	2.64	889.87	15.71	36.70	57.00
	80	63	86	71	1.74	0.95	1796	3.26	967.28	17.98	38.40	59.00
	100	64	86	77	1.86	0.95	1794	3.66	1020.59	20.37	39.80	61.00
B30 + Natural gas	40	61	79	59	1.45	0.95	1792	1.96	820.93	12.60	32.70	45.00
	60	62	81	61	1.63	0.95	1795	2.59	914.58	16.67	33.00	51.70
	80	62	83	67	1.77	0.95	1792	3.13	983.61	17.99	36.00	55.80
	100	61	86	75	2.04	0.95	1795	3.67	1153.88	20.74	32.40	56.00

n/a Not available

Figure 4 shows NO_x emissions as a function of engine load. From the analysis of this figure we can see that the mixture diesel, biodiesel and gas, presented pollutant emissions rates inferior to those of commercial diesel fuel. In all situations for increased power we have higher NO_x emission. Most of the studies reported that the use of biodiesel caused an increase on the emissions of NO_x when used in diesel engines [19]. On the other hand, in the dual mode, by using diesel and natural gas as fuel, experiments were done by Uma et al. [40] where the lowest rates in the emissions of NO_x were obtained, when compared to pure diesel. Since that 85 % of the fuel used in this research is natural gas, we can say that a similar result was obtained. By comparison among the biodiesel blends, we can see that B10 presented lower NO_x emissions.

Figure 5 illustrates the emissions of NO in the experiments done with the mixtures of diesel, biodiesel and natural gas. A better quality was verified in the emission rates in the mixtures biodiesel blends (B10, B20 and B30) and natural gas, when compared with the commercial diesel. This is due to the predominance of natural gas in the composition of the fuel and the low percentage of biodiesel used, which has provoked a reduction in the emissions of NO. There is a clear tendency of increase in the emissions as the power is increased.

Papagiannakis and Hountalas [41], using a four-cycle diesel engine, working at a speed from 1,000 to 3,000 rpm, and with varied loads (40, 60, and 80 % of the full engine load) reported a percentage of NO equal to 0.05 %, for a condition of 2,000 rpm, using 80 % of engine load and also diesel substitution rate of 80 %.

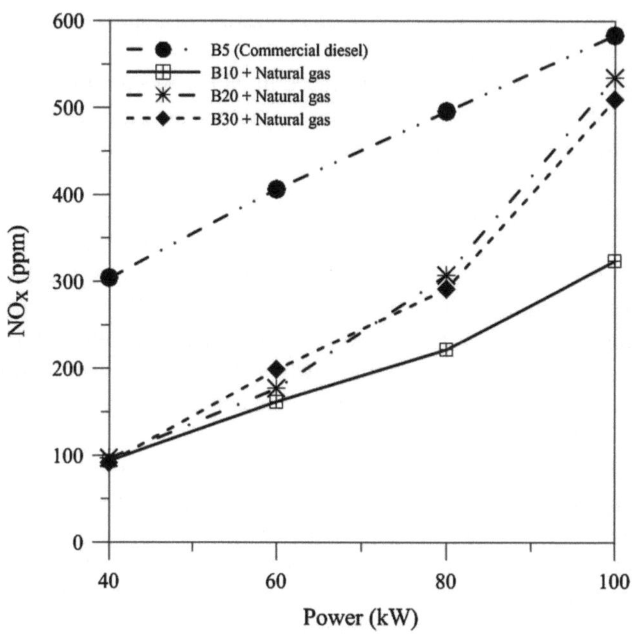

Fig. 4 NO_x emissions of a diesel engine (1,800 rpm) as a function of power

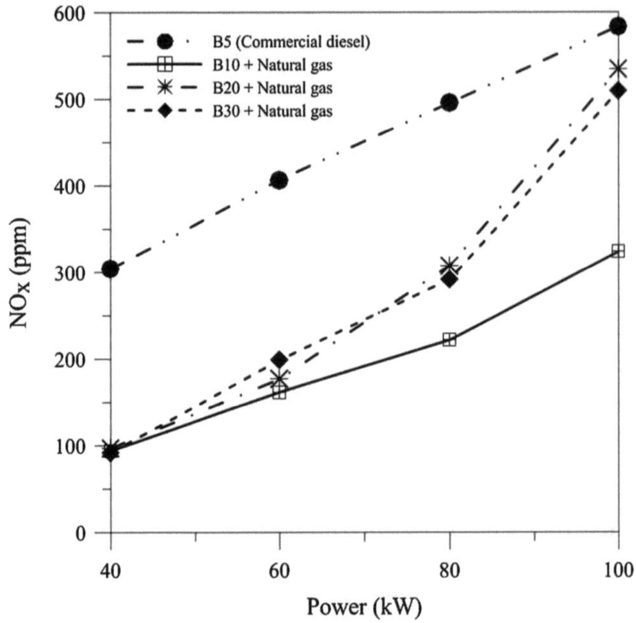

Fig. 5 NO emissions of a diesel engine (1,800 rpm) as a function of power

For a condition of engine speed 1,800 rpm and engine load 140 kW, Mansour et al. [25], report a NO_x ($NO + NO_2$) concentration of approximately 0.05 %, for an unknown substitution rate.

According to Karim [26], a low decrease in NO_x concentration is expected for low fuel-air ratio. With further enriching of the gas-air ratio, NO_x concentration increases with decreasing rates compared to diesel operation, until the effective flammability limit is reached.

Lapuerta et al. [42] carried out an experimental study to verify the influence of the degree of unsaturation of biodiesel fuels on NO_x and PM emissions from a direct-injection diesel engine. According to these authors as the biodiesel fuel became more unsaturated, NO_x emissions increased and PM emissions decreased.

Carbon monoxide (CO) comes mainly from an incomplete combustion. The greatest amount of fuel in the mixture air/fuel, known as a rich mixture, usually generates an incomplete combustion. When the combustion is complete, the CO is converted in CO_2 [43].

According to related research, it is a common tendency to reduce the emissions of CO, when the diesel is substituted by the biodiesel [19]. On the other hand, experiments done on the dual mode, using diesel and natural gas with a replacement rate varying from 67 to 86 %, has presented high levels of CO [34]. Figure 6 illustrates the emissions of CO obtained in each experiment done. It was verified that by using a mixture of biodiesel blends and natural gas, the CO emissions has

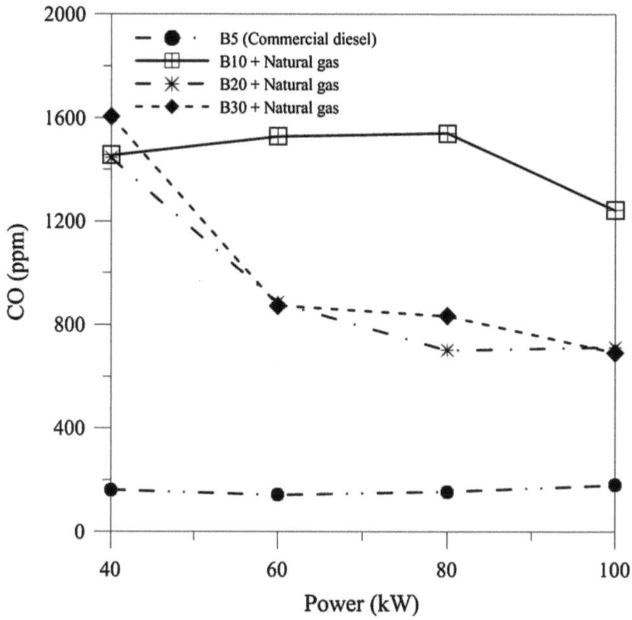

Fig. 6 CO emissions of a diesel engine (1,800 rpm) as a function of power

decreased as the power was increased. In all cases, the emissions were higher than the originated from engines operating with only commercial diesel (B5).

Shenghua et al. [44], testing a WD 615-64 super feed diesel engine, with maximum power of 164 kW at 2,200 rpm and specific consumption of 228 g/kWh, have reported that the CO emissions of the engine operating at 1,000 rpm increased when the methane concentration (gas) was higher, reaching values of approximately 0.2 %, for a percentage of gas equal to 84 %.

Henham and Makkar [45], testing a 2-cylinder, four-cycle Lister Peler LPWS2 engine, with indirect injection, have reported CO concentrations of approximately 0.35 % for a mixture with 55 % of methane and 45 % of diesel, and 0.04 % for the case of pure diesel, both at 2,000 rpm and torque 40 Nm.

The carbon dioxide is known as the main component of the emissions which negatively contributes to the climate alterations in the planet. It is originated mainly from the combustion of fossil fuels in industries and in the transportation system. The transportation system on roads contributes with a high rate of CO_2 emission, which is confirmed by different authors [46]. These authors also reported that the use of biodiesel results in less emission of CO_2, when compared to the commercial diesel.

With the experiments realized herein it has confirmed that the mixtures of B10, B20, B30 and natural gas have presented lower CO_2 emissions when compared to the emissions of commercial diesel, as can be seen on Fig. 7. The level of emissions increase as a more powerful engine is used.

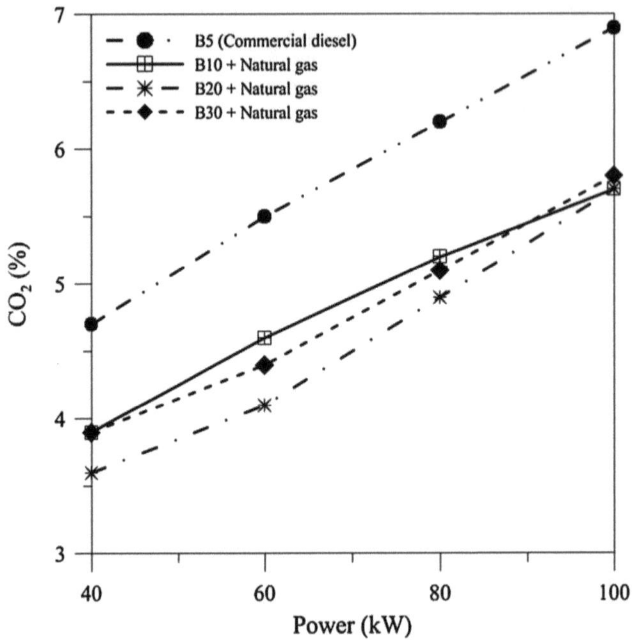

Fig. 7 CO_2 emissions of a diesel engine (1,800 rpm) as a function of power

Experiments made by Costa et al. [34], using diesel and natural gas, state that for the highest (86 %) and in the lowest (67 %) replacement rates, the CO_2 emissions have presented better quality (lower values) when compared to standard diesel fuel emissions.

This parameter provides some information on the quality of the combustion, indicating higher efficiency for the highest level of this component and a decrease in the level of CO emissions. Higher percentages of CO_2 in the exhaust gas suggest higher fuel oxidation rates at constant engine speed, more release of heat for power conversion and enhanced combustion as more CO is converted to CO_2.

Mansour et al. [25] report concentrations of $CO_2 \approx 4.5$ % at 1,800 rpm and 140 kW for unknown substitution rates.

The industrial processes are the main artificial sources responsible for the emissions of sulfur dioxide in the atmosphere which contributes directly to the greenhouse effect. According to Kegl [47] low amount of sulfur in the fuel composition contributes to the reduction of the wearing out of the engine components, besides minimizing the level of particulates and reducing the emissions of sulfur oxides released in the atmosphere. The releasing of SO_2 directly depends upon the percentage of sulfur content present in the fuel.

The emissions of SO_2 illustrated in Fig. 8, indicate that the mixtures of diesel, biodiesel and natural gas has presented inferior quality to that originated by

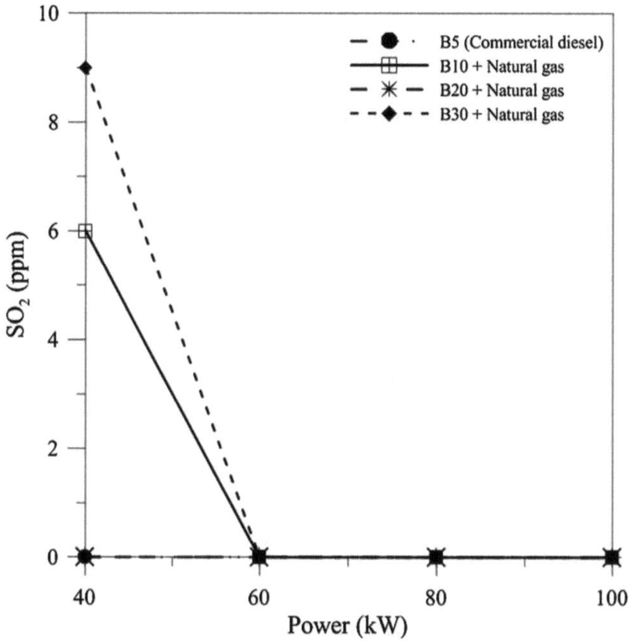

Fig. 8 SO_2 emissions of a diesel engine (1,800 rpm) as a function of power

commercial diesel (B5). Further, the levels of emissions decreased when increased power is used.

According to Uma et al. [40] there will occur a decrease in the level of emission of this pollutant in the dual mode, due to the expected decrease in the concentration of sulfur in the diesel-gas natural mixture.

The exhaust gas analysis of a dual fuel engine normally indicates that appreciable proportions of the fuel gas can survive the combustion processes. Hydrocarbon emissions increase due to several factors, including quenched, lean combustion, wall wetting, cold starting and poor mixture preparation [48]. Heywood [49] reports values of HC emissions (as C1) that can reach 3,000 ppm or 0.3 %, depending on the engine type and work conditions, in agreement with the approximate values of 0.2 % at 2,000 rpm, 80 % of load and 80 % of substitution rate reported by Papagiannakis and Hountalas [41], and Mansour et al. [25], for conditions of 1,800 rpm, 140 kW and unknown diesel substitution rate.

According to literature, the emissions of hydrocarbons reduce when the diesel is substituted by the biodiesel [46, 50]. In a research done by Costa et al. [34] using diesel fuel and natural gas, the C_xH_y emissions of a diesel engine using standard diesel presented lower rates than the emissions when a mixture of diesel and natural gas was used in the power of 40 and 80 kW.

5 Concluding Remarks

In this chapter, theoretical and experimental studies in a dual-fuel engine (natural gas, diesel and biodiesel) have been explored. Interest in this type of physical problem is motivated by its importance in many practical situations, especially to electrical energy generation. Here, the performance and emissions characteristics of a commercial diesel engine operating with natural gas, diesel and biodiesel injection were investigated, and the viability of the use of diesel engine to operate in dual mode with natural gas and biodiesel was verified.

Based on the results, the following conclusions were obtained: The engine was operated in a satisfactory form and a replacement rate of more than 85 % had been reached without presenting any abnormality such as detonation phenomenon; the emissions of NO_x ($NO + NO_2$), NO and CO_2 originated from the mixtures of B10, B20, B30 and natural gas, have presented a better quality when compared to commercial diesel (B5), having presented a tendency to increase as the power is increased, however the emissions of CO and SO_2 originated from the mixtures of B10, B20, B30 and natural gas, have presented inferior quality, when compared with commercial diesel (B5).

Acknowledgments The authors are grateful to Brazilian research agencies CNPq, CAPES, FINEP and FAPEAL for the financial support, to the CETENE (Brazil) for supplying the biodiesel, to the QUALITEX (Brazil) for provides the analysis of the fuels, to the authors of the references in this paper that helped in our understanding of this complex subject, and to the Editor by the opportunity given to present our research in this book.

References

1. Demirbas, A.: Biodiesel: A Realistic Fuel Alternative for Diesel Engines. Springer, London (2008)
2. Atabani, A.E., Silitonga, A.S., Badruddin, I.A., Mahlia, T.M.I., Masjuki, H.H., Mekhilef, S.: A comprehensive review on biodiesel as an alternative energy resource and its characteristics. Renew. Sust. Energ. Rev. **16**, 2070–2093 (2012)
3. Lacour, S., Chinese, T., Alkadee, D., Perilhon, C., Descombes, G.: Energy and environmental balance of biogas for dual-fuel mobile applications. Renew. Sust. Energ. Rev. **16**, 1745–1753 (2012)
4. Panwar, N.L., Kaushik, S.C., Kothari, S.: Role of renewable energy sources in environmental protection: A review. Renew. Sust. Energ. Rev. **15**, 1513–1524 (2011)
5. Sidibé, S.S., Blin, J., Vaitilingom, G., Azoumah, Y.: Use of crude filtered vegetable oil as a fuel in diesel engines state of the art: Literature review. Renew. Sust. Energ. Rev. **14**, 2748–2759 (2010)
6. Escobar, J.C., Lora, E.S., Venturini, O.J., Yañez, E.E., Castillo, E.F., Almazan, O.: Biofuels: Environment, technology and food security Renew. Sust. Energ. Rev. **13**, 1275–1287 (2009)
7. Murugesan, A., Umarani, C., Subramanian, R., Nedunchezhian, N.: Bio-diesel as an alternative fuel for diesel engines—A review. Renew. Sust. Energ. Rev. **13**, 653–662 (2009)
8. Sahoo, B.B., Sahoo, N., Saha, U.K.: Effect of parameters and type of gaseous fuel on the performance of dual-fuel gas diesel engines: A critical review. Renew. Sust. Energ. Rev. **13**, 1151–1184 (2009)

9. Ejaz, M., Shahid, Y.J.: A review of biodiesel as vehicular fuel. Renew. Sust. Energ. Rev. **12**, 2484–2494 (2008)
10. Speight, J.G.: Handbook of Petroleum Product Analysis. Wiley, New Jersey (2002)
11. Pedersen, K.S., Christensen, P.L.: Phase Behavior of Petroleum Reservoir Fluids. CRC Taylor & Francis, Boca Raton (2007)
12. BP Statistical review of World energy. http://www.bp.com/statisticalreview (2012). Accessed 10 Jan 2013
13. ANP—National Agency of Petroleum, Natural gas and Biofuels. http://www.anp.gov.br (2013). Accessed 10 Jan 2013
14. Purcella, G.: Do it Yourself Guide to Biodiesel. Ulysses Press, Berkeley (2008)
15. Benjumea, P., Agudelo, J.R., Agudelo, A.F.: Effect of the degree of unsaturation of biodiesel fuels on engine performance, combustion characteristics, and emissions. Energ. Fuels **25**, 77–85 (2011)
16. Canakci, M., Hosoz, M.: Energy and energy analyses of a diesel engine fuelled with various biodiesels. Energ. Sour. Part B **1**, 379–394 (2006)
17. Fazal, M.A., Haseeb, A.S.M.A., Masjuki, H.H.: Biodiesel feasibility study: An evaluation of material compatibility; performance; emission and engine durability Renew. Sust. Energ. Rev. **15**, 1314–1324 (2011)
18. Sekmen, P., Yilbaşi, Z.: Application of energy and energy analyses to a engine using biodiesel fuel. Math. Comput. Appl. **16**(4), 797–808 (2011)
19. Xue, J., Grift, T.E., Hansen, A.C.: Effect of biodiesel on engine performances and emissions. Renew. Sust. Energ. Rev. **15**, 1098–1116 (2011)
20. Gupta, K.K., Rehman, A., Sarviya, R.M.: Bio-fuels for the gas turbine: A review Renew. Sust. Energ. Rev. **14**, 2946–2955 (2010)
21. Kegl, B.: Effects of biodiesel on emissions of a bus diesel engine. Bioresour. Technol. **99**, 863–873 (2008)
22. Nabi, M.N., Akhter, M.S., Zaglul Shahadat, M.M.: Improvement of engine emissions with conventional diesel fuel and diesel–biodiesel blends. Bioresour. Technol. **97**, 372–378 (2006)
23. Lapuerta, M., Armas, O., Rodnguez-Fernandez, J.: Effect of biodiesel fuels on diesel engine emissions. Prog. Energ. Comb. Sci. **34**, 198–223 (2008)
24. Mbarawa, M., Milton, B.E., Casey, T.C.: Experimental and modeling of natural gas combustion ignited by a pilot diesel fuel spray. Inter. J. Therm. Sci. **40**, 927–936 (2001)
25. Mansour, C., Abdelhamid, B., Abdelkader, A., Françoise, G.: Gas-diesel (dual-fuel) modeling in diesel engine environment. Inter. J. Therm. Sci. **40**, 409–424 (2001)
26. Karim, G.A.A.: Review of combustion processes in the dual engine—The gas diesel engine. Progr. Energ. Comb. Sci. **6**, 277–285 (1980)
27. Papagiannakis, R.G., Rakopoulos, C.D., Hountalas, D.T., Rakopoulos, D.C.: Emission characteristics of high speed, dual fuel, compression ignition engine operating in a wide range of natural gas/diesel fuel proportions. Fuel **89**, 1137–1406 (2010)
28. Papagiannakis, R.G., Kotsiopoulos, P.N., Zannis, T.C., Yfantis, E.A., Hountalas, D.T., Rakopoulos, C.D.: Theoretical study of the effects of engine parameters on performance and emissions of a pilot ignited natural gas diesel engine. Energy **35**, 1129–1138 (2010)
29. Carlucci, A.P., de Risi, A., Laforgia, D., Naccarato, F.: Experimental investigation and combustion analysis of a direct injection dual-fuel diesel-natural gas engine. Energy **33**, 256–263 (2008)
30. Misra, R.D., Murthy, M.S.: Straight vegetable oils usage in a compression ignition engine—A review. Renew. Sust. Energ. Rev. **14**, 3005–3013 (2010)
31. No, S.-Y.: Inedible vegetable oils and their derivatives for alternative diesel fuels in CI engines: A review. Renew. Sust. Energ. Rev. **15**, 131–149 (2011)
32. Misra, R.D., Murthy, M.S.: Blending of additives with biodiesels to improve the cold flow properties, combustion and emission performance in a compression ignition engine—A review. Renew. Sust. Energ. Rev. **15**, 2413–2422 (2011)

33. Costa, Y. J. R.: Energetic and exergetic analyzes of internal combustion engine operating with diesel and natural gas mixture. Ph.D. Thesis, Brazil: Process Engineering, Center of Sciences and Technology, Federal University of Campina Grande (2007) (In Portuguese)
34. Costa, Y.J.R., Lima, A.G.B., Grilo, M.B., Bezerra Filho, C.R., Lima, A.M.N.: Exhaust emissions characteristics: An experimental study on diesel engine operated with mixtures of diesel and natural gas. Bra. J. Pet. Gas **2**(1), 36–44 (2008)
35. Costa, Y.J.R., Lima, A.G.B., Bezerra Filho, C.R., Lima, L.A.: Energetic and exergetic analyses of a dual-fuel diesel engine. Renew. Sust. Energ. Rev. **16**, 4651–4660 (2012)
36. Medeiros, M. A. O., Araújo, A. S., Fernandes, N. S.: Comparative study of the physicochemical properties of diesel fuel in the states of Paraiba and Rio Grande do Norte. In: 2° Brazilian Congress de R&D in Oil and Gas. Rio de Janeiro (2003) (In Portuguese)
37. PBGAS. Report of the properties of natural gas distributed in state of Paraiba, Brazil. http://www.pbgas.pb.gov.br (2006). Accessed 15 Dec 2006 (In Portuguese)
38. Hoekman, S.K., Robbins, C.: Review of the effects of biodiesel on NO_x emissions. Fuel Process. Technol. **96**, 237–249 (2012)
39. Song, H., Tompkins, B.T., Bittle, J.A., Jacobs, T.J.: Comparisons of NO emissions and soot concentrations from biodiesel-fuelled diesel engine. Fuel **96**, 446–453 (2012)
40. Uma, R., Kandpal, T.C., Kishore, V.V.N.: Emission characteristics of an electricity generation system in diesel alone and dual fuel modes. Biomass Bioenerg. **27**, 195–203 (2004)
41. Papagiannakis, R.G., Hountalas, D.T.: Experimental investigation concerning the effect of natural gas percentage on performance and emissions of a dual fuel diesel engine. Appl. Therm. Eng. **23**, 353–365 (2003)
42. Lapuerta, M., Armas, O., Rodrıguez-Fernandez, J.: Effect of the degree of unsaturation of biodiesel fuels on NO_x and particulate emissions. SAE Int. J. Fuels Lubrif. **1**, 1150–1158 (2009)
43. Nabi, M.N., Rahman, M.M., Akhter, M.S.: Biodiesel from cotton seed oil and its effect on engine performance and exhaust emissions. Appl. Therm. Eng. **29**, 2265–2270 (2009)
44. Shenghua, L., Longbao, Z., Ziyan, W., Jiang, R.: Combustion characteristics of compressed natural gas/diesel dual-fuel turbocharged compressed ignition engine. Proc. Inst. Mech. Eng. Part D: J. Automob. Eng. **217**(9), 833–838 (2003)
45. Henham, A., Makkar, M.K.: Combustion of simulated biogas in a dual-fuel diesel engine. Energ. Convers. Manag. **39**, 2001–2009 (1998)
46. Ozsezen, A.N., Canakci, M., Turkcan, A., Sayin, C.: Performance and combustion characteristics of a DI diesel engine fueled with waste palm oil and canola oil methyl esters. Fuel **88**, 629–636 (2009)
47. Kegl, B.: Influence of biodiesel on engine combustion and emission characteristics. Appl. Energ. **88**, 1803–1812 (2011)
48. Nwafor, O.M.I., Rice, G.: Combustion characteristics and performance of natural gas in high speed indirect injection diesel engine. Renew Energ **5**, 841–848 (1994). Part I
49. Heywood, J.B.: Internal Combustion Engine Fundamentals. McGraw-Hill, New York (1998)
50. Godiganur, S., Murthy, C.S., Reddy, R.P.: Performance and emission characteristics of a Kirloskar HA394 diesel engine operated on fish oil methyl esters. Renew. Energ. **35**, 355–359 (2010)

Thermoeconomic Evaluation of Biomass Conversion Systems

Sergio Usón, Wojciech J. Kostowski and Jacek Kalina

Abstract The chapter presents the application of thermoeconomics to the analysis of a sample biomass energy conversion system. Thermoeconomics is aimed at the analysis, optimization and diagnosis of energy conversion systems; it is based on the concept of exergy which is constantly gaining more popularity among engineers and researchers, and the concept of cost. The case study chosen for demonstrating the usefulness of the methodology concerns an integrated biomass gasification and utilization system with an atmospheric fluidized bed gasification unit and a gas turbine unit fed with the produced gas. Definition of fuel and product flows allows one to identify key components responsible for conversion of the largest amounts of exergy. Further concepts of thermoeconomics, such as exergy cost and its formation process allow one to determine the quality of each component of the plant. Attention is paid to the explanation of methodology rather than to particular technological issues, which are explained in the cited references, and are also subject of other chapters of this book.

1 Introduction

Thermoeconomics (a term coined by Tribus and Evans [1]) is a method of analysis and optimization of energy systems in which the flow of *energy* in industrial systems is traced from the perspective of the 2nd law of thermodynamics. From the 1st

S. Usón (✉)
CIRCE—Universidad de Zaragoza, Mariano Esquillor Street 15, 50018 Zaragoza, Spain
e-mail: suson@unizar.es

W. J. Kostowski · J. Kalina
Institute of Thermal Technology, Silesian University of Technology, Konarskiego Street 22, 44100 Gliwice, Poland
e-mail: wojciech.kostowski@polsl.pl

J. Kalina
e-mail: jacek.kalina@polsl.pl

G. Ferreira (ed.), *Alternative Energies*, Advanced Structured Materials 34,
DOI: 10.1007/978-3-642-40680-5_4,

law of thermodynamics we learn that energy is conserved and can neither be created nor destroyed, it can only be converted from one form into another. From the 2nd law of thermodynamics we learn that not all directions of energy conversion are possible, for instance, a friction force applied to a moving object converts its kinetic energy into heat, while a reverse process in not possible. This observation suggests that kinetic energy is a more valuable form of energy than heat; in other words, the 2nd law introduces a concept of quality, while the 1st law only deals with energy quantity. Besides exergy, thermoeconomic analysis is based on the concepts of purpose and efficiency, that leads to the idea of cost.

In the first section of the chapter, fundamentals of exergy analysis are presented. Afterwards, an example of renewable energy conversion systems is presented and exergy of its flows are obtained. Then, thermoeconomic methodology is presented and, finally, applied to this example.

2 Exergy Analysis

In common knowledge, the term *energy* is used in a somewhat confusing way. On the one hand, it is known that energy is conserved: it cannot be destroyed, but can be transformed from one form into another. On the other hand, the term *energy sources* is commonly used, although it is not correct from the point of view of physics. Our intuition tells us that whenever we speak about *energy sources*, we are actually thinking about sources of *usable energy*, i.e. about sources that are practically applicable to meet the consumers' demand for electricity, heat and cold.

The term *usable energy sources* can be defined more precisely by means of a special function *exergy*, defined by Rant [2] and further developed by Szargut [3]. The word exergy is of hybrid origin and derives from the latin *ex* and the greek *ergon*, that means *from the work*. Exergy is expressed in units of energy and has the following features:

1. it tells us what happens with the work supplied at the beginning of the process (this feature explains the name of the function);
2. it tells us, how much work can a given substance perform while brought to the equilibrium with the environment;
3. in each irreversible process (e.g. friction), part of the exergy is lost; therefore exergy does not meet the conservation law;
4. it allows us to determine the quality of a process; the higher exergy loss, the worse is the process;

From the second feature results that in contrary to a popular belief, actually *exergy*, not energy, determines the *ability to generate work*. A substance can carry a large amount of energy, but be unable to generate work: this is the case of heat rejected in a condenser in a thermal power plant. On the other hand, a substance can have low energy, but be able to generate work: this is the case of compressed air in a compressed air energy storage unit. In processes realized below ambient temperature, exergy shows the actual direction of processes: in a domestic refrigerator, exergy is supplied with electricity, is converted into exergy of a cold working fluid, and finally, it is supplied to the refrigerated space.

Rules for calculating exergy have been presented by Szargut [4] and for the most common components can be summarized as follows:

Exergy of mechanical or electric energy equals that energy:

$$B_{el} = E_{el} \quad \text{and} \quad B_m = E_m \tag{1}$$

Exergy of heat is calculated as a maximum work that could be performed by the Carnot cycle at the given temperature level:

$$B = Q\left(1 - \frac{T_0}{T}\right) \tag{2}$$

where T_0 is the temperature of the environment, and T is the temperature of the body rejecting heat. **Physical exergy** of a substance (per unit mass) is calculated as

$$b = h - h_0 - T_0(s - s_0) \tag{3}$$

where h is enthalpy and s is entropy of the substance, both depending on its thermal parameters.

Chemical exergy of a substance results from the difference in its composition with relation to the composition of a reference environment. The chemical exergy of fuels is determined based on the following concepts:

1. *Reference species for a chemical element*—substances containing the given element which most commonly appears in the environment. Each element has one reference species. For example, the reference species for carbon, nitrogen, oxygen and hydrogen are CO_2, N_2, O_2 and H_2O respectively, since these substances commonly appear in the Earth's crust and atmosphere;
2. *Standard chemical reaction for a substance*—a chemical reaction involving only reference species and the considered substance; frequently it is the reaction of oxidation;
3. *The standard chemical exergy of a substance*—result of exergy balance of a standard chemical reaction for that substance (see Eq. 2.7 in [4] for details).

We should underline the difference in methodology of determining chemical energy and chemical exergy of fuels. Chemical energy of fuels lower heating value LHV is determined experimentally and the method does not require any knowledge on fuel composition. On the contrary, chemical exergy is determined for each component separately based on tables of the standard chemical exergy. For solid organic fuels, calculation of chemical exergy is difficult, since these fuels contain many complex and partially unknown components. Therefore an approximate method proposed by Styrylska and Szargut, based on analogy with the chemical exergy of pure organic substances, is used. Results of the approximate method for common fuels are given in Table 1.

Table 1 Ratio of chemical exergy to the lower heating value calculated by the approximate method of Styrylska and Szargut [4]

Fuel	b_{ch}/LHV
Hard coal	1.09
Lignite	1.17
Coke	1.06
Wood	1.15
Liquid hydrocarbon fuels	1.07
Natural gas (high methane)	1.04
Coke-oven gas	1.00
Blast furnace gas	0.98

As can be concluded from Table 1, the chemical exergy is close to the lower heating value, but in general is not equal to it.

In the case of complete combustion, the reaction products (exhaust gases) contain only the reference species. Their chemical exergy is relatively low and results only from the change of the components' concentration compared to the environment:

$$b_{ch} = T_0 R \sum_{i=1}^{N} x_i \ln \frac{x_i}{x_{i,0}} \tag{4}$$

Chemical exergy of non-combustible products was calculated by considering ideal theoretical machines converting the mentioned change in concentration into work [5].

3 Biomass Plant Case Study

The presented case study concerns a Biomass Integrated Gasification Combined Heat and Power Plant. This chapter deals with a chosen plant configuration among several possible options for integrating various types of gasifiers with various types of gas turbine plants, analyzed by Kalina [6]. Therefore, it should be treated as an example for illustrating the thermoeconomic methodology, and not as a representation of a typical plant. Scheme of plant is presented in Fig. 1.

In the following description references to Fig. 1 are made in the following form: flux No. *n* is denoted as F.*n*, and devices are denoted with numbers and abbreviations listed in Table 3.

Raw biomass (spruce wood in form of chips) consists of 60 % dry matter and 40 % water. The dry mater consists of 98.5 % organic combustible substance (51.2 % carbon C, 6.1 % hydrogen H, 40.9 % oxygen O, 0.3 % of nitrogen N) and 1.5 % ash. Wet biomass (F.1) is supplied to the drying unit (11.BDU), where it is dried to the level of 10 % water content (F.2). Subsequently, it is supplied to atmospheric bed gasifier (12.AFB), fed with air (F36) under pressure of 150 kPa (abs) and temperature 374 °C. Raw gas leaving the gasifier (F.3) contains combustible components: hydrogen, carbon monoxide, methane as well as inert components

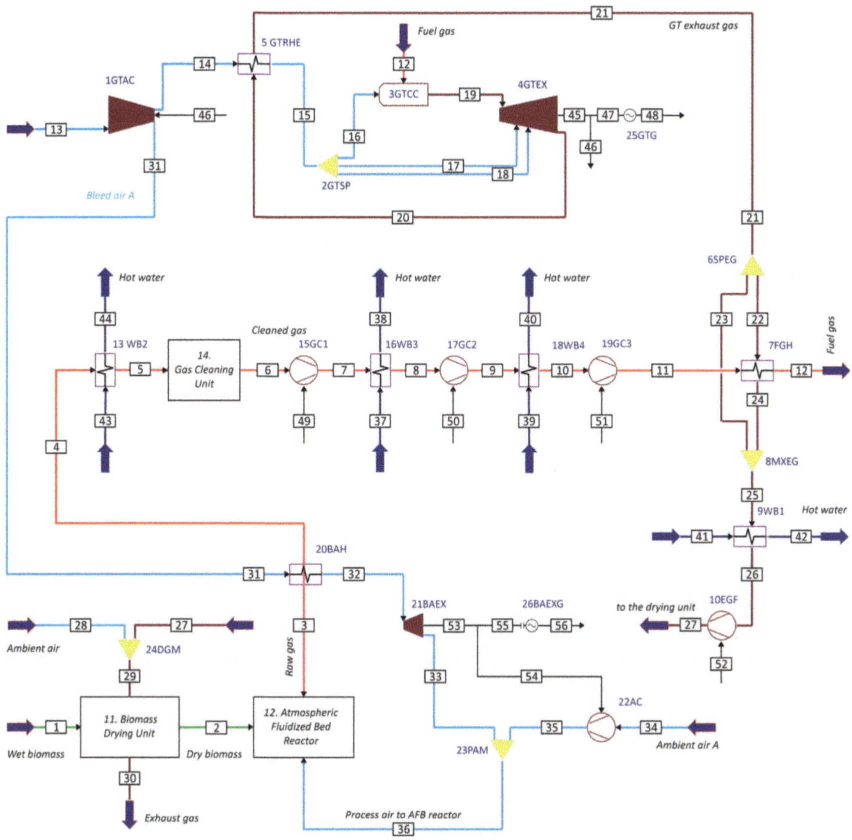

Fig. 1 Scheme of the AFB biomass gasification plant integrated with Mercury TM50 recuperated gas turbine

(see Table 2). The gas is then supplied to the cleaning unit (14.GCU) for removal of solid components. Solid products leaving the gasifier and the cleaning unit comprise ash and unconverted char. Details concerning the modelling of the gasification process are given in [7]. Table 2 presents predicted characteristics of raw and product gas from the considered technology.

The cleaned gas (F.6) is supplied to the gas turbine through a series of compressors (D.15, D.17 and D.19) which raise its pressure to 1.18 MPa. Inlet enthalpy of the raw gas (F.4) as well as heat rejected between compression stages is used to produce hot water (fluxes 43–44, 37–38, 39–40). Finally, the compressed cleaned gas is heated in a heat recovery heat exchanger (D.7) and supplied to the combustion chamber of the gas turbine (D.3).

Gas turbine exhaust gas (F.20) is used in a regenerative heat exchanger (D.5) to heat the main flux of air supplied to the combustion chamber. Subsequently, it is used to heat the fuel gas (as mentioned above), and to generate hot water (F41–42). Eventually, it is supplied by a fan (10.EGF) to the biomass drying unit (11.BDU).

Table 2 Predicted characteristics of raw product gas and gas turbine fuel gas for the atmospheric fluidized bed gasification technology [6]

Molar share, %	Raw gas	Cleaned (product) gas
CO	15.96	17.09
H_2	12.00	12.85
CH_4	3.10	3.31
C_2H_4	0.64	0.69
CO_2	11.78	12.62
H_2O	11.91	5.91
N_2	43.82	46.93
Ar	0.56	0.60
Tars (kg/kg)	0.107	0.00
LHV, kJ/m_n^3	5,009	5,139

The BDU is also partially fed with ambient air to reduce the partial pressure of water.

Part of the flux of air compressed in the gas turbine module is extracted (bleed air, F.31) and supplied to the AFB reactor. This solution allows one to increase the temperature of air for the gasification process. The mechanical exergy of bleed air bound to its pressure (974 kPa) is utilized in the bleed air expander 21.BAEX, which in turn drives the ambient air compressor 22.AC. Excess energy drives the generator BAEXG.

A list of system devices is set in Table 3, and the parameters of fluxes are listed in Table 4. Fluxes of mechanical and electric energy are listed in Table 5.

4 Thermoeconomic Analysis

Thermoeconomic analysis is based on the combination on the concept of exergy (2nd law of thermodynamics) with the concept of cost (economics), and aims at the analysis, optimization and diagnosis of energy conversion systems and, in general, of energy and materials intensive processes. Thermoeconomics [1] goes further than exergy analysis by introducing the concept of purpose through the definition of efficiency. Devices within a man designed installation use resources (fuel, F) to produce useful effects (products, P). Accordingly, if purpose is introduced, exergy balance becomes:

$$F = P + I + L \tag{5}$$

The previous equation states that part of the resources are transformed into useful flows and another part is either destroyed within component boundary (irreversibility, I) or crosses this boundary and is dissipated in the environment (losses, L). All terms in the previous equations are expressed in exergy. At this point it should be commented the interest of the analysis of wastes [8], that will not

Table 3 Components of the considered AFB gasification plant. GT—Gas turbine

ID	Symbol	Description
1	GTAC	GT Air Compressor
2	GTSP	GT air Splitter
3	GTCC	GT Combustion Chamber
4	GTEX	GT EXpander
5	GTRHE	GT Regenerative Heat Exchanger
6	SPEG	Splitter of GT Exhaust Gas
7	FGH	Fuel Gas Heater
8	MXEG	MiXer for GT Exhaust Gas
9	WB1	Water Boiler 1
10	EGF	Exhaust Gas Fan
11	BDU	Biomass Drying Unit
12	AFB	Atmospheric Fluidized Bed reactor
13	WB2	Water Boiler 2
14	GCU	Gas Cleaning Unit
15	GC1	Gas Compressor 1
16	WB3	Water Boiler 3
17	GC2	Gas Compressor 2
18	WB4	Water Boiler 4
19	GC3	Gas Compressor 3
20	BAH	Bleed Air Heater
21	BAEX	Bleed Air EXpander
22	PAC	Process Air Compressor
23	AM	Process Air Mixer
24	DGM	Drying Gas Mixer
25	GTG	GT Generator
26	BAEXG	Generator driven by BAEX

be considered in the formulation presented here. It is important to make clear that neither fuel corresponds necessarily to input nor product is always output. For instance, a turbine uses expansion of steam to produce power: accordingly, its fuel is the exergy drop of the steam (input minus output) and its product is the actual produced power. Besides, a heat exchanger uses the enthalpy drop of a hot stream to increase the temperature of a cold flow: for this reason, its fuel is the exergy drop of the hot stream whereas its product is the exergy increment of the cold one.

Efficiency (η) of a component is its ability to transform resources (fuel) into useful streams (products); accordingly, it is defined as the quotient between these terms:

$$\eta = \frac{P}{F} \tag{6}$$

Since exergy is not conserved, this magnitude decreases through the productive process due to irreversibility and losses. For this reason, for each energy unit of useful product leaving the system, more than one unit of exergy resources are

Table 4 Thermal parameters and exergy of fluxes of the considered AFB gasification plant

ID	Description	p (kPa)	T (°C)	$\bar{m}$ (kg/s)	$\bar{B}_{ph}$ (kW)	$\bar{B}_{ch}$ (kW)	$\bar{B}$ (kW)
1	Wet biomass	–	15	1.56	0	20,234	20,234
2	Dry biomass	–	75	1.04	0	20,208	20,208
3	Raw gas	140	920	2.79	2,056	12,034	14,090
4	Raw gas	140	737	2.79	1,481	12,034	13,515
5	Raw gas	140	75	2.79	104	12,034	12,138
6	Clean gas	125	40	2.64	55	12,399	12,454
7	Clean gas	264	126	2.64	291	12,399	12,690
8	Clean gas	264	40	2.64	239	12,399	12,638
9	Clean gas	558	77	2.64	439	12,399	12,838
10	Clean gas	558	40	2.64	424	12,399	12,823
11	Clean gas	1,180	81	2.64	625	12,399	13,024
12	Clean gas	1,180	300	2.64	894	12,399	13,294
13	Air	101	15	16.65	0	0	0
14	Air	974	308	14.87	4,172	0	4,172
15	Air	944	634	14.87	7,322	0	7,322
16	Air	944	634	12.57	6,190	0	6,190
17	Air	944	634	1.37	675	0	675
18	Air	944	634	0.93	458	0	458
19	Comb gas	916	1,208	15.25	15,813	471	16,284
20	Exhaust gas	108	655	17.55	6,000	447	6,447
21	Exhaust gas	98	389	17.55	2,477	447	2,924
22	Exhaust gas	98	389	5.00	706	127	833
23	Exhaust gas	98	389	12.55	1,771	320	2,091
24	Exhaust gas	98	239	5.00	291	127	419
25	Exhaust gas	98	347	17.55	2,026	447	2,474
26	Exhaust gas	98	150	17.55	394	447	841
27	Exhaust gas	101	155	17.55	472	447	919
28	Air	101	15	2.58	0	0	0
29	Drying gas	101	140	20.13	441	421	862
30	Drying gas wet	101	68	20.65	93	432	525
31	Air	974	308	1.78	499	0	499
32	Air	974	700	1.78	974	0	974
33	Air	150	400	1.78	318	0	318
34	Air	101	15	0.15	0	0	0
35	Air	150	58	0.15	5	0	5
36	Air	150	374	1.93	315	0	315
37	Water	500	30	10.00	20	0	20
38	Water	500	40	10.00	47	0	47
39	Water	500	30	10.00	20	0	20
40	Water	500	38	10.00	39	0	39
41	Water	1,000	50	10.00	91	0	91
42	Water	1,000	137	10.00	861	0	861
43	Water	1,000	50	8.00	73	0	73
44	Water	1,000	124	8.00	567	0	567

Table 5 Fluxes of mechanical or electric exergy of the AFB plant mechanical exergy equals energy

ID	Type	$\bar{E} = \bar{B}$ (kW)
45	Mechanical	10,390
46	Mechanical	5,012
47	Mechanical	5,378
48	Electric	5,146
49	Electric*	278.4
50	Electric*	257.9
51	Electric*	255.8
52	Electric*	89.51
53	Mechanical	592.5
54	Mechanical	6.66
55	Mechanical	585.8
56	Electric	556.5

*Indicates that conversion to mechanical energy is included in the corresponding compressor

needed. This leads to the concept of exergy cost [9], which are the units of exergy needed for producing a given flow. This concept represents the connection between thermodynamics and economics. A similar concept is the cumulative exergy consumption, developed by Szargut and Morris [10]. If exergy is used to assess economic costs of the flows of a given plant (considering fuel, investment, and operation and maintenance), exergoeconomic costs are obtained [11, 12].

Based on the Exergy Cost Theory and a compact matrix formulation, Torres developed Symbolic Thermoeconomics, which is the methodology used here and that will be explained in next section [13]. The same author extended the methodology to consider waste flows [8], but this formulation will not be presented here. Some other relevant publications on thermoeconomic methodologies are [14–19].

4.1 Symbolic Thermoeconomics Fundamentals

The first step in the application of the methodology is to define the productive structure. Components of this structure are not necessarily the same as those of the physical structure, and are connected among them and with the environment by exergy flows. These flows are defined according to productive purposes and, thus, they are not necessarily physical flows (they are sometimes combination of them).

Components of productive structure are numbered from 1 to n, whereas number 0 corresponds to the environment. The part of the product (P) of component i that becomes part of the fuel (F) of component j is represented as E_{ij}. Accordingly:

$$F_i = \sum_{j=0}^{n} E_{ji} \tag{7}$$

$$P_i = \sum_{j=0}^{n} E_{ij} \tag{8}$$

It should be noted that summations in equations above include component 0 because there are also productive flows connected with the environment.

Two possible representations of the system can be considered: from product back to fuel (PF) or from fuel to product (FP).

PF representation is based on the unit exergy consumptions, κ_{ij}, defined as the number of units of exergy that each component requires from the other components to obtain a unit of its product:

$$\kappa_{ij} = \frac{E_{ij}}{P_j} \tag{9}$$

The sum of all unit exergy consumptions in a component is the inverse of the exergy efficiency (η) of that component:

$$k_j = \sum_{i=0}^{n} \kappa_{ij} = \frac{F_j}{P_j} = \frac{1}{\eta_j} \tag{10}$$

The exergy cost of a flow E_{ij} is the amount of exergy resources needed to produce that flow and it is represented by E^*_{ij}. The unit exergy cost of a flow is the quotient between its exergy cost and its exergy.

$$k^*_{ij} = \frac{E^*_{ij}}{E_{ij}} \tag{11}$$

Besides, costs of fuel and of product of each component are represented as F^* and P^*, respectively.

Considering that, in each component, cost of the product must equal cost of fuel, it can be demonstrated that the unit exergy cost of the product of all components of a given system can be calculated as:

$$\boldsymbol{k}^*_P = \left(\boldsymbol{U}_D - \langle \boldsymbol{KP} \rangle^t\right) \boldsymbol{\kappa}_e \tag{12}$$

where KP is a matrix containing elements κ_{ij} and $\boldsymbol{\kappa}_e$ is the vector of unit exergy consumptions of external resources (κ_{0i}).

4.2 Decomposition of Exergy Cost According to Irreversibility

Besides the values of unit exergy cost of flows, it is useful to analyze how these costs are formed. For this reason, it is interesting to consider an ideal system without irreversibility. In such a system, cost of the product of each component would be equal to its exergy; in other words, unit exergy cost would be equal to unity. In a real system, irreversibility does exist. Accordingly, it is interesting to quantify how much irreversibility in each component causes increments of the cost of each flow. A methodology for obtaining this was developed in [13] that is summarized below.

The first step is to consider FP representation of the system. It is based on distribution coefficients y_{ij}, which indicate the proportion of the production of the j_{th} component used as resource for the i_{th} component:

$$y_{ij} = \frac{E_{ji}}{P_j} \tag{14}$$

After some transformations, it was demonstrated that the cost of the products of all components can be decomposed as:

$$\boldsymbol{P}^* = \boldsymbol{P} + (\boldsymbol{U_D} - \langle \boldsymbol{FP} \rangle)^{-1} \boldsymbol{I} \tag{15}$$

where $\boldsymbol{P}^*$ is the vector containing the exergy costs of the products of all plant components, $\boldsymbol{P}$ is the vector of these products, $\boldsymbol{U_D}$ is the identity matrix, $\langle \mathbf{FP} \rangle$ is a matrix containing elements y_{ij} and $\boldsymbol{I}$ is the vector containing irreversibility and losses of all components. The previous equation indicates that the cost of the products is equal to its exergy plus a summation of terms due to irreversibility appearing in the different components. Once exergy cost decomposition has been obtained, unit exergy cost decomposition can be calculated by dividing it into exergy of products.

5 Thermoeconomic Analysis of the ASF Biomass Plant

In this section, thermoeconomic analysis is applied to the example of biomass gasification plant described in Sect. 3. The first step is to define components of productive structure and to identify their fuel and product, which is done in Table 6. Most components of the productive structure were already present in the physical one. However, some aggregation of components has been made in order to integrate heat exchangers producing hot water without later use (i.e. this heat is finally released to the environment), and also mixers and splitters in the nearest component. It should be noted that, in some cases, this heat may have a use, and its release to the environment causes loss in global system efficiency, as it will be discussed later. Fuel and product definition has been made considering the purpose of the components. Heat exchangers use decrease in exergy of hot stream for increasing exergy of the cold stream; compressors consume electricity for increasing the exergy of an stream, combustion chamber uses exergy of fuel for increasing exergy of gas, and expander uses exergy decrease of flue gases for producing power. It should be noted that flows with exergy equal to zero (air from the atmosphere) have not been considered. Besides, increment of exergy of water in heat exchangers cooling gas does not appear because this effect is not used in other components and,

Table 6 Components of the productive structure

ID	Description	Components of the physical structure	Fuel	Product
1	Biomass drying unit	11BDU	B1 + B27 − B30	B2
2	Gasification unit	12AFB, 23PAM	B2 + B35 + B33	B3
3	Air compressor	22AC	B54	B35
4	Heat exchanger	20BAH	B3 − B4	B32 − B31
5	Expander	21BAEX	B32 − B33	B54 + B55
6	Generator	26BEAEXG	B55	B56
7	Gas cleaning unit	14GCU, 13WB2	B4	B6
8	Gas compressor	15GC1	B49	B7 − B6
9	Gas compressor	17GC2, 16WB3	B50	B9 − B7
10	Gas compressor	19GC3, 18WB4	B51	B11 − B9
11	Heat exchanger	7FGH, 6SPEG, 8MXEG	B21 − B25	B12 − B11
12	Exhaust gas fan	10EGF, 9WB1	B52 + B25 − B30	B27 − B30
13	Turbine air compressor	1GTAC	B46	B14 + B31
14	Turbine regenerator	5GTRHE, 2GTSP	B20 − B21	B16 + B17 + B18 − B14
15	Combustion chamber	12GTCC	B12	B19 − B16 − B30
16	Expander	4GTEX	B17 + B18 + B19 − B20	B46 + B47
17	Generator	25GTG	B47	B48
18	Power mixer		B48 + B56	B49 + B50 + B51 + B52 + B57

thus, it is a loss. Finally, a power mixer not present in the physical structure has been included in order to have a single flow with the net power generated in the plant (this is flow 57 that has also been added).

With the identification of fuel and product for the productive structure, the first results can be obtained. They are listed in Table 7. After the numerical values of fuel and product of each component, its difference (irreversibility + losses) is listed. Percentage of this irreversibility related to the plant fuel is also calculated. It can be seen that the highest irreversibility + losses appears in the gasifier, followed by combustion chamber. It is also important to note the relatively high values of gas cleaning unit (7) and gas fan (12). The last column shows exergy efficiency of all components. The worse value correspond to gas fan (12) that integrates gas cooler, and this result can be explained because gas is cooled down to low temperature (and thus low exergy) whereas exergy of water heated in the process is not used but released to the environment. The last row summarizes results for the whole plant: it consumes 20,234 kW of biomass and produces 4,821 kW of electricity, which corresponds to an exergy efficiency of 23.83 % (in other words, 76.17 % of consumed fuel is lost in irreversibilities and losses.

Analysis performed up to now allows to assess efficiency of the different components and of the total system, but does not takes into account interactions between components and, thus, does not allow to calculate cost. To do that, it is necessary to build the fuel-product table. This table contains elements E_{ij} and each

Table 7 Fuel, product irreversibility and efficiency of the components

ID	Description	Fuel (kW)	Product (kW)	Irrev + losses (kW)	Irrev + losses (%)	Efficiency (%)
1	Biomass drying unit	20,628	20,208	420	2.08	97.96
2	Gasification unit	20,531	14,090	6,441	31.83	68.62
3	Air compressor	6.66	5.46	1.20	0.01	81.93
4	Heat exchanger	575	474.3	100.7	0.50	82.48
5	Expander	655.8	592.5	63.34	0.31	90.34
6	Generator	585.8	556.5	29.3	0.14	95.00
7	Gas cleaning unit	13,515	12,454	1,061	5.24	92.15
8	Gas compressor	278.4	236	42.4	0.21	84.77
9	Gas compressor	257.9	148	109.9	0.54	57.39
10	Gas compressor	255.8	186	69.8	0.34	72.71
11	Heat exchanger	450	270	180	0.89	60.00
12	Exhaust gas fan	2,039	394	1,645	8.13	19.33
13	Turbine air compressor	5,012	4,671	340.6	1.68	93.20
14	Turbine regenerator	3,523	3,151	372.4	1.84	89.43
15	Combustion chamber	13,294	9,569	3,725	18.41	71.98
16	Expander	10,970	10,390	579.6	2.86	94.72
17	Generator	5,378	5,146	232	1.15	95.69
18	Power mixer	5,703	5,703	0	0	100
	Whole plant	20,234	4,821	15,413	76.17	23.83

row of this table represent how product of each component is distributed among the other components and the environment; besides, each column indicates the origin of the fuel of each component (from other components and from the environment).

Table 8 contains a generic fuel-product table for the analyzed example. By using this table, elements E_{ij} can be obtained from the values of exergy flows of the physical structure (B_i). For instance the product of component 2 (B_3) is partially consumed by component 4 (B_3–B_4) and partially by component 7 (B_4). Definition of the table is not difficult when the product of a single component is consumed by several components or when the fuel of a single component is produced by several components. However, a problem appears when the product of more than one component is consumed simultaneously by more than one component. In this situation, the total products have to be added and then distributed proportionally to the fuel of each component, by using suitable factors (r). For instance, the product of heat exchanger (4) and turbine air compressor (13) is consumed simultaneously by gasification unit (3) and expander (5). The problem is solved by introducing factors r_1 and r_2 that distribute proportionally these products of 4 and 13:

$$E_{4,2} = (B_{32} - B_{31}) \cdot r_1 \tag{16}$$

$$E_{4,5} = (B_{32} - B_{31}) \cdot r_2 \tag{17}$$

$$E_{13,2} = B_{31} \cdot r_1 \tag{18}$$

$$E_{13,5} = B_{31} \cdot r_2 \tag{19}$$

where r_1 and r_2 are calculated as:

$$r_1 = \frac{B_{33}}{B_{32}} \tag{20}$$

$$r_2 = \frac{B_{32} - B_{33}}{B_{32}} \tag{21}$$

A similar situation appears in the gas turbine, because the product of compressor (13), regenerator (14) and combustion chamber (15) is fuel for two heat exchangers (11,12), the regenerator (14) and the expander (16):

$$E_{13,11} = B_{14} \cdot r_3 \tag{22}$$

$$E_{13,12} = B_{14} \cdot r_4 \tag{23}$$

$$E_{13,14} = B_{14} \cdot r_5 \tag{24}$$

$$E_{13,16} = B_{14} \cdot r_6 \tag{25}$$

$$E_{14,11} = (B_{16} + B_{17} + B_{18} - B_{14}) \cdot r_3 \tag{26}$$

Table 8 Generic fuel product table

	F1	F2	F3	F4	F5	F6	F7	F8	F9	F10
P0	B_1	0	0	0	0	0	0	0	0	0
P1	0	B_2	0	0	0	0	0	0	0	0
P2	0	0	0	$B_3 - B_4$	0	0	B_4	0	0	0
P3	0	B_{35}	0	0	0	0	0	0	0	0
P4	0	$E_{4,2}$	0	0	$E_{4,5}$	0	0	0	0	0
P5	0	0	B_{54}	0	0	B_{55}	0	0	0	0
P6	0	0	0	0	0	0	0	0	0	0
P7	0	0	0	0	0	0	0	0	0	0
P8	0	0	0	0	0	0	0	0	0	0
P9	0	0	0	0	0	0	0	0	0	0
P10	0	0	0	0	0	0	0	0	0	0
P11	0	0	0	0	0	0	0	0	0	0
P12	$B_{27} - B_{30}$	0	0	0	0	0	0	0	0	0
P13	0	$E_{13,2}$	0	0	$E_{13,5}$	0	0	0	0	0
P14	0	0	0	0	0	0	0	0	0	0
P15	0	0	0	0	0	0	0	0	0	0
P16	0	0	0	0	0	0	0	0	0	0
P17	0	0	0	0	0	0	0	0	0	0
P18	0	0	0	0	0	0	0	B_{49}	B_{50}	B_{51}
TOTAL	$B_{21} + B_{27} - B_{30}$	$B_2 + B_{33} + B_{35}$	B_{54}	$B_3 - B_4$	$B_{32} - B_{33}$	B_{55}	B_4	B_{49}	B_{50}	B_{51}

(continued)

Table 8 (continued)

	F11	F12	F13	F14	F15	F16	F17	F18	F0	Total
P0	0	0	0	0	0	0	0	0	0	B_1
P1	0	0	0	0	0	0	0	0	0	B_2
P2	0	0	0	0	0	0	0	0	0	B_3
P3	0	0	0	0	0	0	0	0	0	B_{35}
P4	0	0	0	0	0	0	0	0	0	$B_{31} - B_{31}$
P5	0	0	0	0	0	0	0	0	0	$B_{54} + B_{55}$
P6	0	0	0	0	0	0	0	B_{56}	0	B_{56}
P7	0	0	0	0	B_6	0	0	0	0	B_6
P8	0	0	0	0	B_7-B_6	0	0	0	0	$B_7 - B_6$
P9	0	0	0	0	B_9-B_7	0	0	0	0	$B_9 - B_7$
P10	0	0	0	0	B_{11}-B_9	0	0	0	0	$B_{11} - B_9$
P11	0	0	0	0	B_{12}-B_{11}	0	0	0	0	$B_{12} - B_{11}$
P12	0	0	0	0	0	0	0	0	0	$B_{27} - B_{30}$
P13	$E_{13,11}$	$E_{13,12}$	0	$E_{13,14}$	0	$E_{13,16}$	0	0	0	$B_{14} + B_{31}$
P14	$E_{14,11}$	$E_{14,12}$	0	$E_{14,14}$	0	$E_{14,16}$	0	0	0	$B_{16} + B_{17} + B_{18} - B_{14}$
P15	$E_{15,11}$	$E_{15,12}$	0	$E_{15,14}$	0	$E_{15,16}$	0	0	0	$B_{19} - B_{16} - B_{30}$
P16	0	0	B_{46}	0	0	0	B_{47}	0	0	$B_{46} + B_{47}$
P17	0	0	0	0	0	0	0	B_{48}	0	B_{48}
P18	0	B_{52}	0	0	0	0	0	0	B_{57}	$B_{49} + B_{50} + B_{51} + B_{52} + B_{57}$
TOTAL	$B_{21} - B_{25}$	$B_{52} + B_{25} - B_{30}$	B_{46}	$B_{20} - B_{21}$	B_{12}	$B_{17} + B_{18} + B_{19} - B_{20}$	B_{47}	$B_{48} + B_{56}$	B_{57}	

Table 9 Fuel-Product table

	F1	F2	F3	F4	F5	F6	F7	F8	F9	F10	F11	F12	F13	F14	F15	F16	F17	F18	F0	Total
P0	20,234	0	0	0	0	0	0	0	0	0	0	0	0	0	0	0	0	0	0	20,234
P1	0	20,208	0	0	0	0	0	0	0	0	0	0	0	0	0	0	0	0	0	20,208
P2	0	0	0	575	0	0	13,515	0	0	0	0	0	0	0	0	0	0	0	0	14,090
P3	0	5.46	0	0	0	0	0	0	0	0	0	0	0	0	0	0	0	0	0	5.46
P4	0	1,54.8	0	0	319.4	0	0	0	0	0	0	0	0	0	0	0	0	0	0	4,74.3
P5	0	0	6.66	0	0	585.8	0	0	0	0	0	0	0	0	0	0	0	0	0	592.5
P6	0	0	0	0	0	0	0	0	0	0	0	0	0	0	0	0	0	556.5	0	556.5
P7	0	0	0	0	0	0	0	0	0	0	0	0	0	0	12,454	0	0	0	0	12,454
P8	0	0	0	0	0	0	0	0	0	0	0	0	0	0	236	0	0	0	0	236
P9	0	0	0	0	0	0	0	0	0	0	0	0	0	0	148	0	0	0	0	148
P10	0	0	0	0	0	0	0	0	0	0	0	0	0	0	186	0	0	0	0	186
P11	0	0	0	0	0	0	0	0	0	0	0	0	0	0	270	0	0	0	0	270
P12	394	0	0	0	0	0	0	0	0	0	0	0	0	0	0	0	0	0	0	394
P13	0	163.0	0	0	336.4	0	0	0	0	0	111.1	481.4	0	870.1	0	2709	0	0	0	4671
P14	0	0	0	0	0	0	0	0	0	0	83.93	363.5	0	657.1	0	2046	0	0	0	3,151
P15	0	0	0	0	0	0	0	0	0	0	254.9	1104	0	1,996	0	6,214	0	0	0	9,569
P16	0	0	0	0	0	0	0	0	0	0	0	0	5,012	0	0	0	5,378	0	0	10,390
P17	0	0	0	0	0	0	0	0	0	0	0	0	0	0	0	0	0	5,146	0	5,146
P18	0	0	0	0	0	0	0	278.4	257.9	255.8	0	89.51	0	0	0	0	0	0	4,821	5,703
TOTAL	20,628	20,531	6.66	575	655.8	585.8	13,515	278.4	257.9	255.8	450	2,039	5,012	3,523	13,294	10,970	5,378	5,703	4,821	

$$E_{14,12} = (B_{16} + B_{17} + B_{18} - B_{14}) \cdot r_4 \tag{27}$$

$$E_{14,14} = (B_{16} + B_{17} + B_{18} - B_{14}) \cdot r_5 \tag{28}$$

$$E_{14,16} = (B_{16} + B_{17} + B_{18} - B_{14}) \cdot r_6 \tag{29}$$

$$E_{15,11} = (B_{19} - B_{16} - B_{30}) \cdot r_3 \tag{30}$$

$$E_{15,12} = (B_{19} - B_{16} - B_{30}) \cdot r_4 \tag{31}$$

$$E_{15,14} = (B_{19} - B_{16} - B_{30}) \cdot r_5 \tag{32}$$

$$E_{15,16} = (B_{19} - B_{16} - B_{30}) \cdot r_6 \tag{33}$$

where the distribution factors r_3, r_4, r_5 and r_6 are calculated as:

$$r_3 = \frac{B_{21} - B_{25}}{B_{17} + B_{18} + B_{19} + B_{30}} \tag{34}$$

$$r_4 = \frac{B_{25} - B_{30}}{B_{17} + B_{18} + B_{19} + B_{30}} \tag{35}$$

$$r_5 = \frac{B_{20} - B_{21}}{B_{17} + B_{18} + B_{19} + B_{30}} \tag{36}$$

$$r_6 = \frac{B_{17} + B_{18} + B_{19} - B_{20}}{B_{17} + B_{18} + B_{19} + B_{30}} \tag{37}$$

Table 10 Unit exergy cost of the products of the components

ID	Description	Unit exergy cost of the product (k_P*)
1	Biomass drying unit	1.3885
2	Gasification unit	2.0709
3	Air compressor	4.6519
4	Heat exchanger	2.5106
5	Expander	3.8116
6	Generator	4.0123
7	Gas cleaning unit	2.2473
8	Gas compressor	4.9512
9	Gas compressor	7.3138
10	Gas compressor	5.7722
11	Heat exchanger	6.3700
12	Exhaust gas fan	19.8599
13	Turbine air compressor	4.3294
14	Turbine regenerator	4.2738
15	Combustion chamber	3.4520
16	Expander	4.0352
17	Generator	4.2171
18	Power mixer	4.1972

Table 11 Unit exergy cost decomposition

Component		P	I1	I2	I3	I4	I5	I6	I7	I8	I9	I10	I11	I12	I13	I14	I15	I16	I17	I18
1	Biomass drying unit	1	*0.0288*	**0.1232**	0.0000	0.0009	0.0003	0.0001	*0.0208*	0.0008	0.0022	0.0014	0.0035	**0.1128**	0.0063	0.0073	*0.0729*	0.0063	0.0009	0.0000
2	Gasification unit	1	0.0430	**0.6592**	0.0001	0.0038	0.0004	0.0002	0.0330	0.0013	0.0034	0.0022	0.0056	*0.1683*	0.0108	0.0116	*0.1158*	0.0107	0.0014	0.0000
3	Air compressor	1	0.0966	**1.4809**	*0.2207*	0.1493	0.1324	0.0008	*0.1874*	0.0075	0.0194	0.0123	0.0318	*0.3781*	0.1077	0.0658	*0.6578*	0.0969	0.0065	0.0000
4	Heat exchanger	1	0.0521	**0.7992**	0.0001	*0.2170*	0.0005	0.0002	0.0400	0.0016	0.0041	0.0026	0.0068	*0.2040*	0.0131	0.0140	*0.1404*	0.0129	0.0017	0.0000
5	Expander	1	0.0791	**1.2134**	0.0002	0.1224	0.1084	0.0007	*0.1535*	0.0061	0.0159	0.0101	0.0260	*0.3098*	0.0883	0.0539	*0.5389*	0.0794	0.0053	0.0000
6	Generator	1	0.0833	**1.2773**	0.0002	0.1288	0.1142	0.0534	*0.1616*	0.0064	0.0167	0.0106	0.0274	*0.3261*	0.0929	0.0567	*0.5674*	0.0835	0.0056	0.0000
7	Gas cleaning unit	1	0.0466	**0.7154**	0.0001	0.0042	0.0005	0.0002	*0.1210*	0.0014	0.0037	0.0024	0.0060	*0.1826*	0.0118	0.0126	*0.1257*	0.0116	0.0015	0.0000
8	Gas compressor	1	0.1028	**1.5762**	0.0003	0.0246	0.0155	0.0072	*0.2596*	*0.1900*	0.0269	0.0171	0.0440	*0.4024*	0.0833	0.0911	*0.9115*	0.1418	0.0568	0.0000
9	Gas compressor	1	0.1518	**2.3283**	0.0004	0.0364	0.0229	0.0106	*0.3835*	0.0153	*0.7823*	0.0252	0.0651	*0.5944*	0.1231	0.1346	*1.3464*	0.2095	0.0839	0.0000
10	Gas compressor	1	0.1198	**1.8375**	0.0003	0.0287	0.0181	0.0083	*0.3027*	0.0121	0.0314	*0.3952*	0.0513	*0.4691*	0.0972	0.1062	*1.0626*	0.1653	0.0662	0.0000
11	Heat exchanger	1	0.1322	**2.0278**	0.0004	0.0139	0.0033	0.0015	*0.3420*	0.0137	0.0354	0.0225	*0.7247*	*0.5177*	0.1031	0.1200	*1.2007*	0.0995	0.0115	0.0000
12	Exhaust gas fan	1	0.4122	**6.3223**	0.0012	0.0461	0.0129	0.0057	*1.0650*	0.0426	0.1103	0.0701	0.1807	**5.7880**	0.3220	0.3738	*3.7392*	0.3227	0.0452	0.0000
13	Turbine air compressor	1	0.0899	**1.3782**	0.0003	0.0095	0.0023	0.0010	*0.2324*	0.0093	0.0241	0.0153	0.0394	*0.3519*	0.1430	0.0816	*0.8160*	0.1275	0.0078	0.0000
14	Turbine regenerator	1	0.0887	**1.3605**	0.0003	0.0093	0.0022	0.0010	*0.2294*	0.0092	0.0238	0.0151	0.0389	*0.3476*	0.0692	*0.1987*	*0.8056*	0.0668	0.0077	0.0000
15	Combustion chamber	1	0.0717	**1.0989**	0.0002	0.0075	0.0018	0.0008	*0.1853*	0.0074	0.0192	0.0122	0.0314	*0.2806*	0.0241	0.0261	**0.6507**	0.0278	0.0063	0.0000
16	Expander	1	0.0838	**1.2846**	0.0002	0.0088	0.0021	0.0009	*0.2166*	0.0087	0.0224	0.0143	0.0368	*0.3280*	0.0653	0.0760	**0.7605**	0.1188	0.0073	0.0000
17	Generator	1	0.0875	**1.3425**	0.0003	0.0092	0.0022	0.0010	*0.2264*	0.0090	0.0235	0.0149	0.0384	*0.3427*	0.0682	0.0795	*0.7949*	0.1242	0.0527	0.0000
18	Power mixer	1	0.0871	**1.3361**	0.0002	0.0209	0.0131	0.0061	*0.2201*	0.0088	0.0228	0.0145	0.0373	*0.3411*	0.0707	0.0772	*0.7726*	0.1202	0.0481	0.0000

Once a generic fuel-product table has been defined, the table with numerical values can be obtained by considering the actual values of exergy of the flows of the physical structure. This table appears in Table 9.

Once the fuel product table has been obtained, it is possible to calculate the unit exergy consumptions and, finally, the unit exergy costs of the products of all plant components. These unit exergy costs are listed in Table 10, where is interesting to see how unit exergy cost increases as productive process is developed: biomass drying unit (1), gasification unit (2), heat exchanger (4), gas cleaning unit (7), combustion chamber (15), expander (16) and finally power mixer (18). Unit exergy cost of the product of the latter is 4.1972, which is the inverse of the plant exergy efficiency. It should be noted that this increment of unit cost through the productive process is not completely true because of recirculations. In fact, unit exergy costs of the products of compressors consuming electricity are higher than the cost of electricity, because it is necessary to consider the additional irreversibility of these compressors; this is especially relevant in the fan of flue gases, characterized by a very low efficiency.

It is interesting to analyze how the unit exergy cost presented above is formed due to irreversibility of the different components. This decomposition is presented in Table 11. In order to simplify the interpretation of the table, values with average relevance (i.e. between 5 % and 25 % of the increment of unit exergy cost) are written in italics whereas most important values (more than 25 %) are in bold. Each row shows how the unit exergy cost of each device is formed due to irreversibility of all components. It can be seen how gasification unit (2) and combustion chamber (15) have strong impact on all components. Gas fan (12) and gas cleaning unit (7) also have a relevant contribution. Last but not least, it can be appreciated a kind of "diagonal" of higher values, pointing out the contribution that irreversibility of each component has on the cost of the product of the same component.

6 Conclusions

It is interesting to organize conclusions in two parts: one related to case study and other general.

6.1 Case Study Conclusions

The complete analysis performed yields the following case-related conclusions:

1. The gasification unit is a major source of irreversibility. This results from complex and irreversible phenomena, including diffusion, mixing, heat transfer and chemical reactions, involved in the gasification process. A comparison with

other type of gasification units, such as Pressurized Fluidized Bed (PFB) or Allothermal Fast Internally Circulated Fluidized Bed (FICFB) can be done by interested readers based on data presented in [6]. It should be stressed that exergy flows and irreversibility related to the gasification unit can be calculated based on the 'black-box' approach, i.e. knowing the composition, flow and thermal parameters of fluxes incident to the reactor. At this level of analysis, it is not necessary to perform exergy analysis of the complex processes inside the reactor.

2. Major irreversibility is bound to the combustion chamber and other components of the gas turbine. Exergy losses due to combustion and mixing with excess air are related to the maximum temperature that flue gases can have at expander input, that is a major issue in the development of gas turbines with higher efficiencies. In general, since gas turbines are delivered as closed systems of devices, the potential of improvement consists in the choice between different commercially available turbines rather than in improving a particular solution.
3. Gas cooling in gas cleaning (7) and flue gas cooling before the fan (12) are also relevant sources of inefficiency. This effect can be reduced by using exergy of produced hot water instead of dissipated it in the environment. With this change, the system would operate as cogeneration plant. Actually, recovery of this heat was proposed in the design of the installation. This is the most important opportunity for improving exergy efficiency of the system, although a demand of low temperature heat is needed.
4. The drying process does not increase the exergy of feedstock biomass, even though it is energy-consuming. However that, consumption would be higher in the case of a dedicated dryer instead of the integrated version proposed.
5. Despite the improvement potential commented above, it should be noted that the system analyzed obtain good efficiency figures, mainly due to the proper integration among components. These good characteristics of the configuration would have been highlighted if the analysis had been also applied to a similar system not so integrated.

6.2 General Conclusions

The presented methodology can be used as an advanced tool for process design engineers. It is a rigorous and systemic approach able to both assess efficiency of individual components but also to consider interactions among them. Once sources of inefficiency have been identified and quantified, it is much easier to improve the design by substituting devices or changing plant configuration, of course taking into account limitation in technology and economic-related issues.

As can be observed in Table 7, particular plant devices differ in the quantity of processed exergy, and in the quality of processing. The amount of exergy processed by 'key' plant devices is by one or two orders of magnitude higher than in the case of 'secondary' devices. The distinction between 'key' and 'secondary' is of course arbitrary. In the presented case study, five components: biomass drying unit, gasification unit, gas cleaning unit, combustion chamber and expander process more than 10,000 kW of fuel exergy each. On the other hand, some devices (e.g. gas compressors 8–10) process less than 300 kW of fuel exergy. It is easy to conclude that any improvement (or deterioration) in the performance of key devices has a substantial impact on the total plant efficiency, while attempts to improve secondary components will fail to enhance the system performance, in spite of a perhaps high investment cost.

It should be noted that detecting 'key' and 'secondary' devices and formulating the above conclusions is much easier if the discussed fuel-product approach is applied. In a conventional engineering approach, thermodynamic and economic performance of the system is calculated for the entire system, since the investor generally asks about basic economic indicators such as NPV or IRR. This approach is correct, but it lacks the transparency and fails to analyze components independently. Following this overall approach, an inexperienced process engineer could e.g. suggest investing money in better gas compressors, supposing that these components are important since they are working in the fuel supply line. On the contrary, a process engineer familiar with the fuel-product approach is immediately able to indicate which devices are worth further improvements.

On the other hand, it may happen that efficiency of a 'secondary' device can be easily improved (e.g. by changes in operation). In this situation, the method is also useful because it provides a complete picture of the system, not only focusing on the more relevant components.

The presented methodology allows one to identify possible *potential of improvement*, expressed by irreversibility, and to focus on plant devices having the highest potential. Part of this irreversibility may be easily avoided (e.g. by changes in operation or small modifications in plant configuration), other part can be avoided by increasing investment in components, and another part is unavoidable due to the physical nature of the process [11]. Finding the optimum value is a question of economics and current market conditions.

Finally, it should be stressed that the presented methodology is only one of the possible applications of the 2nd law of thermodynamics, dedicated to intrinsic characteristics of the plant and its components. Another extensive field of research is bound to the cumulative- and life cycle calculus, accounting to the consumption of resources and generation of wastes not only within the analyzed plant in the phase of its operation, but in all related chains of resources extraction, transportation and conversion in the entire life cycle of the installation. An example of such an analysis for the biomass gasification plant analyzed in this chapter, and also for several alternative plants, is shown in [20], based on the methodology of the thermoecological cost (TEC) given by Szargut and Stanek [21].

Both areas of research are based on the 2nd law of thermodynamics, and both aim at improving the performance of processes for a better utilization of available renewable and non-renewable energy resources in our Planet.

References

1. Tribus, M., Evans, R.B.: A contribution to the theory of thermoeconomics. Technical report, Report No. 62–63; Department of Engineering, UCLA: Los Angeles, CA, USA 1962
2. Rant, Z.: Exergie, ein neues Wort für technische Arbeitsfähigkeit (Exergy: a new word for 'technical work ability'). Forschung auf dem Gebiet des Ingenieurwesens **22**, 36–37 (1956)
3. Szargut, J.: Die Exergiebilanz der Wärmeprozesse in der Hüttenindustrie (Exergy balance of thermal processes in the steel industry). Freiberger Forschungshefte Nr B 68:81-103, Berlin 1962
4. Szargut, J.: Exergy method: Technical and ecological applications. WIT Press, Southampton (2005)
5. Szargut, J.: Analiza termodynamiczna i ekonomiczna w energetyce przemysłowej (Thermodynamic and economic analysis in industrial energy engineering, in Polish). WNT Press, Warsaw (1983)
6. Kalina, J.: Comparative analysis of alternative configurations of the Mercury 50 recuperated gas-turbine-based biomass integrated gasification combined heat and power (BIGCHP) plant. Energy Fuels **26**, 6452–6465 (2012)
7. Kalina, J.: Modelling of fluidized bed biomass gasification in the quasi-equilibrium regime for preliminary performance studies of energy conversion plants. Chem. Process Eng. **32**, 73–89 (2011)
8. Torres, C., Valer, A., Rangel, V., Zaleta, A.: On the cost formation process of the residues. Energy **33**, 144–152 (2008)
9. Valero, A., Lozano, M.A., Muñoz, M.: A general theory of exergy saving. Part I: On the exergetic cost. In: Computer Aided Engineering and Energy System, vol 3: Second Law Analysis and Modelling. ASME, New York, 2–3:1–8 (1986)
10. Szargut, J., Morris, D.: Cumulative exergy consumption and cumulative degree of perfection of chemical processes. Int. J. Energy Res. **11**, 245–261 (1987)
11. Tsatsaronis, G., Park, M.H.: On avoidable and unavoidable exergy destructions and investment costs in thermal systems. Energy Convers. Manage. **43**, 1259–1270 (2002)
12. Bejan, A., Tsatsaronis, G., Moran, M.: Thermal design and optimization. Wiley, New York (1996)
13. Torres, C.: Symbolic thermoeconomic analysis of energy systems. In: Frangopoulos, C.A. (ed.) Exergy, energy system analysis and optimization, Encyclopedia of life support systems (EOLSS), EOLSS Publishers, Oxford 2006
14. El Sayed, Y., Evans R.: Application of the thermoeconomic approach to the analysis and optimization of a vapour compression desalting system. J. Eng. Power. 17–26 (1970)
15. Frangopoulos, C.A.: Thermoeconomic functional analysis and optimization. Energy, vol. **7**, 563–571 (1986)
16. Tsatsaronis, G.: Thermoeconomic analysis and optimization of energy systems. Progress in Energy Combustion Science, vol. **19**, 227-257 (1993)
17. von Spakovsky, M.R., Evans R.B.: Engineering functional analysis. Parts I, II. J. Energ. Resour. ASME **115**, 86–99 (1993)
18. Valero, A., Torres, C.: Thermoeconomic analysis. In: Frangopoulos, C.A. (ed.) Exergy, energy system analysis and optimization, Encyclopedia of life support systems (EOLSS), EOLSS Publishers, Oxford (2006)
19. El-Sayed, Y.M.: The thermoeconomics of energy conversions. Elsevier, Oxford (2005)
20. Stanek, W., Czarnowska, L., Kalina, J.: Thermo-ecological evaluation of biomass integrated gasification gas turbine based cogeneration technology. Proceedings of ECOS 2012—the 25th International Conference on Efficiency, Cost, Optimization, Simulation and Environmental Impact of Energy Systems, June 26–29, Perugia, Italy (2012)
21. Szargut, J., Stanek, W.: Fuel part and mineral part of the thermoecological cost. Int. J. Thermodyn. **15**, 187–190 (2012)

Effect of Transitional Turbulence Modelling on a Straight Blade Vertical Axis Wind Turbine

K. M. Almohammadi, D. B. Ingham, L. Ma and M. Pourkashanian

Abstract The flow around straight blade vertical axis wind turbines is typically complex at low tip speed ratios (TSR < 2). In this paper, the turbulence models which are based on the assumption of fully developed turbulent flow, such as S-A, RNG κ-ε and SST κ-ω have been investigated in comparison to the SST transitional model (both with and without curvature correction) to account for the laminar-turbulence transition. The investigation is based on the 2D unsteady Reynolds averaged Navier–Stokes (URANS) equations using a sliding mesh technique. It has been found that applying turbulence models based on the assumption of fully developed flow shows significant differences in velocity magnitude if the flow is under stall condition or wake effect compared to the transitional model. Also, the predicted flow structure in the vicinity of the stalled airfoils using different types of turbulence models is found to be different compared to the un-stalled airfoils where no significant differences in the flow field have been observed. In the wake region, the flow varies less significantly compared to the stalled airfoils.

K. M. Almohammadi (✉)
CFD Centre, ETII, Faculty of Engineering, University of Leeds, Leeds LS2 9JT, UK
e-mail: Doctorkmsa@gmail.com

K. M. Almohammadi
Mechanical Engineering, Taibah University, Madinah, Saudi Arabia

D. B. Ingham · L. Ma · M. Pourkashanian
CFD Centre, ETII, Faculty of Engineering, University of Leeds, Leeds LS2 9JT, UK
e-mail: d.b.ingham@leeds.ac.uk

L. Ma
e-mail: L.Ma@leeds.ac.uk

M. Pourkashanian
e-mail: fue6mtz@leeds.ac.uk

G. Ferreira (ed.), *Alternative Energies*, Advanced Structured Materials 34,
DOI: 10.1007/978-3-642-40680-5_5,

1 Introduction

Harnessing wind energy using different types of turbine configurations has become increasingly more investigated. It is a free and clean alternative source of energy that could reduce the dependence on fossil fuels that are rapidly depleting. Wind turbines are generally classified into two categories, namely vertical axis wind turbines (VAWT) and horizontal axis wind turbines (HAWT).

In urban regions, the vertical axis wind turbine can perform better aerodynamically due to its independence on the wind direction compared to the horizontal axis wind turbines [1–5]. Wind turbines are mainly either lift driven turbines or drag driven turbines. In 1920, the maximum theoretical efficiency of the lift driven turbines was found to be 59.3 %, known as the Betz limit [6], referring to the German physicist Albert Betz, whereas the maximum theoretical efficiency for the drag based turbines is only 29.6 % [7]. The lift based vertical axis wind turbines come in different blade shapes, namely straight or curved blades. Due to the simplicity of the straight blade shape and its aerodynamic performance in urban regions [8], it has been under researchers focus in recent years both numerically and experimentally. The accurate modelling, by the means of computational fluid dynamics, requires profound understanding of the modelling aspects, in particular the turbulence modelling, which have become mandatory if the modelling becomes the basis of the turbine optimization [9–11].

A key parameter to accurately model the fluid flow in the vicinity of a straight blade vertical axis wind turbine is the appropriate choice of the turbulence model. This becomes more important when the turbine operates in complex flow structures and at low tip speed ratios (TSR < 2), which is the ratio of the rotational speed of the blades to the upwind air speed [12, 13].

There are many different turbulence models available but none are universally applicable [14, 15]. The understanding of the underlying physics of the modelled problem is one of the most important issues involved in the modelling process. Therefore, this paper focuses on examining the effect of considering the transitional effect that is accounted for by the turbulence model applied, namely the SST Transitional, compared to the turbulence models that assume the flow to be fully developed, such as the Spalart–Allmaras [16], RNG κ-ε [17], and SST κ-ω models [18]. This kind of turbine develops a complex nonlinear phenomenon, known as dynamic stall which may be affected by the assumption of the fully developed flow that is the basis of the most widely used turbulence models. Also, additional modifications of the transitional turbulence model, namely the correction factor, are investigated.

2 Turbulence Modelling of Vertical Axis Wind Turbines

The attempts to understand and model the complex flow in the vicinity of the straight blades of the vertical axis wind turbines started by developing different models such as the momentum, cascade, and vortex based models. These models are not directly appropriate and require modifications and assumptions for modelling vertical axis wind turbines due to the non-linearity and the complexity of the flow. Additional models need to be implemented to account for the dynamic stall and rotational effects. Therefore, computational fluid dynamics based on the Navier–Stokes equation is preferred due to the inclusion of the viscous flow effects in the turbulence modelling. The most widely used technique of modelling is based on the Reynolds averaging of the fluctuating transport variables of the Navier–Stokes equations (RANS) [19, 20]. The general form of the Navier–Stokes equation is given by

$$\frac{\partial}{\partial t}(\rho\phi) + \nabla.(\rho V\phi) = \nabla.(\Gamma\nabla\phi) + S \tag{1}$$

There are different families of turbulence models that are based on the Navier–Stokes equations, such as the one-equation (Spalart–Allmaras), two-equations (ReNormalisation Group (RNG) κ-ε, and SST κ-ω), and four-equations (SST Transitional) models. The Spalart–Allmaras was originally designed for aerospace applications and is appropriate for low Reynolds number flows with adverse pressure gradients in the boundary layers [10, 16], and therefore it could be a candidate for use in vertical axis wind turbine modelling. The model is based on the modelling of one transport variable, namely the turbulent eddy viscosity. It is important to note that the turbulence length scale, which is required for the turbulence dissipation rate computations, is not computed in this model. It has been reported that this model responds poorly to rapid changes in the flows [10]. However, wind turbines experience rapid changes in the flow at high tip speed ratios but the low computational effort and the operation at relatively low tip speed ratios make this model attractive. Therefore the results produced using this model have to be compared with the results obtained from more advanced models for wind turbines.

The second family of turbulence models is the two equation models based on κ and ε. There are many variations of this model, such as the standard κ-ε, RNG κ-ε, realizable κ-ε, and low Re κ-ε models. Among these models, the RNG κ-ε has been shown to have a superior performance and it is claimed to be a universal turbulence model [21]. In this model, the small scale turbulence is determined statistically and incorporated into the Navier–Stokes equations using a random forcing function. These small scales are instantaneously removed from the governing equations and their effects are accounted for in the large eddies. However, the turbulence dissipation rate (ε) in many applications fails to accurately predict flows that include adverse pressure gradients and recirculations [22–24] and these effects normally occur in wind turbine flows and this results in stall flow conditions. Wilcox [25] developed a new model, namely the Shear Stress Transport (SST) κ-ω model, based on the specific turbulence dissipation rate (ω) as a remedy for near wall flows and an improved version of the model was introduced by Menter and Langtry [26, 27] to account for the shear stresses. This model has been shown to produce good agreement with experimental data in many applications [11].

However, the laminar-to-turbulent transition is not accounted for in these models and this could play a significant factor in predicting the wind turbine flow. The boundary layers near the airfoils are always in transitional state due to the continuously changing Reynolds number, and therefore it is essential to include transitional modelling in the selected turbulence model. A candidate for the turbulence model that accounts for laminar-to-turbulent transition is the SST Transitional model which is a four equation model. This model is based on the SST κ-ω model and shows improved results compared to the experimental data in many applications [28–30].

3 Laminar-Turbulent Transition

The importance of applying the transitional model to analyse complex flow structures in low Reynolds flows have been reported in many applications. It is identified that there are five main categories of transition, namely natural transition, bypass transition, separated flow transition, wake induced transition and reverse transition [31]. More than one transition category may be present in straight blade vertical axis wind turbine flows. However, this is an open research question that needs to be properly investigated. Factors such as blade shape, angle of attack, Reynolds number, free stream turbulence intensity play an important role in determining the transition type [32].

The experimental analysis of the flow transition is expensive despite the development of the advancement in the equipment employed due to the difficulties in controlling the affecting parameters and, therefore, the modelling of transitional flows is a viable alternative. There are many methods incorporated within the RANS turbulence models to capture transition. A standard method that has been widely employed in the industry is the e^N method [33, 34] and this has been employed by Lian and Shyy [35]. It was necessary to develop a model that accounts for all factors that affect the transition process, even though this coupling has shown good agreement with experimental results in many applications. Recent methods are based on the concept of the correlation being more comprehensive compared to the e^N method, such as the model employed in this study, namely the SST transition model. Walters and Leylek [36] developed another transitional model which also based on the concept of correlations, namely $k - k_L - \omega$ transitional model. Both transitional model that are based on correlations have been shown to be in good agreement to the results obtained experimentally in many applications [37–40]. In 1990, Durbin [41] has developed $v^2 - f$ model in order to overcome the near anisotropy. This model is found to be very sensitive to the free stream length scale and predicted transitional flow with good accuracy level. However, comparing the transitional model computationally requires enormous computing power and out of the scope of this study.

The SST transition model has been successfully tested in turbomachinery and aerodynamic applications. The model has been investigated by Benini and Ponza on predicting the airfoil coefficients and the results for the drag have been shown to be about 20 % different to the experimental data [42]. Further, their results were found to be highly sensitive to the upstream turbulence levels.

The high sensitivity of the transition process makes it important to investigate the flow around the straight blade vertical axis wind turbine in operating conditions. This allows the flow in the vicinity of the blades upstream of the turbine to interact and affect the flow in the vicinity of the blades downstream of the turbine. Therefore, the cases that have been tested on the airfoils, for the purpose of vertical axis wind turbines, need to be more thoroughly investigated.

There are only a few attempts at applying the transitional model to straight blade vertical axis wind turbines instead of analysing the airfoils designed for wind turbines. Castelli applied the SST transitional model at different tip speed ratios [43]. However, the angle of attack in this study is set to a maximum of 10 degrees in order to avoid the stall condition which is shown by Wang [1, 44] to be highly sensitive to the transitional process.

Another important issue is the computational scheme and the grid employed for modelling the transitional flow. It has been shown that the stability of the calculations is dependent on the type and the quality of the cells employed for the computations [45]. These have not been properly investigated in the literature for straight blade vertical axis wind turbines. Higher-order schemes, such as the MUSCAL scheme, with a low residual of, say, the order of 10^{-7} need to be further investigated.

4 Computational Scheme

4.1 Computational Domain and Grid

A 2D cross section of the experiment performed by Bravo et al. [46, 47] has been created. The turbine consists of three straight blades and the specifications are shown in Table 1.

The computational domain is created as two regions, namely a stationary region and a rotating region. The stationary region is created with a rectangular shape of 87.5 cm (length) × 50 cm (width) which has been tested for size independency by Maclaren [48]. The rotation region is created to be of a circular shape of radius 17.5 cm and centred 8.75 m in length and 10 m in width from the inlet which is selected for the left hand side of the rectangle. A pressure outlet boundary

Table 1 Details of the straight blade VAWT investigated by Bravo et al. [20, 21]

Blades	Cord (m)	Trailing edge curve (mm)	Diameter (m)	Height (m)
NACA 0015	0.4	40	2.5	3.0

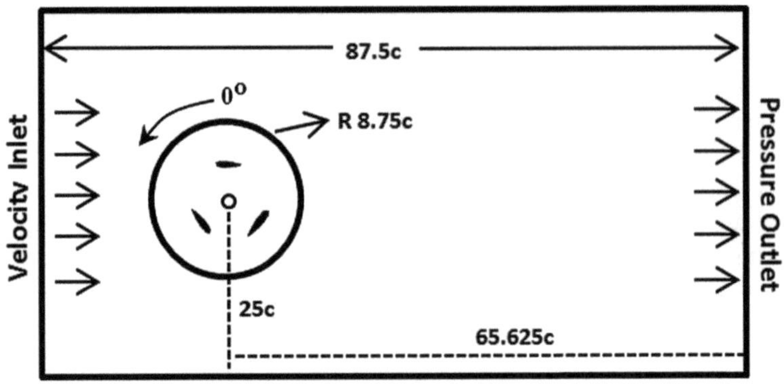

Fig.1 Schematic of the domain of the investigated straight blade vertical axis wind turbine

condition is selected for the right hand side of the rectangle. The other two sides of the rectangle are selected as wall boundary conditions. This is illustrated in Fig. 1.

The contact surface between the stationary and the rotating regions is selected as the interface in order to apply the sliding mesh technique. The domain is created using the commercial software GAMBIT with 15 layers of structured quadrilateral cells (with a y^+ value of about (1) near the airfoil boundary and unstructured quadrilateral cells are employed elsewhere, as recommended by Almohammadi [49] for enhancing the stability and consistency of the computations. This results in a mesh with 235,388 cells which is a relatively fine mesh, and therefore it is used for all the computations presented in this paper based on the assumption that the errors produced by the same mesh will be the same since the turbulence models applied are all URANS (Unsteady Reynolds Averging Navier–Stokes) models.

The inlet velocity is set to be the same as in the experimental investigation, namely 10 m/s with an estimated turbulence level of 0.5 % since it is a wind tunnel test. The tip speed ratio (rotational speed/upwind speed) is fixed at 1.75 in order to compare the performance of the selected turbulence models. To operate at a tip speed ratio 1.75 then the rotational speed of the turbine is set to 14 rad/s.

4.2 Numerical Simulations

A pressure based solver, namely the SIMPLE algorithm, is applied and it is one of the segregated algorithms where the velocity and the pressure are coupled and iteratively solved by the finite volume method using the commercial software FLUENT. On the cell boundaries, the gradients are computed using the least squares method. The computation is initiated using the first-order upwind scheme for the spatial discretization and the first-order implicit scheme for the transient discretization. Then, after obtaining a few successive convergences of the residuals, which are set for all the variables to be 10^{-5}, the spatial discretization is set to be the second-order upwind scheme. The number of time steps per revolution is set to be 500, 1,000, 2,500 and 5,000 in order to compare the torque profile as the time stepping changes. Since the transient formulation is set to be a first-order implicit scheme, the smallest number of time steps per revolution, namely 5,000 steps, has been applied for the computations proposed in this paper in order to minimize the transient discretization errors.

The turbine completes one full rotation in 5,000 steps and at each time step the torque produced by the three airfoils is averaged from the previous 5,000 steps. As the turbine rotates, the residuals convergence and the relative difference in the averaged torque are monitored during the computations. The solution is considered to have converged when the relative difference in the averaged torque at all-time steps is less than 2 %, i.e.

$$Convergence = \frac{\sum_{i=1}^{N} (Torque)_i}{N} \leq 0.02 \tag{2}$$

where N is the total number of time steps applied in the numerical computations and i is the index variable for the time steps. The performance of the turbine is obtained by multiplying the averaged torque obtained from the calculations by the rotational speed of the turbine during computations.

5 Results and Discussion

The 2D model of the turbine takes about 8–10 full rotations before a satisfactory convergence is achieved. After obtaining the converged solution for all the investigated turbulence models, it was clear that the turbine peaks in power coefficient when the blades are at azimuthal angles of about 90°–115°, 210°–225° and 330°–345°, whereas it troughs in power coefficient at about 30°, 150° and 270°. The peak and trough locations are found to be consistent to what have been found in the literature [50–52]. It has been observed from the flow visualization of the velocity magnitude between the angles 90°–180° that the flow around the blade passing through this region produces vortices and the flow becomes more complex. By monitoring the flow field characteristics while the turbine rotates during the computations, the azimuthal positions, namely 0°, 120°, and 240°, have been found to be representative of the flow field development of the investigated straight-blade vertical axis wind turbine.

To compare the performance of the turbulence models, it was essential to select the azimuthal angle where the blade is exposed to the wake and stall condition. Therefore, azimuthal angles of 0°, 120° and 240°, where the blades 1, 2 and 3 are located, respectively, have been selected in order to compare the flow field contours for the velocity magnitude. At this azimuthal position, blade 1 is exposed to the almost undisturbed upwind flow, blade 2 is under stall condition, and blade 3 is under the wake effect. The contours of the velocity field are compared for the three blades at the selected azimuthal positions for the Spalart–Allmaras, RNG κ-ε, SST κ-ω, and SST transitional models. The contour levels of the velocity magnitude presented in all figures are fixed to a maximum velocity value of 40 m/s, which is represented by the red colour and a 5 m/s reduction between each level with different contour colours ending with the blue colour in order to compare the flow in the vicinity of the airfoils.

The profile of the velocity approaching the turbine is assumed to be uniform with a predefined turbulence intensity level of 0.5 set at the inlet of the computational domain. Similar velocity profile is development by the investigated turbulence models in the region between the inlet of the computational domain and the turbine upstream. It is important to note that the convection of the flow structure from the upstream region to the downstream region of the turbine is not considered in this study. Also, the comparison of wake structure of the flow behind the turbine is not produced.

It is clear from Fig. 2 that the flow and the stream lines around blade 1 are similar for all the turbulence models. The flow at this stage is not complex, and therefore all the turbulence models produce similar flow field structures. The main reason is that the angle of attack that the upwind makes with the airfoil is relatively small. Also, the local Reynolds number near the blade surface is relatively small.

At the position of blade 2, namely azimuthal angle 120°, the turbulence models produce significant flow field differences as seen in Fig. 3. This is mainly because the blade is under stall condition, which is a complex non-linear phenomenon. The turbulence models designed for fully developed flows, namely the Spalart–Allmaras, RNG κ-ε and SST κ-ω models, show good agreement for the stream lines and the contours of the velocity magnitude flow field. When the transition is taken into account in the SST transitional model, a significant difference is observed on both the stream lines and the contours of the velocity magnitude field. The location and shape of the vortex have been remarkably changed and this suggests that including the transitional effect in the modelling of the SB-VAWT is a crucial decision. The flow separation that causes the vortex formation, and later the shedding to the wake, occurs near the blade surface which is mostly in transition during the turbine operation, and as a result the transitional effect cannot be ignored in modelling this kind of applications.

The general features of the flow become similar as the blade proceeds to the azimuthal angle 240° for blade 3, as shown in Fig. 4. Despite this, it is clear that on including the transition in the modelling then more vortices are captured in the flow.

Since the turbine is in continuous rotation, it is important to investigate the curvature correction of the stream lines, which modifies the production of the turbulence in the turbulence model. The investigated models, apart from the RNG κ-ε model, are not sensitive to the rotation of the system and therefore the stream lines are not corrected during computations.

In Fig. 5, it is clear that including the curvature correction to the SST transitional model is important since the flow field and the vortices have been

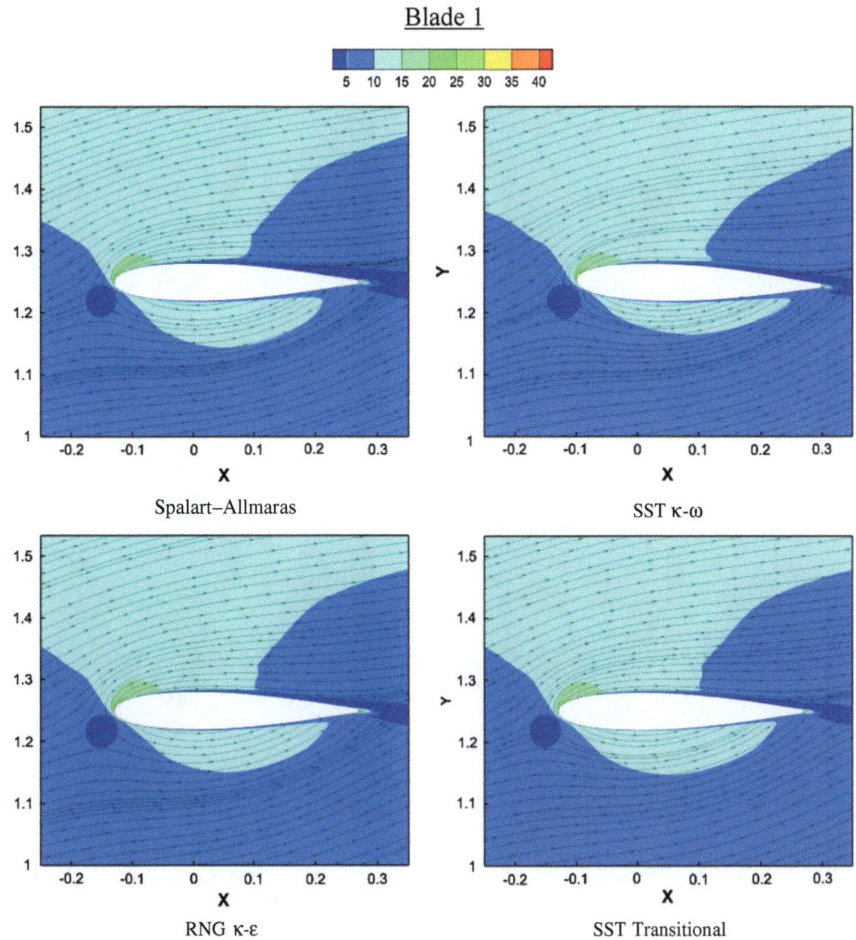

Fig. 2 Contours of the velocity field (m/s) and stream lines of blades 1 of the SB-VAWT at TSR 1.75 for four turbulence models at azimuthal angle 0°

significantly changed when the flow is complex. At the position of the blade 1, it appears that the flow and the stream lines are similar since the rotational effect of the system and the complexity of the flow are relatively small. However, at the position of blade 2, the effect of the curvature correction is clearly seen. The vortex formation is affected by the rotation of the system and therefore it needs to be modified accordingly. It is important to highlight that the turbulence models that are designed for fully developed flows show similarity in the flow field and the stream lines. For blade 3, the general flow features are similar. However, the strength of vortex predicted when applying the curvature correction is higher due to the correction of the streamlines during the calculations for the turbine rotation.

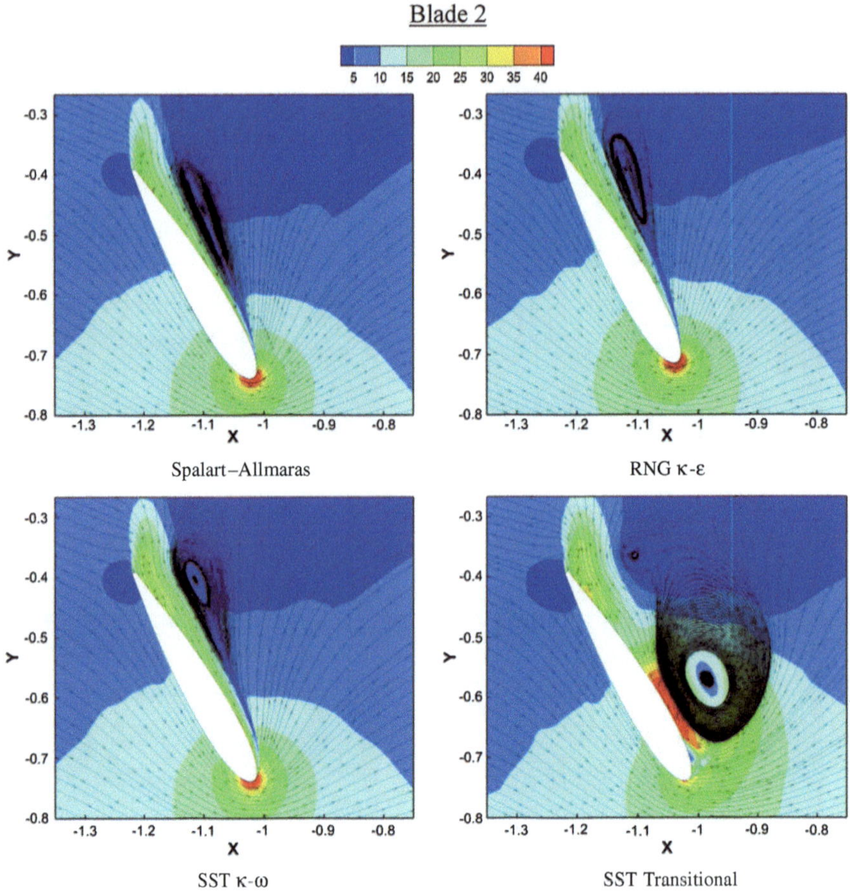

Fig. 3 Contours of the velocity field (m/s) and stream lines of blades 2 of the SB-VAWT at TSR 1.75 for four turbulence models at azimuthal angle 120°

Including the transition in the modelling captures the vortices and these are not predicted when the transitional modelling is not applied. The curvature correction is also important when the blade is under the wake effect or stall condition. It is important to note that the difference in the flow field and the stream lines are not as significant as when the blade is under stall condition. The toque rose needs to be compared in order to closely examine the effect of the transition modelling with and without curvature correction to the development of the flow during turbine.

Figure 6 shows the torque roses generated during the turbine operation at TSR 1.75. The torque at each azimuthal angle is calculated based on the converged torque produced during the computation by all the blades which represent the overall turbine performance during the turbine rotation. The torque produced by

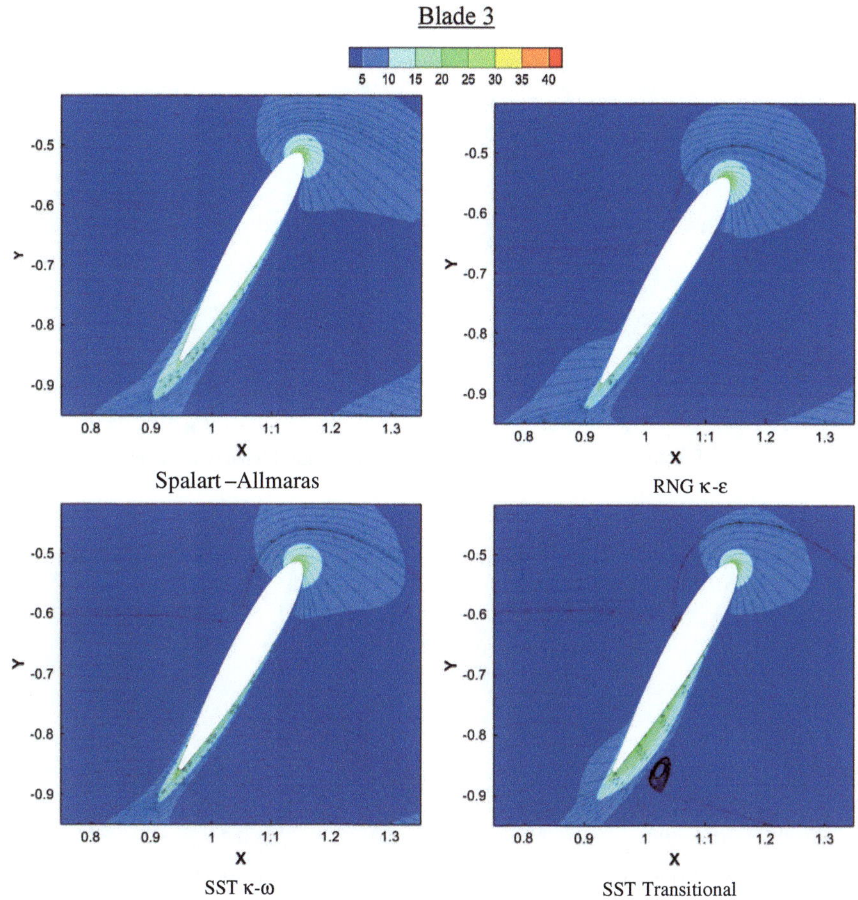

Fig. 4 Contours of the velocity field (m/s) and stream lines of blades 3 of the SB-VAWT at TSR 1.75 for four turbulence models at azimuthal angle 240°

the turbine increases between the angles 30°–65°, 150°–185° and 270°–305°, in other words, none of the blades are in the stall condition, and the turbulence models applied predict similar torque values. At the beginning of the turbine rotation, blades 1, 2, and 3 are at azimuthal angles of 0°, 120°, and 240°, respectively. The torque values predicted by the different turbulence models at this position are not similar for the transitional and the non-transitional models. This is mainly because blade 2, at stall condition, and blade 3 are recovering from the stall and under wake effects. The significant difference in the predicted torque of the turbine occurs when at least one of the blades is under stall condition or wake effect.

The turbine rotates and reaches the position where the blades 1, 2, and 3 are at azimuthal angles 30°, 150°, and 270°, respectively. At this position, all the

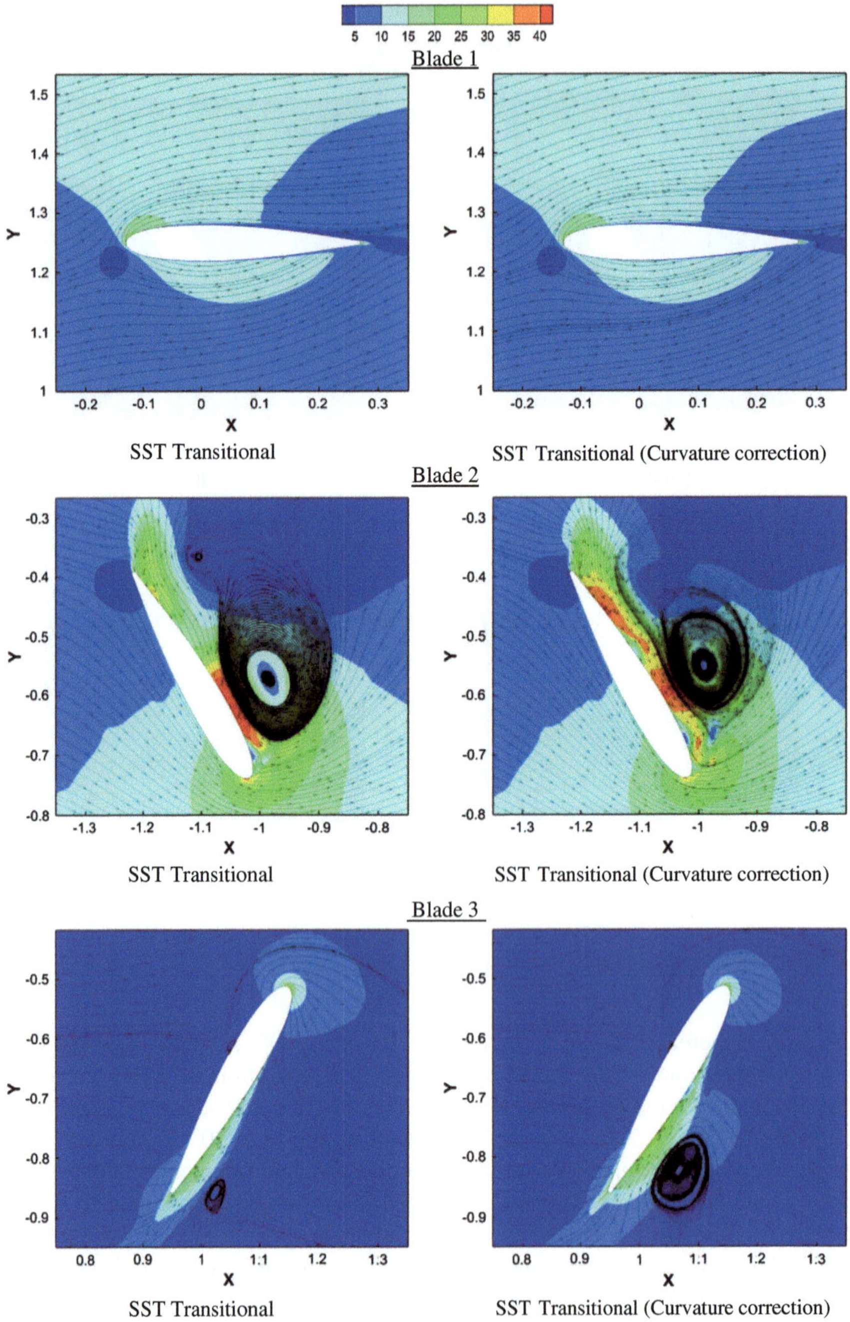

Fig. 5 Curvature correction effect on velocity field (m/s) using transitional model for blade 1, 2, and 3 at azimuthal angle of 0°, 120°, and 240°, respectively

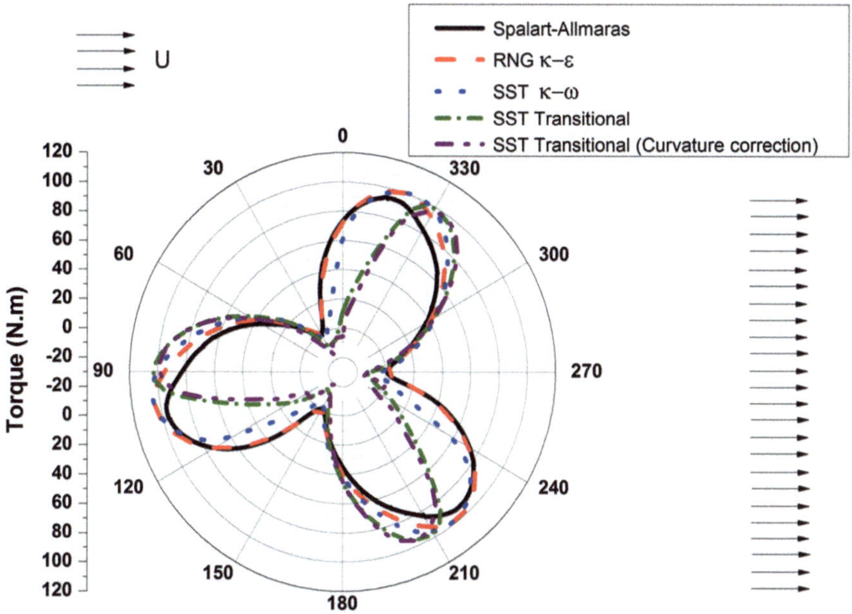

Fig. 6 Torque roses for different turbulence models at TSR 1.75

turbulence models investigated produced similar torque values. A possible reason for this is that all the blades are either under light stall or no stall is present. Another possible reason is that the predicted torque value is already small. It is interesting to note that the turbulence models designed for fully developed flows predict similar results, even in stall condition, despite the changes in the flow field and stream lines. When the transition, both with and without curvature correction, is taken into account then the predicted torque is significantly changed compared to the non-transitional models.

After the position where the blades 1, 2, and 3 are at the azimuthal angles 65°, 185°, and 305°, respectively, the predicted torque value begins to deviate for different turbulence models. However, the transitional model both with and without the curvature correction, predict almost the same torque values until the turbine reaches the position where the torque peaks, namely when blade 1 is at the azimuthal angle 90°. When blade 1 passes the azimuthal angle 90°, the torque drops due to the loss in the lifting force on the blade which is mainly caused by the stall.

When blade 1 travels from the azimuthal angle 90°–150°, blade 2 and blade 3 travel from the azimuthal angles 210°–270° and 330°–30°, respectively. At these azimuthal angles, the turbine recovers from the severe effect of the stall condition and the wake effect. It is apparent that the stall is predicted earlier by the transitional model compared to non-transitional model, namely the Spalart–Allmaras,

RNG κ-ε and SST κ-ω models. The transitional model, both with and without curvature correction, shows a slight difference in the predicted torque value.

It is important to highlight that even though the flow field and the stream lines produced by the transitional model both with and without curvature correction, are significantly different, the overall effect on the turbine performance, namely the predicted torque, is significant. However, this is only true for TSR 1.75 and different tip speed ratios need to be investigated in order to see whether similar conclusions may be obtained.

Examining the boundary layer near the blades at different azimuthal angles may be the key to understanding how the investigated turbulence models vary. Therefore, the normalized velocity magnitude profile is generated, as seen in Fig. 7, at a specified normalized distance for blades 1, 2 and 3 at azimuthal angles 0°, 120°, and 240°, respectively. Six different stations have been selected to compare the results obtained. The velocity magnitude profiles at these stations, namely 1–6, are presented at x = 0.25 and up to a distance of 2.5 % of the chord length (0.025 cm) from the airfoil surface.

At stations 1 and 2 on blade one, the change in velocity magnitude is not significant in both stations and this is mainly due to the domination of the attached flow near the airfoil surface and the absence of separation, as seen in Fig. 8. The small difference in the prediction of the profile of the normalized velocity in the viscous sub layer near the blade, namely the distance from d/c = 0.0025 to

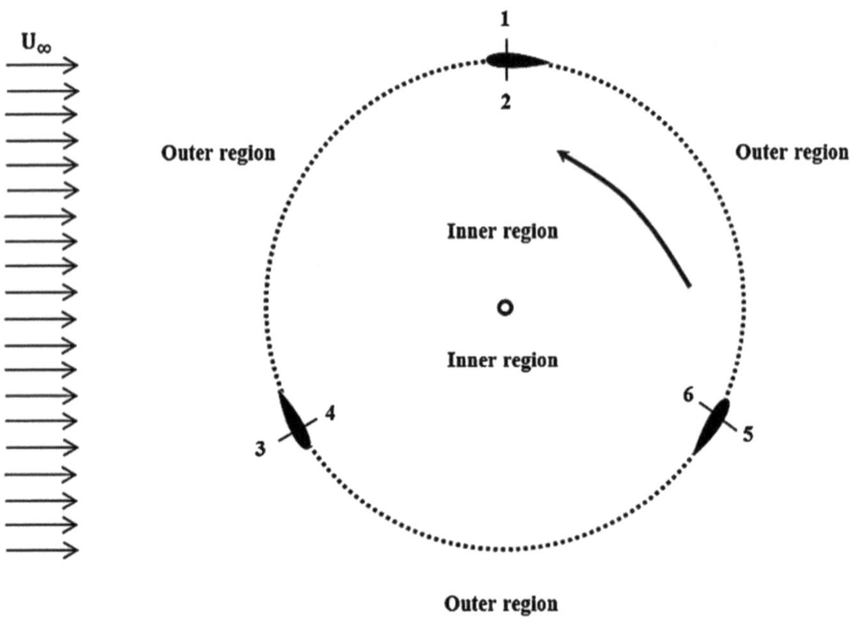

Fig. 7 Schematic of the investigated stations 1-6 of straight blade vertical axis wind turbine (SB-VAWT)

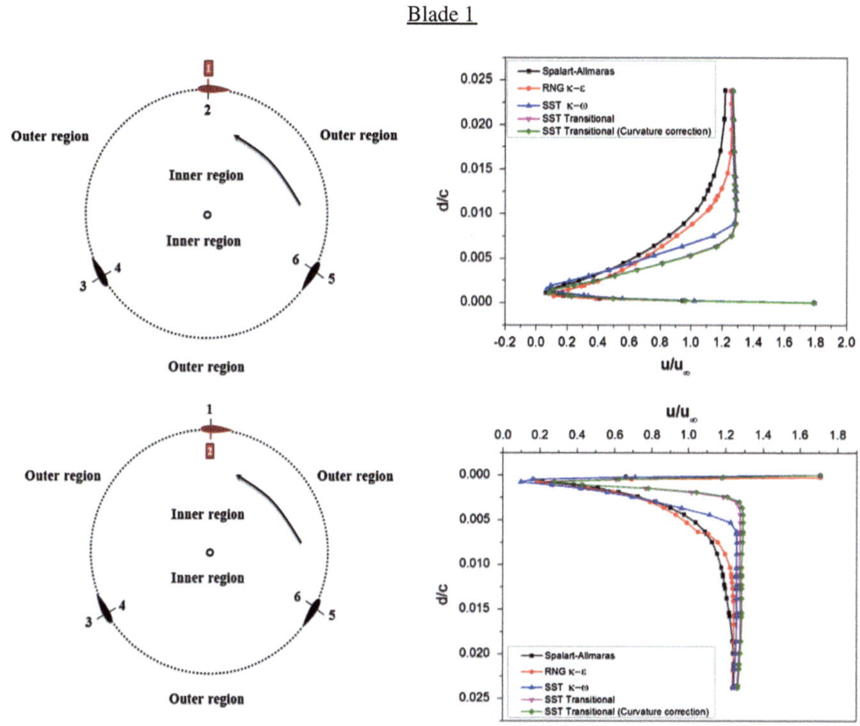

Fig. 8 Normalized velocity magnitude profile verses normalized distance from the airfoil surface at x/c = 0.25 for blades 1 at stations 1 and 2

d/c = 0.0175, is mainly due to the difference in the assumptions that each model is based on. At low Reynolds number, and in the viscous sub layer, this may be a major source of error in the development of the fluid flow and subsequently the prediction of the turbine performance.

However, stations 3 and 4 are under stall condition and the performance of the turbulence models is clearly observed in Fig. 9. At station 3, despite the flow being attached and there are no vortices near the blade surface, the profile of the normalized velocity is not similar for the investigated turbulence models. It is important to note that the transitional model, both with and without curvature correction, predict similar normalized velocity profiles compared to the non-transitional models. At station 4, the modelling of the flow becomes highly sensitive to the turbulence model applied. The profile of the normalized velocity is not quantitatively similar. It is important to note that the curvature correction at this station significantly affects the development of the flow because of the effect of turbine rotation, and as a result the wake structure is substantially corrected.

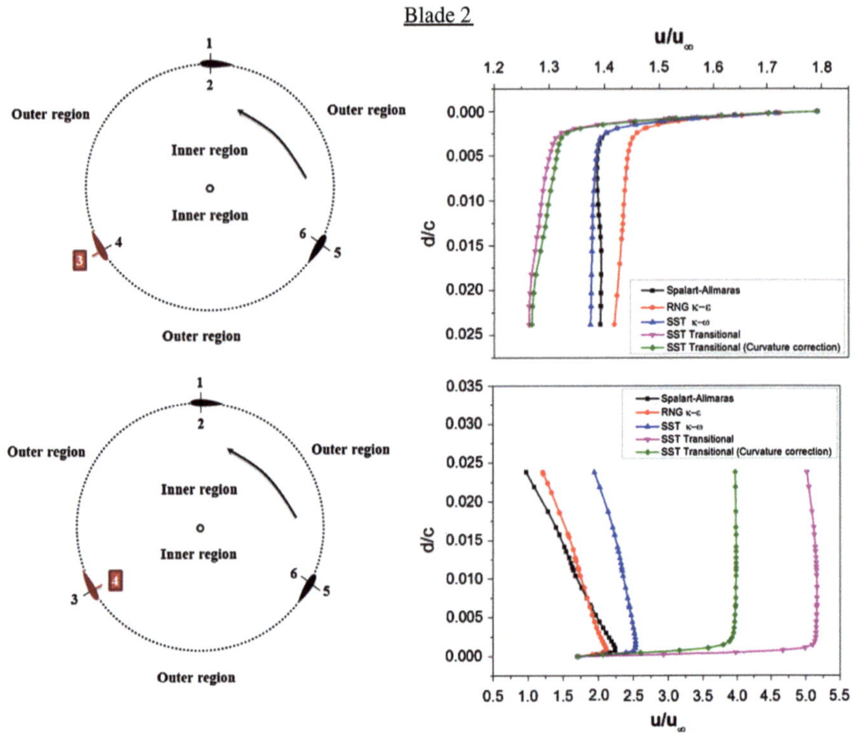

Fig. 9 Normalized velocity magnitude profile verses normalized distance from the airfoil surface at x/c = 0.25 for blades 2 at stations 3 and 4

The disturbance to the flow, and the vortices created upstream of the turbine, is convected downstream of the turbine. The structure of the flow is dependent on the flow development which is based on the turbulence model applied. For station 5, the flow is under these structures, namely the wake effect as shown in Fig. 10. It is observed that the turbulence models predict different normalized velocity profiles. It is interesting to note that the trend of the normalized velocity profiles is not similar. A possible reason for this is that by this stage the effect of the complexity of flow upstream becomes clear on the normalized velocity profiles.

At stations 6, the flow has recovered and becomes attached, and also there is no separation present. Therefore, the predicted normalized velocity profiles for all the employed turbulence models are similar. The viscous layer at this stage is thin, and therefore the turbulence model predictions are almost the same.

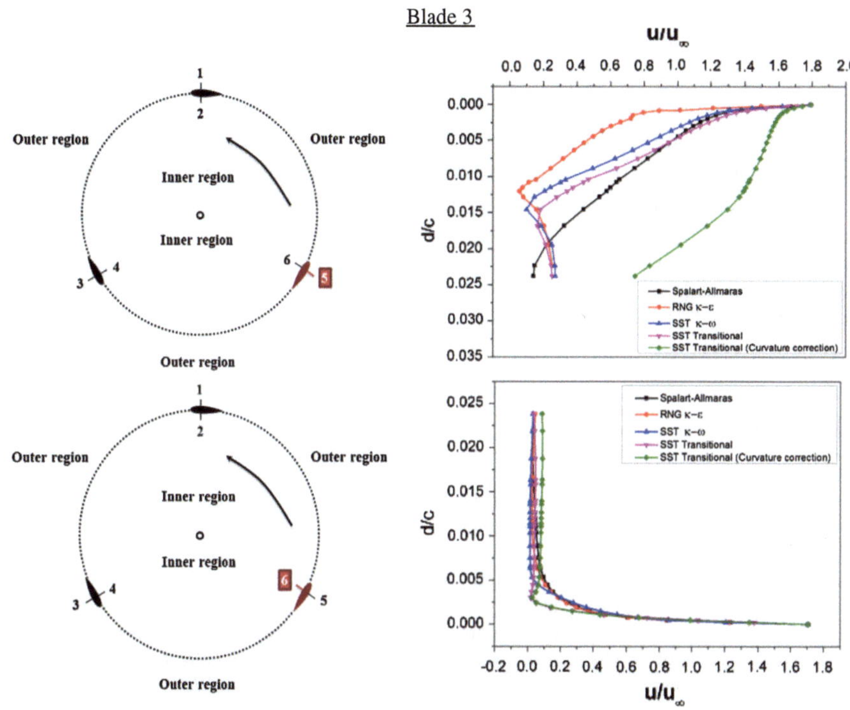

Fig. 10 Normalized velocity magnitude profile verses normalized distance from the airfoil surface at x/c = 0.25 for blades 3 at stations 5 and 6

6 Conclusions

In this paper, four different turbulence models have been investigated for predicting the flow field, stream lines, normalized velocity magnitude profiles and torque roses for a straight blade vertical axis wind turbine. The domain is meshed using quadrilateral cells and the mesh is fixed for all the turbulence models applied. The three turbulence models are based on the assumption of fully developed flow, namely the Spalart–Allmaras, RNG κ-ε, and SST κ-ω models, whereas the fourth turbulence model, namely the SST transitional, accounts for transition. A curvature correction for the stream lines during operation has been applied to the SST transitional model.

The visualization of the flow fields, the stream lines, and the normalized velocity magnitude profiles for all the turbulence models that are based on the assumption of the fully developed flows are similar if the flow is attached. When the transition modelling is included then significant changes are observed in the flow field and the stream lines and the trend of the normalized velocity magnitude profile is similar in the turbine upstream. However, it has been found that in the presence of the stall condition and/or the wake effect, it is important to employ a turbulence model that accounts for the flow transition and to apply the curvature correction to the turbulence production term. In the downstream region of the turbine, the trend in the normalized velocity magnitude profiles are not similar. These conclusions may also be deducted from the torque rose.

In conclusion, not including the transition into the turbulence modelling of the straight vertical axis wind turbine operating at low tip speed ratios is very likely to result in unreliable computational fluid dynamics predictions.

Acknowledgments Khaled M. Almohammadi would like to express his gratitude to Taibah University, Kingdom of Saudi Arabia for supporting him to perform his PhD study in the University of Leeds.

References

1. Wang, S., Ingham, D.B., Ma, L., Pourkashanian, M., Tao, Z.: Numerical investigations on dynamic stall of low Reynolds number flow around oscillating airfoils. Comput. Fluids **39**, 1529–1541 (2010)
2. Paraschivoiu, I.: Wind Turbine Design: With Emphasis on Darrieus Concept. Polytechnic International Press, Canada (2002)
3. Sharpe, D.: A theoretical and experimental study of the Darrieus vertical axis wind turbine. Polytechnic School of Mechanical, Aeronautical and Production Engineering (1977)
4. Holme, O.: A contribution to the aerodynamic theory of the vertical-axis wind turbine. In: International Symposium on Wind Energy Systems, Cambridge, England, pp. C4-55–C4-71 (1976)
5. Strickland, J.: A performance prediction model for the Darrieus turbine. In: International Symposium on Wind Energy Systems, Cambridge, UK, pp. C3-39–C3-54 (1976)
6. Bergey, K.: The lanchester-betz limit. J. Energy **3**, 382–384 (1979)
7. Mathew, S.: Wind Energy: Fundamentals, Resource Analysis and Economics. Springer, Berlin (2006)
8. Lv, Y.Z., Jiang, D.X., Jiang, Y.: Numerical simulation on small scale straight-blade and twisted-blade vertical axis wind turbine. Adv. Mater. Res. **455**, 334–338 (2012)
9. Anderson, J., Wendt, J.: Computational Fluid Dynamics. McGraw-Hill, New York (1995)
10. Versteeg, H., Malalasekera, W.: An Introduction to Computational Fluid Dynamics: The Finite Volume Method. Prentice Hall, NJ (2007)
11. Tu, J., Yeoh, G.H., Liu, C.: Computational Fluid Dynamics: A Practical Approach, 1st edn. Butterworth-Heinemann, Burlington (2008)
12. Islam, M., Amin, M., Carriveau, R., Fartaj, A.: Investigation of low reynolds number airfoils for fixed-pitch straight-bladed VAWT, pp. 5–8
13. Islam, M., Ting, D., Fartaj, A.: Aerodynamic models for Darrieus-type straight-bladed vertical axis wind turbines. Renew. Sustain. Energy Rev. **12**, 1087–1109 (2008)
14. Hinze, J.: Turbulence, vol. 13, pp. 741–773. McGraw Hill Book Company, NY (1975)
15. Launder, B., Spalding, D.: Lectures in Mathematical Models of Turbulence. Academic Press, NY (1979)
16. Spalart, P.R., Allmaras, S.R.: A one-equation turbulence model for aerodynamic flows. La recherche aérospatiale **1**, 5–21 (1994)
17. Yakhot, V., Orszag, S., Thangam, S., Gatski, T., Speziale, C.: Development of turbulence models for shear flows by a double expansion technique. Phys. Fluids A **4**, 1510 (1992)
18. Menter, F.R.: Two-equation eddy-viscosity turbulence models for engineering applications. AIAA J. **32**, 1598–1605 (1994)

19. Reynolds, O.: On the dynamical theory of turbulent incompressible viscous fluids and the determination of the criterion. Phil. Trans. R. Soc. **186**, 123–161 (1895)
20. Jackson, D., Launder, B.: Osborne Reynolds and the publication of his papers on turbulent flow. Fluid Mech. **39**, 19–35 (2007)
21. Papageorgakis, G., Assanis, D.N.: Comparison of linear and nonlinear RNG-based k-epsilon models for incompressible turbulent flows. Numer. Heat Transf. Part B Fundam. **35**, 1–22 (1999)
22. Baldwin, B., Lomax, H.: Thin Layer Approximation and Algebraic Model for Separated Turbulent Flows. AIAA Paper, U.S., pp. 78–257 (1978)
23. Kato, M., Launder, B.: Three-dimensional modelling and heat-loss effects on turbulent flow in a nominally two-dimensional cavity. Int. J. Heat Fluid Flow **16**, 171–177 (1993)
24. Ince, N., Launder, B.: Three-dimensional and heat-loss effects on turbulent flow in a nominally two-dimensional cavity. Int. J. Heat Fluid Flow **16**, 171–177 (1995)
25. Wilcox, D.: Turbulence Modeling for CFD. DCW Industries, Inc., CA (1993)
26. Menter, F., Kuntz, M., Langtry, R.: Ten years of industrial experience with the SST turbulence model. Turbul. Heat Mass Transf. **4**, 2003 (2003)
27. Menter, F.R.: Review of the shear-stress transport turbulence model experience from an industrial perspective. Int. J. Comput. Fluid Dyn. **23**, 305–316 (2009)
28. Langtry, R., Menter, F., Likki, S., Suzen, Y., Huang, P., Völker, S.: A correlation-based transition model using local variables—Part II: Test cases and industrial applications. J. Turbomach. **128**, 423 (2006)
29. Menter, F., Langtry, R., Likki, S., Suzen, Y., Huang, P., Völker, S.: A correlation-based transition model using local variables—Part I: Model formulation. J. Turbomach. **128**, 413 (2006)
30. Wang, S., Ma, L., Ingham, D.B., Pourkashanian, M., Tao, Z.: Turbulence modelling of deep dynamic stall at low Reynolds number. Lect. Notes Eng. Comput. Sci. **2**, 184 (2010)
31. Genç, M.S., Karasu, I., Açıkel, H.H., Akpolat, M.T.: An experimental study on aerodynamics of NACA2415 aerofoil at low Re numbers. Exp. Thermal Fluid Sci. **39**, 252–264 (2012)
32. White, F.M.: Viscous Fluid Flow, vol. 2. McGraw-Hill, New York (1991)
33. Cebeci, T., Mosinskis, G.J., Smith, A.M.O.: Calculation of separation points in incompressible turbulent flows. J. Aircraft **9**, 618–624 (1972)
34. Drela, M., Giles, M.: Viscous-inviscid analysis of transonic and low Reynolds number airfoils. AIAA J. **25**, 1347–1355 (1987)
35. Lian, Y., Shyy, W.: Laminar-turbulent transition of a low Reynolds number rigid or flexible airfoii. AIAA J. **45**, 1501–1513 (2007)
36. Walters, D.K., Leylek, J.H.: Computational fluid dynamics study of wake-induced transition on a compressor-like flat plate. J. Turbomach. **127**, 52–63 (2005)
37. Cutrone, L., De Palma, P., Pascazio, G., Napolitano, M.: Predicting transition in two-and three-dimensional separated flows. Int. J. Heat Fluid Flow **29**, 504–526 (2008)
38. Nakamori, I., Ikohagi, T.: Dynamic hybridization of MILES and RANS for predicting airfoil stall. Comput. Fluids **37**, 161–169 (2008)
39. Abu-Ghannam, B., Shaw, R.: Natural transition of boundary layers—the effects of turbulence, pressure gradient, and flow history. J. Mech. Eng. Sci. **22**, 213–228 (1980)
40. Sørensen, N.N.: CFD modelling of laminar-turbulent transition for airfoils and rotors using the γ-model. Wind Energy **12**, 715–733 (2009)
41. Durbin, P.: Near-wall turbulence closure modelling without "damping functions". Theoret. Comput. Fluid Dyn. **3**, 1–13 (1991)
42. Benini, E., Ponza, R.: Laminar to turbulent boundary layer transition investigation on a supercritical airfoil using the $\gamma - \theta$ transitional model. In: 40th Fluid Dynamics Conference and Exhibit, Chicago, Illinois, p. 4289 (2010)
43. Castelli, M.R., Garbo, F., Benini, E.: Numerical investigation of laminar to turbulent boundary layer transition on a Naca 0012 airfoil for vertical-axis wind turbine applications. Wind Eng. **35**, 661–686 (2011)

44. Wang, S., Ingham, D.B., Ma, L., Pourkashanian, M., Tao, Z.: Turbulence modelling of deep dynamic stall at relatively low Reynolds number. J. Fluids Struct. **33**, 191–209 (2012)
45. Almohammadi, K.M., Ingham, D.B., Ma, L., Pourkashanian, M.: CFD sensitivity analysis of a straight-blade vertical axis wind turbine. Wind Eng. **5**, 571 (2012)
46. Kooiman, S., Tullis, S.: Response of a vertical axis wind turbine to time varying wind conditions found within the urban environment. Wind Eng. **34**, 389–401 (2010)
47. Bravo, R., Tullis, S., Ziada, S.: Performance testing of a small vertical-axis wind turbine. In: 21st Canadian Congress of Applied Mechanics, Toronto, Ontario, Canada (2007)
48. McLaren, K.W.: A numerical and experimental study of unsteady loading of high solidity vertical axis wind turbines. Phd, Mechanical Engineering, McMaster University, McMaster (2011)
49. Almohammadi, K.M., Ingham, D., Ma, L., Pourkashanian, M.: CFD modelling investigation of a straight-blade vertical axis wind turbine. In: Presented at the 13th International Conference on Wind Engineering, Amsterdam, Netherland (2011)
50. Ferreira, C.S., Bijl, H., van Bussel, G., van Kuik, G.: Simulating dynamic stall in a 2D VAWT: Modeling strategy, verification and validation with particle image velocimetry data. J. Phys: Conf. Ser. p. 012023 (2007)
51. Camporeale, S.M., Magi, V.: Streamtube model for analysis of vertical axis variable pitch turbine for marine currents energy conversion. Energy Convers. Manage. **41**, 1811–1827 (2000)
52. McLaren, K., Tullis, S., Ziada, S.: Computational fluid dynamics simulation of the aerodynamics of a high solidity, small-scale vertical axis wind turbine. Wind Energy **15**, 349–361 (2011)

Design Optimization of a Vertical Axis Water Turbine with CFD

S. Laín, O. López, B. Quintero and D. Meneses

Abstract This study presents two-dimensional and three-dimensional numerical simulations of a cross-flow vertical-axis marine current turbine (straight-bladed Darrieus type) with particular emphasis on rotor-performance prediction and hydrodynamic characteristics. Numerical investigations of a model turbine (torque coefficient, power coefficient, tangential force coefficient, normal force coefficient and flow behavior) were undertaken using developed computational models. Turbine design was studied using a time-accurate Reynolds-averaged Navier–Stokes (RANS) commercial solver (ANSYS-CFD). A physical transient rotor–stator model with a sliding mesh technique was used to capture change in flow field at a particular time step. A shear stress transport k-ω turbulence model was used to model turbulent features of the flow. Two-dimensional simulations were employed to test the influence of the profile type and thickness not only in the output power coefficient of the turbine but also on the radial force over the turbine shaft, while three-dimensional simulations were used to compute the curve of power coefficient versus tip speed ratio. Moreover, several flow phenomena as the interference between blades and detached tip vortices and the development of von Karman vortices are identified in the simulations. These phenomena are the reason for the decrease of power coefficient in the three-dimensional case regarding the two-dimensional situation.

S. Laín (✉) · B. Quintero
Energetics and Mechanics Department, Universidad Autónoma de Occidente, Cali, Colombia
e-mail: slain@uao.edu.co

B. Quintero
e-mail: bquintero@uao.edu.co

O. López · D. Meneses
Mechanical Engineering Department, Universidad de los Andes, Bogotá, Colombia
e-mail: od.lopez20@uniandes.edu.co

D. Meneses
e-mail: dp.meneses@uniandes.edu.co

G. Ferreira (ed.), *Alternative Energies*, Advanced Structured Materials 34,
DOI: 10.1007/978-3-642-40680-5_6, © Springer-Verlag Berlin Heidelberg 2013

1 Introduction

Exhaustion of fossil fuels resources combined with greenhouse gas negative impact has recently raised the interest for renewable energies. Among them, hydropower takes a particular place because of its huge potential. Numerous micro-hydro power plants, presently build on rivers and canals, contribute also to the global growth of the hydropower production. Beyond this classical exploitation, hydropower is heading toward huge ocean energy potential, especially kinetic energy of tidal currents. Tides are created by the gravitational attraction of the moon and the sun acting on the oceans of the rotating earth, the relative motions causing the surface of the oceans to be raised and lowered periodically. Tidal power can be extracted by impounding a tide with a barrage to recover potential energy, or alternatively by extracting kinetic energy directly from the tidal stream. Actually, harvesting the tidal current energy rather than the tidal head has lower environmental impact.

Water turbines shapes are inspired from wind turbine shapes. Most of them are driven by lift rather than by drag forces. They can be classified depending on the direction of the rotational axis relative to the water flow direction. Axial flow water turbines (AFWT) have their axis of rotation parallel to water stream direction. Other turbines, cross flow water turbines (CFWT) or Darrieus type water turbines, have rotational axis perpendicular to current direction. A vertical-axis turbine is able to extract power from any direction without adjustment.

Development and optimization of tidal turbines require accurate and time-efficient mathematical models. Based on the computational tools available, different models with different computational costs were developed and applied for optimization and analysis purposes. These models range from computationally inexpensive but low in accuracy momentum models, to three-dimensional computational fluid dynamics (CFD) models of turbine with all the physical details taken into account. It can be concluded from a comprehensive literature review that there are two main families of approaches to numerically model a tidal turbine: potential flow codes and CFD codes. With the use of powerful computers and parallel processing technology, CFD simulations are becoming more popular in industrial and academic sectors [1, 2]. Contrary to potential flow codes, CFD simulations do not need any external data (experimental lift and drag) and can include separation from foils and drag induced vortices from turbine's shaft. Also, they are able to simulate dynamic stall phenomenon (although it is not perfect due to the limitations of turbulence models). CFD modelling is also a powerful tool for complex geometries. However, CFD simulations for tidal turbines still suffer from high computational cost and time. The main advantage of CFD is that it allows reproducing physical unsteady flow around turbine using the so-called sliding mesh methodology, wherein relative motion between steady domain and rotor (unsteady domain) is captured by coupling them through an interface, which is updated at each time step and allows conservative interchange of fluxes between both domains. Rotor grid turns at each time step an angle relative to steady domain. At each time step a new solution is calculated. Transient behavior is built by adding solutions at each time step. In this methodology, integral values (torque) must be averaged in a complete revolution. The main disadvantage of CFD regarding potential flow methods is its higher computational cost (CPU time and memory). This is the reason why there are not many publications applying this methodology to CFWT.

There are not many previous studies approaching the simulation of vertical axis machines, considering air or water as fluid, from the same perspective as suggested here. Ferreira [3], employing air as fluid, presented a detailed state of art of different strategies for predicting aerodynamic characteristics of a VAWT. This author performs an exhaustive study about the ability of different turbulence models (Spalart–Allmaras, k-ε, Detached Eddy Simulation, DES, and Large Eddy Simulation, LES) to reproduce the dynamics of the detached vortices from a single airfoil in two dimensions during its trajectory in the half cycle upstream of a VAWT. The best comparisons with experiments are obtained for LES, but the computational cost is prohibitive to think about useful design and evaluations from a practical point of view. This fact is common when LES is applied to industrial relevant flows.

Also using air as fluid, Maître et al. [4] applied the sliding mesh strategy to a two bladed VAWT using Fluent v. 6.0 software combined with the one equation turbulence model Spalart–Allmaras. These authors defined two zones, an outer fixed zone and another inner rotating zone containing the blades. The results were compared with fairly old experimental data [5], obtaining an overprediction for the measured aerodynamic forces on the airfoils. However, in this paper no geometric details of the considered turbine were given.

Using water as fluid, Nabavi [6] performed a two-dimensional very detailed numerical study about hydrodynamic performance of a three-bladed CFWT introduced in a duct, to accelerate flow upstream the turbine. The author compared the two dimensional computational results obtained with Fluent in free flow conditions with own experimental measurements, resulting in an overprediction of the power coefficient. This result is in line with that obtained by Maître et al. [4] because both used Spalart–Allmaras turbulence model. Additionally, Nabavi [6] tried other RANS models (k-ε, k-ω and Reynolds Stress model) obtaining similar qualitative results. However, the authors believe that the use of the Spalart–Allmaras turbulence model in these specific simulations cannot be recommended in general because of the high adverse pressure gradients appearing in vertical CFWT and the lack of streamlines curvature correction of the model.

Dai and Lam [7] also performed a two-dimensional numerical study of three-bladed CFWT using the software ANSYS CFX v. 11, which is extensively employed for numerical simulation of hydraulic turbo machines. In this case, the turbulence model chosen was the two equation model SST (Shear Stress Transport). As in [6], they validate their numerical results versus own experimental measurements. The quantitative results of the validation are, however, only provided in a point, comparing them with the experiments and also with the results obtained by the double multiple stream tube model. As in the former studies, the averaged values of the torque provided by CFD are above the experimental values. In this study, enough information was provided about geometric parameters of turbine, so this configuration has been chosen in present work.

Howell et al. [8] developed a combined experimental and computational study of the aerodynamics and performance of a small (aspect ratio of 4) vertical axis wind turbine straight-bladed Darrieus type with three blades. Two- as well as three-dimensional numerical simulations were performed using the commercial code Fluent in connection with the k-ε RNG turbulence model. Authors found that the power coefficient of the two-dimensional computations was significantly higher than that of the three-dimensional calculations and the experimental measurements, which was attributed to the presence of the over tip vortices in the three-dimensional situation.

This study performs full transient simulations of flow around a CFWT using CFD tools, including underlying turbulence of fluid flow and also viscous effects, without employing tabulated lift and drag data. This reason has motivated the choice of the sliding mesh method, which is the only one that allows describing the real unsteady behaviour of the CFWT blades in the fluid domain. Two dimensional simulations are used to study the influence of selected geometrical parameters of the airfoil on the performance (torque, tangential force and normal force coefficients) of the turbine. The airfoil thickness and symmetry were the selected factors for the present parametric study. Three-dimensional simulations allow identifying several flow phenomena, as detachment of blade tip vortices and vortex pairing shedding, relevant in the fluid dynamics of cross flow water turbines.

2 Vertical Axis Turbine Operation

Hydraulic operation of a water turbine can be characterized by rotor torque M, rotor drag D, rotor angular velocity ω and power output $P = \omega M$. These values can be made dimensionless as:

Tip speed ratio (TSR),

$$\lambda = \omega R / V_0 \tag{1}$$

Torque coefficient,

$$C_m = \frac{M}{\left(\frac{1}{2}\rho V_0^2 R S_{ref}\right)} \tag{2}$$

Power coefficient or efficiency,

$$C_P = \frac{P}{\left(\frac{1}{2}\rho V_0^3 S_{ref}\right)} \tag{3}$$

Drag coefficient,

$$C_D = \frac{D}{\left(\frac{1}{2}\rho V_0^2 S_{ref}\right)} \tag{4}$$

where R represents the maximum radius of rotor, V_0 incident current velocity, ρ water density and S_{ref}, cross section ($S_{ref} = \pi R^2$ for AFWT and $S_{ref} = 2RH$ for CFWT). Solidity is defined as σ = NC/R with N the number of blades and C the chord length. H is blade span.

Turbine blades rotate around vertical axis with rotational velocity $\vec{\omega}$. In 2D cylindrical coordinates (r, θ), the blade local relative velocity **W** corresponding to an incident flow velocity V_0, is given as:

$$\vec{W} = \vec{V}_0 - \vec{\omega} \times \vec{R} \tag{5}$$

When a blade rotates, its angle of attack α (the angle between local relative velocity and chord) changes leading to variable hydrodynamic forces as (see Fig. 1):

$$\alpha = \tan^{-1}\left(\frac{\sin\theta}{\lambda + \cos\theta}\right) \tag{6}$$

Resultant hydrodynamic force acting on the blades is decomposed in two components (normal F_n, perpendicular to chord; and tangential F_t, parallel to chord). Forces values can be inferred from classical computations over an airfoil in an unbounded domain or from available experiments in wind tunnel tests at fixed α and Reynolds number. The tangential force coefficient is the nondimensional form of F_t and it is directly related to the turbine torque. Negative tangential component is responsible for turbine rotation. In the upstream semicircle, with θ increasing from 0° position, tangential force becomes negative and reaches a minimum near $\theta = 90°$ before increasing until $\theta = 180°$. Same behaviour occurs in downstream semicircle between $\theta = 180$ and 360° positions. In vicinity of $\theta = 0$ and 180° positions, blade has a positive tangential component force F_t, opposed to rotational motion. In this configuration of blades, small tip speed ratios lead to large incidence variations during a revolution. In particular, α becomes very large and overtakes the static stall angle of airfoils, about 12–15°. The normal force coefficient is the non-dimensional form of F_n, so is related to cycle loading and fatigue on the turbine shaft. Since the simulations are performed in a X-Y reference frame, forces coefficients (C_x and C_y) are numerically obtained in this system (see Fig. 1). A simple transformation (rotation) is needed between the X-Y coordinate system and the t-n system of the airfoil (see Eq. 7).

$$\begin{aligned} C_t &= C_{tx} + C_{ty} = C_x \cos(\theta) + C_y \sin(\theta) \\ C_n &= C_{ny} - C_{nx} = C_y \cos(\theta) - C_x \sin(\theta) \end{aligned} \tag{7}$$

Real flows around blades in CFWT may differ from above conclusions because of two reasons: (1) Relative flow passing through a CFWT blade is unsteady; and (2) Oncoming far field seen by a blade is not V_0, but some unspecified velocity. Flow field around a Darrieus type CFWT is inherently unsteady and three-dimensional due to dynamic stall phenomenon experienced by a rotating blade and also to the interference of detached vortices from moving blades [5, 9]. Such vortices stay near the generating blade. Therefore, a strong coupling between them

Fig. 1 Coordinate systems

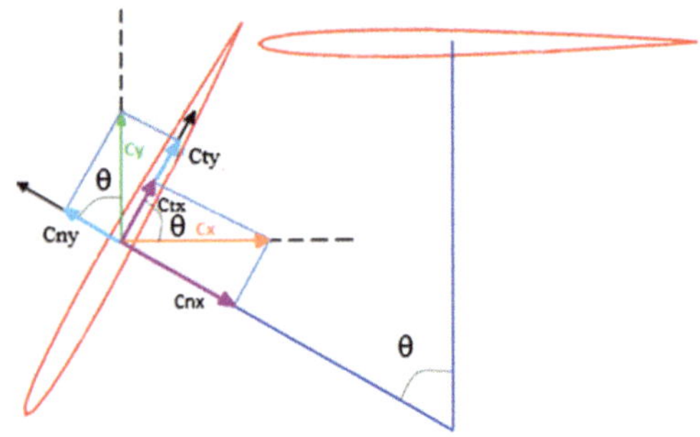

and the flow around the blade increases lift, improving turbine efficiency. However, although the presence of dynamic stall at low tip speed ratios can have a positive impact on the power generation of a wind turbine, the formation of vortices can generate other problems such as vibrations, noise and reduction of fatigue life of the blades due to unsteady forces. Larsen et al. [10] show that dynamic stall is mainly characterized by flow separations at the suction side of the airfoil. This can be summarized in four crucial stages: (1) Leading edge separation starts, (2) Vortex build-up at the leading edge, (3) Detachment of the vortex from leading edge and build-up of trailing edge vortex, (4) Detachment of trailing edge vortex and breakdown of leading edge vortex. The sequence of these four flow events will generate unsteady lift, drag and pitching moment coefficients with a large range of flow hysteresis dependent on the angle of attack [11].

3 Geometrical Configuration and Mesh Generation

CFWT [7] has been chosen in this study due to availability of all geometric data (Diameter D, 900 mm; reference area $S_{ref} = 0.63$ m^2) of the turbine. Span of straight blades (H = 700 mm) are based on symmetric NACA0025 airfoil. Considered case has been 3S2R1 [7] (profile chord, 132.75 mm), resulting in a solidity $\sigma = 0.89$. The turbine rotates with a constant angular velocity of 6.28 rad/s.

In principle, the geometry employed in the simulation was a two-dimensional version of the real three-dimensional turbine. Moreover, neither the supporting arms of the blades nor the shaft have been included as in [7]. The dimensions of the two-dimensional simulation domain were eight rotor diameters in length and five rotor diameters in width, resulting in a blockage ratio of 20 % [7].

The boundary conditions employed in the two-dimensional computations consist of a velocity inlet on the left side, a pressure outlet on the right and two moving walls on top and bottom with the same fluid velocity as the inlet. Moreover, the profiles representing the blades are in the inner part of a rotating ring, which is separated from two steady domains, inner and outer, by two sliding

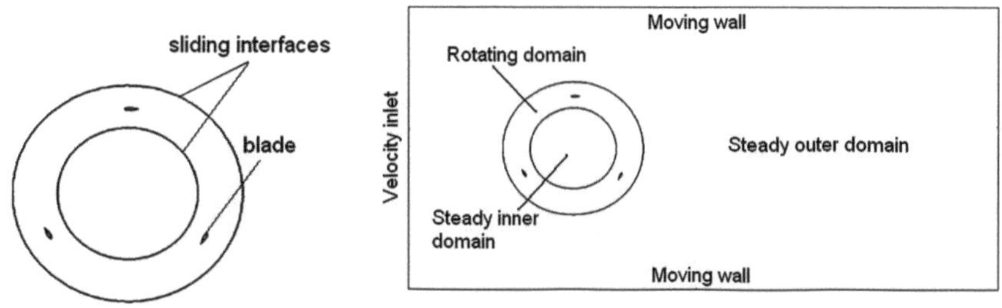

Fig. 2 Geometry employed for the simulated *three-bladed* CFWT described in [7]

interfaces specified as a boundary condition of type sliding mesh. A sketch of the computational domain and boundary conditions are shown in Fig. 2.

In the case of the CFWT, the computational domain consists of a rotating zone (the rotor in a ring-like domain) and a steady zone. The last one includes the water environment outside and inside of the ring-like domain (see Fig. 3).

The computational domain was meshed using a non-structured grid generated with the software GAMBIT. The mesh closest to the profiles must be refined enough to be able to describe with sufficient precision the boundary layer flow. To this end, the created mesh had an O-grid topology based on quads and it is shown in Fig. 4. Outside of this prisms layer, a non-structured grid based on triangles was chosen, keeping an aspect ratio similar to that of the quads [2].

The steady domain was also discretized with a non-structured grid based on triangles (Fig. 3). As it can be observed, the grid node density is higher near the blades than in the rest of the domain. Moreover, due to the complexity of the flow in the turbine wake, also the grid node density is higher downstream than upstream the CFWT.

Obviously, the most interesting zone for the simulation is the ring-like domain, because is here where the flow interacts with the blades, which is responsible for the turbine performance. Again, in this region the grid node density is higher than in the steady domain.

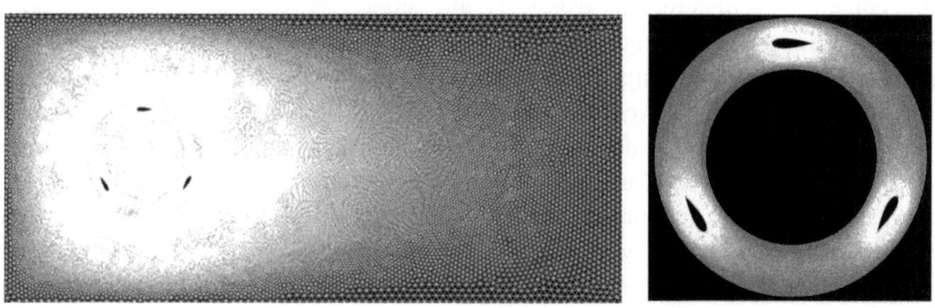

Fig. 3 Details of the generated grid. *Right* detail of the grid in the *rotating ring* containing the profiles

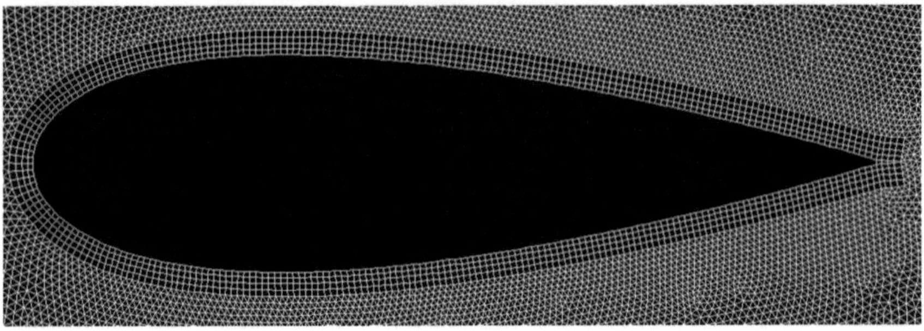

Fig. 4 Detail of the grid near the profile surface showing the prisms layer

4 Numerical Simulation Methodology

Numerical simulations of the straight-bladed CFWT were carried out using the software ANSYS-CFD, based on the finite volume method. Two-dimensional simulations were carried out with Fluent, whereas CFX was employed in three-dimensional simulations due its higher flexibility to deal with complex geometries.

In the unsteady simulation a transient rotor stator model was employed to capture the change of the flow field at a particular time. A moving mesh technique was applied in order to rotate the turbine blades at a constant rotational speed and the Shear Stress Transport (SST) k-ω turbulence model was used to model the turbulent features of the flow. This method is a combination of the k-ε and k-ω models [12]: it uses the k-ω model near the wall and switches to a function of the k-ε model when moving away from the wall closer to the upper limit of the boundary layer. The SST k-ω model has been shown to give superior results for flows with strong adverse pressure gradients such as those appearing in the CFWT flow configuration [11], being able to describe the generation of specific vortices at the leading and trailing edges respectively. The governing equations of the SST k-ω turbulence model are given in Menter [12].

The effectiveness of physical transport within the solver depends not only on the turbulence model but also on the discretization scheme. The diffusive term in the equations is discretized using second order centered differences as usual. However, for the advection term a second order upwind scheme is utilized. The pressure–velocity coupling algorithm chosen has been transient Semi-Implicit Method for Pressure-Linked Equations (SIMPLE). Finally, the time integration is performed by a second order implicit scheme to obtain a good resolution in time.

Typically, the simulation starts with the computation of the steady flow around a fixed position of the turbine blades. From this initial condition, the transient simulation begins, firstly with first order schemes to ease convergence. Once that the total torque on the turbine has reached a quasi-periodic regime, after three or four complete rotor revolutions, the discretization schemes are switched to second order. Finally, the simulation runs during a sufficient number or rotor revolutions in the quasi-periodic regime to extract an average value for the torque, which is used to estimate the turbine performance.

5 Two-dimensional Verification and Validation

The verification process of the CFD simulation implies to perform calculations in different grids varying number of elements, for evaluating convergence of most relevant variables, which in this case has been the non-dimensional torque transferred from fluid to blades, C_m. The number of grid nodes were: coarse, 60,558; medium, 157,130; and fine, 297,302. Results of validation in three different grids for average torque coefficient were: coarse, 0.1200; medium, 0.1399; and fine, 0.14352. Along a complete revolution of blades (at $\lambda = 1.745$), the C_m values obtained for three grids (Fig. 5) are presented. As a result, the medium grid is selected as it represents a compromise between precision and computational cost.

Moreover, a temporal verification study was performed for the medium grid. As a result, a time step of $\Delta t = 5$ ms was adopted, which is the same employed in [7]. Figure 6 shows the torque coefficient obtained with different time steps along a complete revolution of the blades for a tip speed ratio of $\lambda = 1.745$.

After a time of around 6 s, C_m reached a quasi-periodic regime after initial transient. Both maxima and minima of C_m by each blade have higher absolute value than total C_m due to existence of cancellations and compensations of torque among blades, resulting in a total C_m lower than that experienced by a single blade.

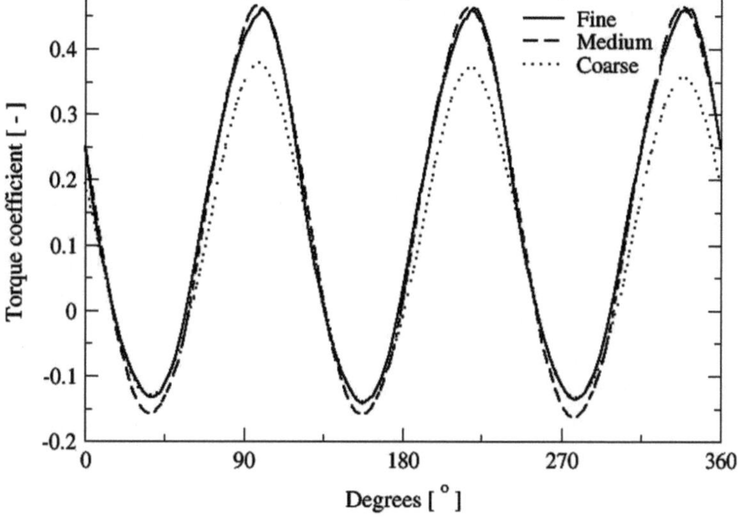

Fig. 5 Torque coefficient at $\lambda = 1.745$ in the grid convergence study

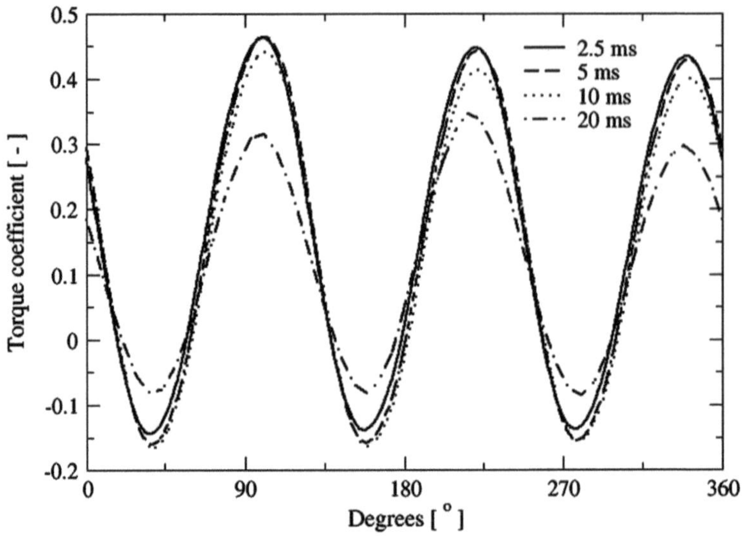

Fig. 6 Torque coefficient for different time steps in the medium grid, $\lambda = 1.745$

The flow field around the turbine is quite complex because, as the blades rotate, high and low velocity zones appear, leading to a detachment of the boundary layer in certain angular positions. Moreover, the flow behind the CFWT is characterized by low velocities which imply a smaller contribution to the total torque than in the upstream region. As a result, in a turn, each blade produces a positive C_m in about a third of revolution, whereas in other two thirds, C_m is slightly negative. After averaging the three coefficients generated by blades, a nearly sinusoidal curve is obtained with three positive maxima at each turn (Fig. 7) and three negative minima, meaning that during a revolution there are periods of time where turbine produces torque on fluid. The number of maxima at each turn equals number of blades of turbine.

The plot of average torque coefficient versus tip speed ratio λ (Fig. 8) shows positive values up to λ close to 2.1, meaning that the fluid is providing torque to the turbine. Beyond $\lambda = 2.1$ the torque coefficient is negative which indicates that the turbine, rotating at constant angular speed, exerts torque on the fluid. This situation appears because a high tip speed ratio implies a high turbine angular speed and, in such case, the kinetic energy contained in the flow is not enough to deliver torque to the CFWT and making it to rotate with the same angular velocity. On the other hand, the curve $C_m(\lambda)$ presents a maximum around $\lambda = 1.35$ and decreases for lower values of the tip speed ratio. This behavior is due to the fact that, for low values of λ, the flow around the blades is separated implying low lift and high drag. As a result the transferred torque from the fluid to the turbine decreases.

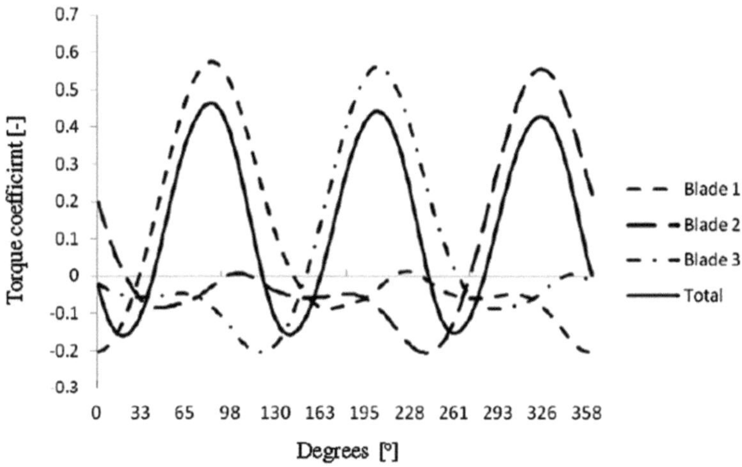

Fig. 7 Generation of torque coefficient by each blade in a revolution

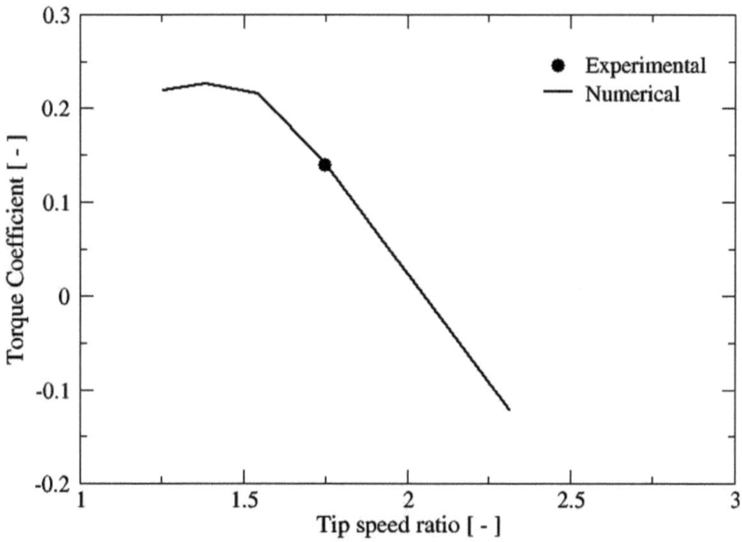

Fig. 8 Average torque coefficient versus tip speed ratio λ

Moreover, as the power coefficient equals the torque coefficient times the tip speed ratio, a behavior of C_P similar to that of C_m is expected. It can be observed in the solid curve of Fig. 9. The maximum power coefficient predicted by CFD is 33 % at $\lambda = 1.6$. Similar C_P curves are found in other CFWT [1, 6]. Unfortunately, Dai and Lam [7] only provide data for a single point with $\lambda = 1.745$, instead of the full $C_P(\lambda)$ curve, which is also shown in Fig. 9.

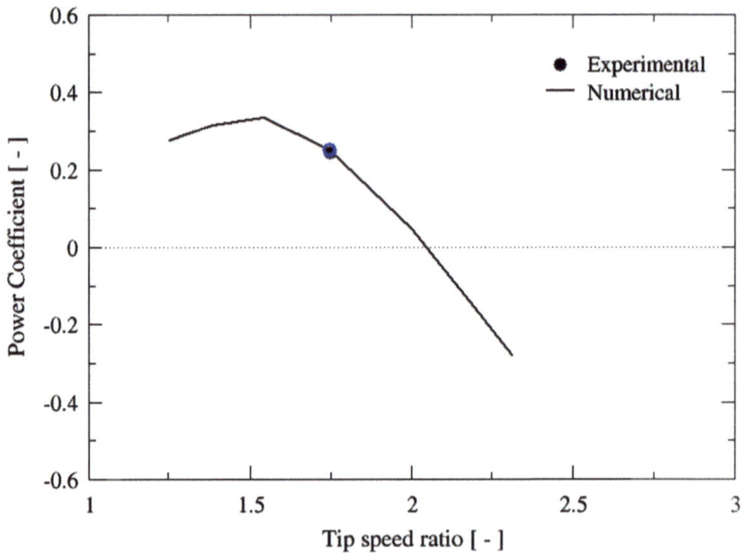

Fig. 9 Average power coefficient versus tip speed ratio λ

6 Two-dimensional Optimization Parametric Studies

For the parametric two-dimensional study, the same characteristics of the case 3S2R1 of Ref. [7] were used and kept constant except for the airfoil geometry which changes between computational experiments. In this case, the commercial software Gridgen was used to generate the mesh and to ensure that all the computational grids had the same topological details. The experimental design (a factorial 3^2) for the parametric study consists in developing a series of tests in which the geometry of the airfoil is changed to determine its influence on the performance of the turbine. The airfoils used here are NACA 4-digit series, which are characterized by its camber and thickness [13]. These two geometric variables of the airfoil profile were used as factors, each one with 3 levels as shown in Table 1. Figure 10 shows the different profiles used in the present parametric study, in total 9 different meshes were generated for each computational experiment proposed for the parametric study.

A Gridgen script written in Glyph language was created in order to ensure that the mesh generated is similar for all the computational experiments. The mesh generation starts by defining the airfoil profile with the maximum thickness (t) as a fraction of the chord, the maximum camber (m) also as a fraction of the chord and the location of the maximum camber (p) as a tenth of the chord. With these parameters the lower and upper surfaces of the airfoil are created and divided in a fixed number of elements. Then, a hyperbolic mesh is generated from the profile in order to define the boundary layer region. The script allows controlling the size of

Table 1 Factors and levels for the design

Factor	Level		
	Low	Medium	High
Thickness (as percentage of c)	6	12	15
Camber	00XX	24XX	44XX

the first element, growth rate, the size of the boundary layer region and the number of cells on it.

The size of the first element and the growth rate are important to capture the physics of the flow close to the airfoil (Fig. 11 left). Depending on the radio and the number of blades of the turbine, the airfoil and the boundary layer mesh is copied and pasted in the correct location. Then the rotating domain and the steady inner domain are created and meshed with triangular elements. The growth rate between the boundary layer region and the rotating domain is also specified in order to achieve a smooth transition between the structured and unstructured meshes (Fig. 11 right). Finally, the steady outer domain is generated and meshed with triangular elements. The growth rate in the free stream direction can be also controlled and specified. The different parameters in the mesh generation were fixed and their values were achieved with a grid convergence study.

Integral results based on moment, tangential force and normal force coefficients were statistically analyzed using the commercial software Minitab. Moment coefficient (C_m) was computed in the turbine shaft showing a similar behavior and magnitude to Fig. 7. The influence of thickness and camber on the average C_m for each airfoil profile is shown in Fig. 12. It is clear that the thickness is the most important parameter in the turbine performance at this point of operation ($\lambda = 1.745$). As the airfoil becomes thicker, better performance is achieved. However, C_m does not improve significantly for airfoils thicker than 15 % of its chord. In order to confirm the observed influence of the thickness an Analysis Of Variance (ANOVA) [14] was performed, in which the *p* value found for the thickness is below 0.05 while for the camber was 0.193. Camber seems to have an important influence when thin airfoils are used, but it does not make a difference as the airfoil thickness increases. Figure 13 shows the evolution of C_m for the thicker airfoils with different camber, in which it is observed that there is no significant difference in the averaged C_m. Nevertheless, NACA4415 shows a smaller negative C_m region which is desirable for smooth turbine operation.

The tangential force coefficient (C_t) has also a cyclic behavior related to C_m. Since the behavior of C_t is similar for all the blades, only one blade whose motion starts at $\theta = 0°$ will be analyzed. Figure 14 shows the evolution of C_t for one blade with different camber. It is clear that NACA0015 airfoil presents the deepest minimum which is a desired condition since the performance increases, but it is also observed that between 210 and 40° this airfoil does not produce torque. As the camber increases the contribution of the airfoil in the total torque improves (between 250 and 360°). This observation supports the argument that camber

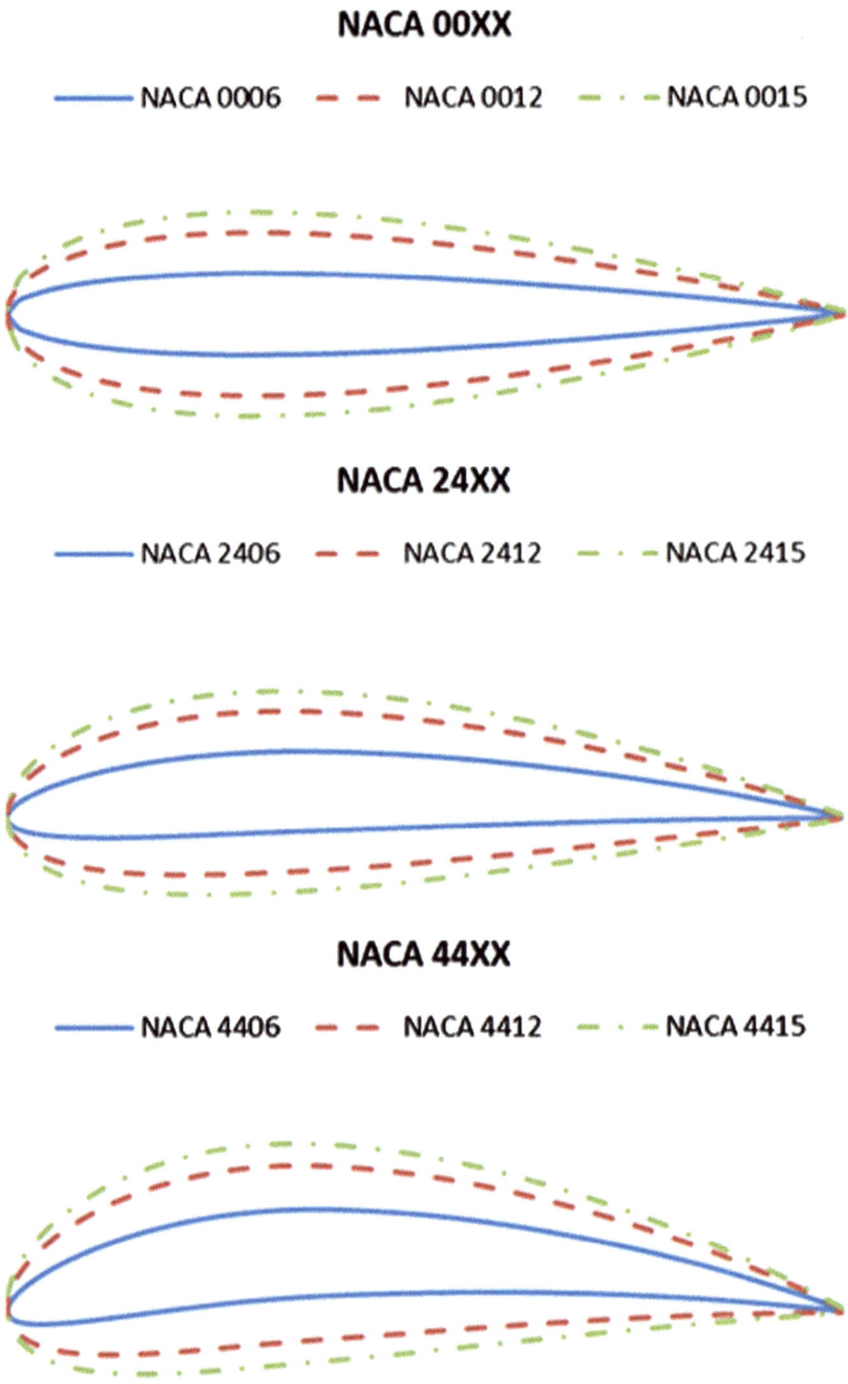

Fig. 10 NACA airfoil profiles used in the parametric study

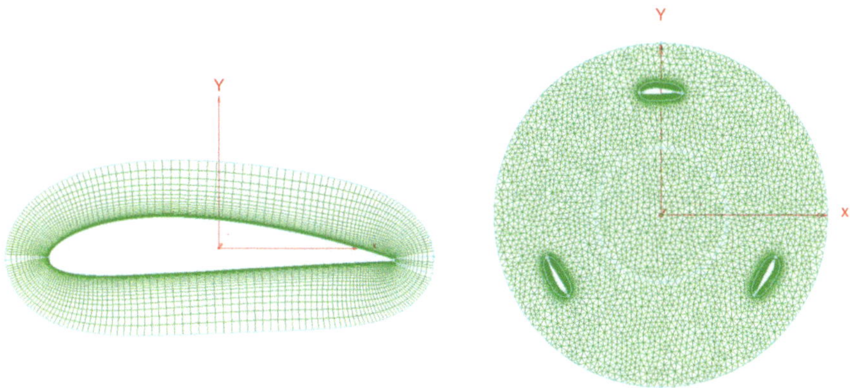

Fig. 11 Generation of the hyperbolic mesh around the profile (*left*) and mesh in the *inner* domain (*right*)

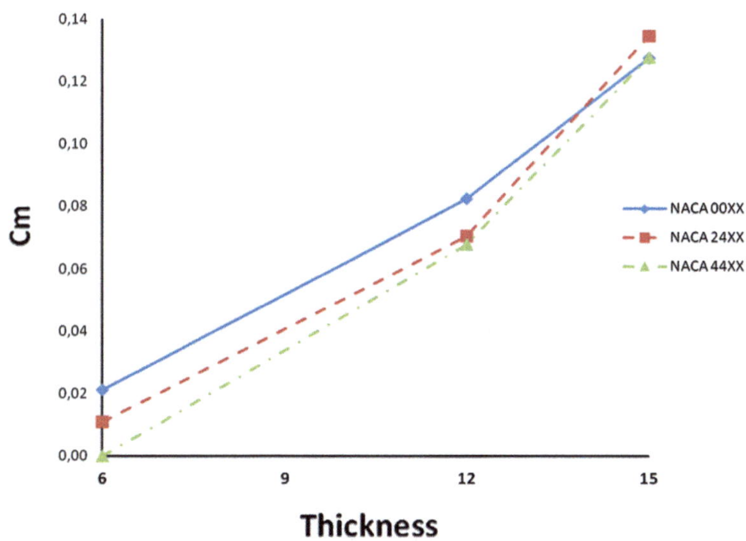

Fig. 12 Influence of symmetry/asymmetry and thickness on C_m

could improve a smooth turbine operation, since during one rotation of the turbine at least one of the blades is always producing negative tangential force. Average values of C_t for one blade are computed and compared (Fig. 15), it is clear that the thickness is the most important factor that influences the tangential force. It is also observed that the camber drives the averaged tangential force into more negative values in particular for thicker airfoils. In order to confirm the observed influence of the thickness an ANOVA was performed, in which the p-value found for the thickness is below 0.05 while for the camber was 0.41.

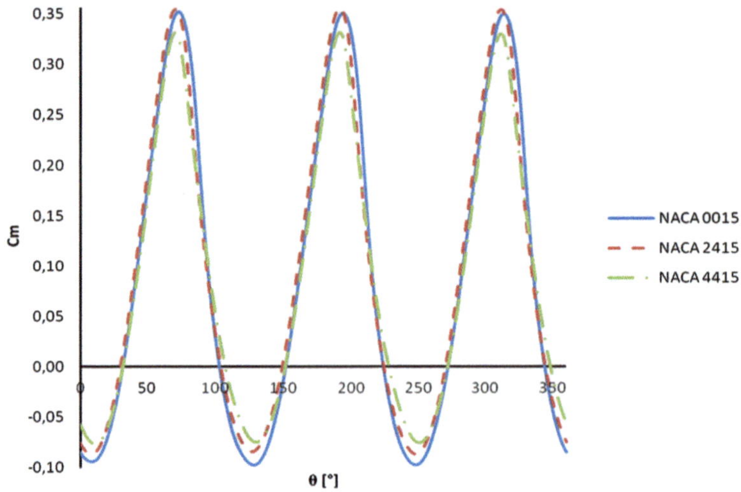

Fig. 13 C_m comparison for different airfoil camber

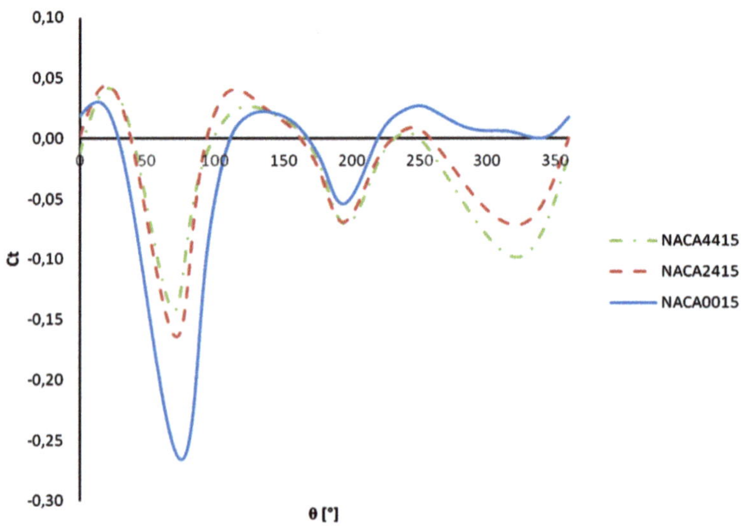

Fig. 14 C_t comparison for one blade for three different camber

The normal force coefficient (C_n) has also a cyclic behavior. Similar to the analysis performed to C_t, only one blade whose motion starts at $\theta = 0°$ will be studied. Figure 16 shows the evolution of C_t for one blade with different camber. It is observed that the curve is shifted towards zero as the camber increases. The influence of the airfoil thickness and camber is shown in Fig. 17, in this case the camber is the factor that primary influence the normal force. This observation is

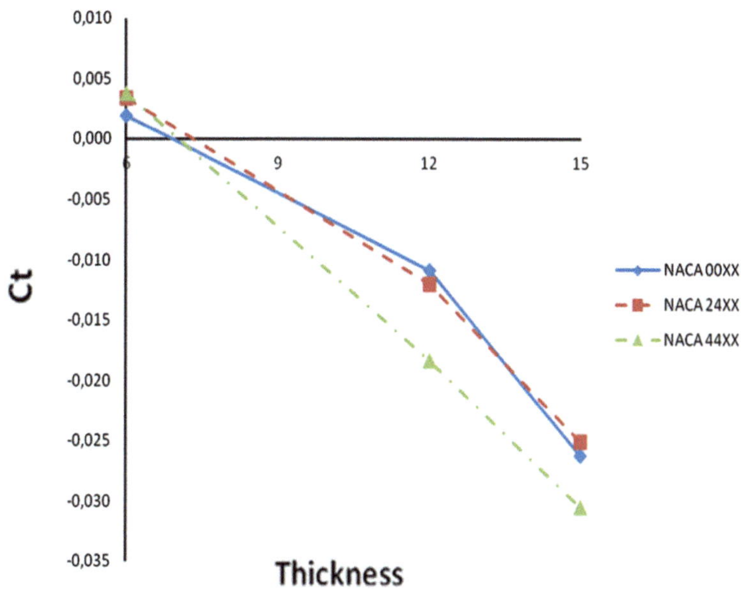

Fig. 15 Influence of symmetry/asymmetry and thickness on C_t

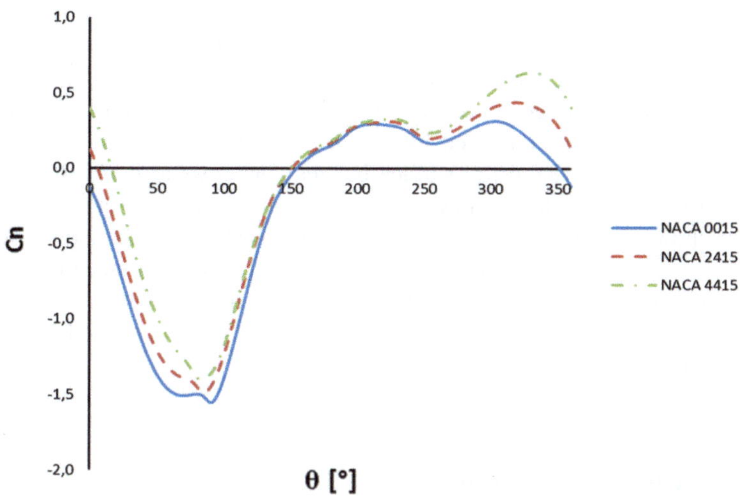

Fig. 16 C_n comparison for one blade for three different cambers

confirmed with an ANOVA, in which the p-value found for the camber is below 0.05 while for the thickness was 0.47. The average normal force is related to the resultant force on the turbine shaft. From this analysis, it is concluded that a cambered profile could improve the resistance of a specific turbine design to fatigue failure.

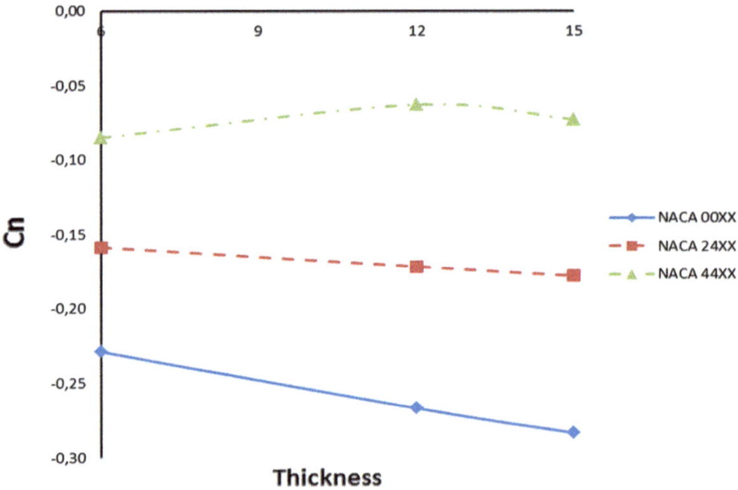

Fig. 17 Influence of symmetry/asymmetry and thickness on C_n

Figure 18 shows the evolution of the vorticity field (magnitude) during one cycle of the turbine with the NACA4415 airfoil. When the cycle starts ($\theta = 0°$) a vortex forms in the trailing edge of the blade. At 90°, a leading edge vortex starts to form and grow in the lower surface of the airfoil. This observation is consistent with the increase in C_n since the lower surface of the airfoil is exposed to a very low pressure while the upper surface of the airfoil has a high pressure. At $\theta = 125°$ the leading edge vortex and the trailing edge vortex joint and it is clear that the wake is being convected by the free stream flow. When $\theta = 180°$ the vortex detaches from the airfoil surface but remains close to it. As the vortex is convected by the free stream flow, its strength diminishes and interacts with the blade as shown in Fig. 18e. It is also observed that at this angle another leading edge vortex starts to be formed but in this case in the upper surface of the airfoil. At the end of the cycle, a very strong wake is formed which is convected downstream. Moreover, the airfoil interacts with the wake generated by the previous profile (low pressure region).

7 Three-dimensional Simulations

The geometry employed in the three-dimensional simulation is a simplified version of real turbine. Therefore, as in the 2D simulations, neither supporting arms of blades nor shaft have been included [7]. As in the two-dimensional computations, a sliding mesh was used to model the rotor geometry. Moreover, a symmetry plane was used at rotor mid span to reduce the computational requirements. Dimensions

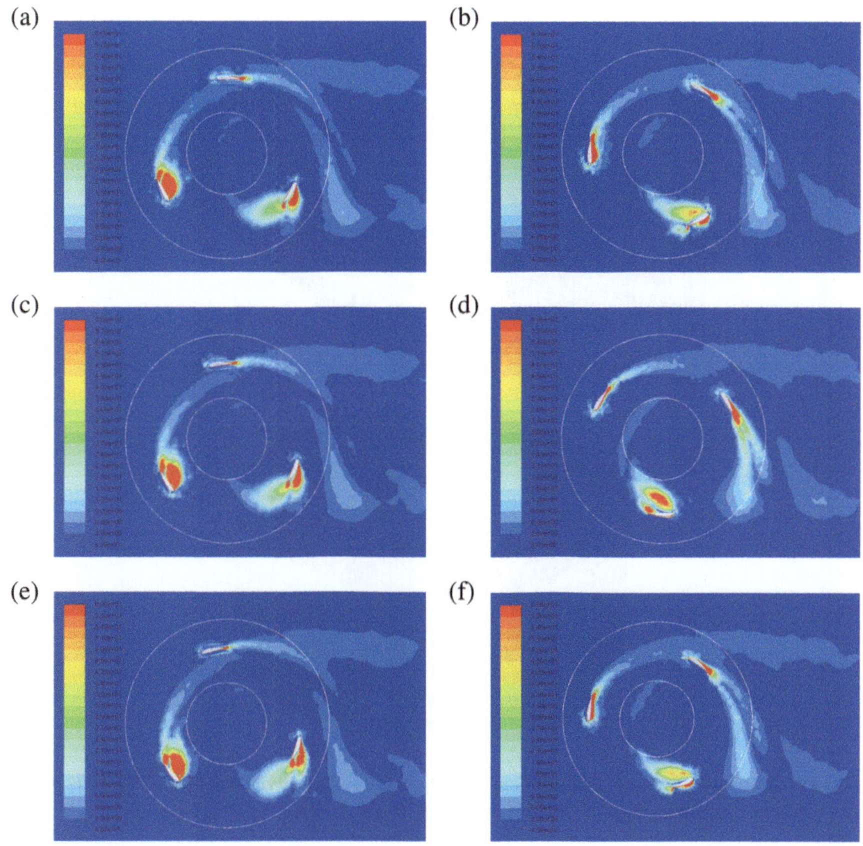

Fig. 18 Vorticity field (magnitude in 1/s) evolution in one cycle. **a** $\theta = 0°$, **b** $\theta = 90°$, **c** $\theta = 125°$, **d** $\theta = 180°$, **e** $\theta = 250°$, **f** $\theta = 325°$

of computational domain are: length (y), 20D; width (x), 13D; height (z), 3.5D, where D is the length of the diameter of the rotor.

Following the standard notation of CFD, boundary conditions employed in the computations (Fig. 19) consist of a velocity inlet on south side, a pressure outlet on north side and free slip walls on west, east as well as on top side; south side was specified as a symmetry plane at rotor mid span. The blades are in inner part of a rotating ring, which is separated from two steady domains (inner and outer) by two sliding interfaces specified as a boundary condition of type sliding mesh. In case of CFWT, computational domain consists of a rotating zone (rotor in a ring-like domain) and a steady zone, which includes water environment outside and inside of ring-like domain (Fig. 19 bottom).

The three-dimensional computational domain was meshed using a non-structured grid generated with the software ICEM CFD. As in the two-dimensional case, the mesh closest to profiles must be refined to describe with sufficient

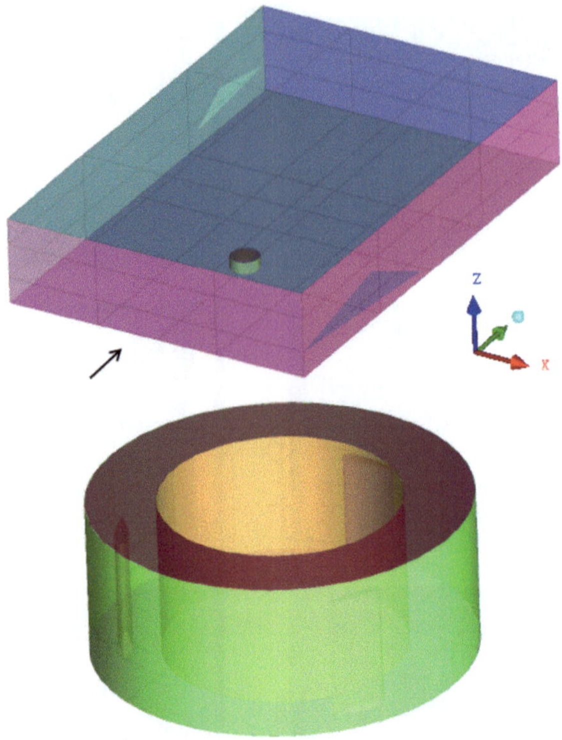

Fig. 19 Schematics of the three-dimensional geometry considered for the CFWT. *Top* total computational domain. *Bottom* rotating domain with the blades

precision the boundary layer flow. The created mesh had an O-grid topology based on prisms. Outside of this prisms layer, a non-structured grid based on tetras was chosen, keeping an aspect ratio similar to that of prisms (Fig. 20). Grid node density is higher near blades than in rest of the domain.

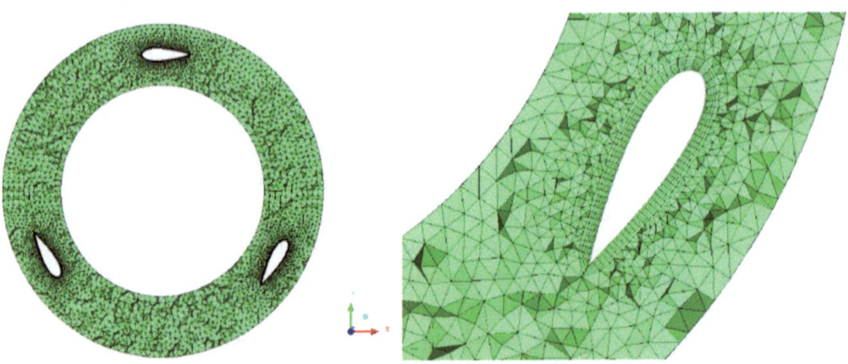

Fig. 20 Detail of the 3D mesh in the ring-like domain (*left*) and around the blade (*right*) showing the prisms layers

Steady domain was also discretized with a non-structured grid based on tetras. Moreover, due to complexity of flow in turbine wake, also grid node density is higher downstream than upstream CFWT. The most interesting zone for simulation is the ring-like domain, where flow interacts with blades, which is responsible for turbine performance. Also, in this region, the grid node density is higher than in the steady domain. In the final mesh the number of elements was around 1.370000.

Same as in the two-dimensional simulations, the employed turbulence model was SST [12] in combination with second-order discretization of all spatial terms of the partial differential equations and using a second order implicit scheme for the time integration with a time step of 5 ms.

Figure 21 presents the obtained results for the power coefficient versus tip speed ratio in the present 3D transient simulations together with the 2D results shown previously and the experimental point of Dai and Lam [7]. As expected, the 3D results are below the 2D values because the additional losses due to the tip blade vortices, which are naturally absent in the 2D computations (see below). The trailing tip vortices are associated with the generation of lift in three-dimensions and they are unavoidable side-effect of it. Such trailing vortices induce a downward component of the fluid velocity in the neighborhood of the blade reducing the angle of attack and creating a component of the drag, known as induced drag. Such drag component reduces the effective torque that fluid delivers to the turbine rotor and therefore, also its performance.

Similarly to the discussion given in the two-dimensional case, the plot of average power coefficient versus tip speed ratio λ (Fig. 21) shows positive values up to λ close to 2.1, meaning that the fluid is providing torque to the turbine. Beyond $\lambda = 2.1$ the power coefficient is negative which indicates that the turbine, rotating at constant angular speed, delivers work to the fluid. This situation appears because a high tip speed ratio implies a low incident velocity and, in such case, the kinetic energy contained in the flow is not enough to deliver power to the CFWT to keep constant angular velocity. On the other hand, the curve $C_p(\lambda)$ presents a maximum around $\lambda = 1.55$ and decreases for lower values of the tip speed ratio.

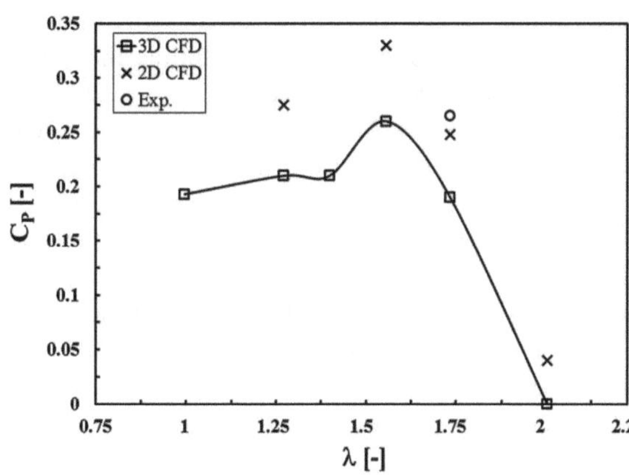

Fig. 21 Power coefficient versus tip speed ratio curve for the considered turbine. 3D versus 2D results

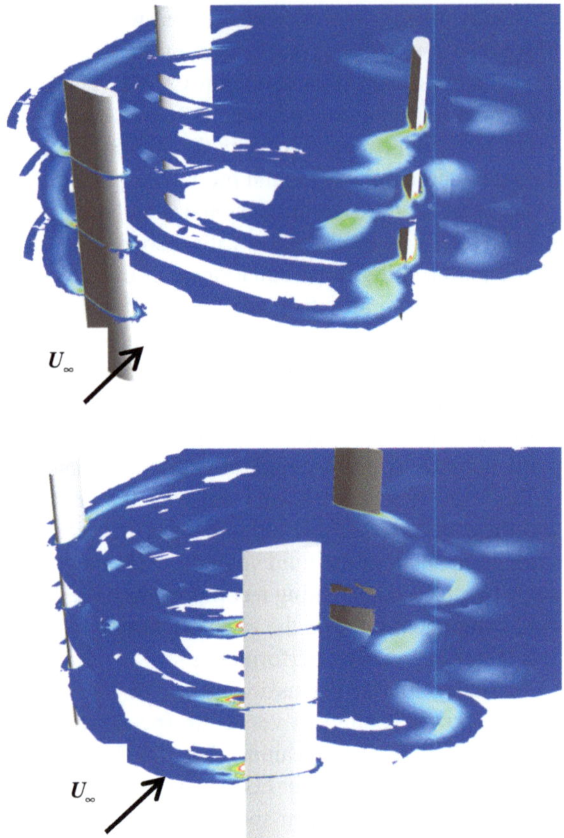

Fig. 22 Detail of the flow around the blades in two different angles in a revolution with a phase lag of 60°. Vortex shedding of the blades trailing edge is clearly observed

This behavior is due to the fact that, for low values of λ, the flow around the blades is separated (dynamic stall) implying low lift and high drag. As a result, the transferred power from the fluid to the turbine decreases.

In order to illustrate some important flow physics occurring in the three-dimensional simulations, Figs. 22 and 23 show, in isometric perspective, the vortex shedding along the blade, including the blade tip trailing vortices. In these, it can be clearly seen that those vortices travel downstream eventually interacting with the other blades. Moreover, in Fig. 22 the so-called von Karman vortex street of the detached vortices from the blade trailing edge can be identified. The von Karman vortex street denotes a repeating pattern of swirling vortices caused by the unsteady separation of the fluid flow from the blades. Over a certain range or Reynolds numbers, eddies are shed continuously from both sides of the blades forming rows of vortices in their wakes. When a single vortex is shed, an asymmetrical flow pattern forms around the blade and changes the pressure distribution. This means that the alternate shedding of vortices can create periodic lateral forces on the blade, causing it to vibrate. If the vortex shedding frequency is similar to the

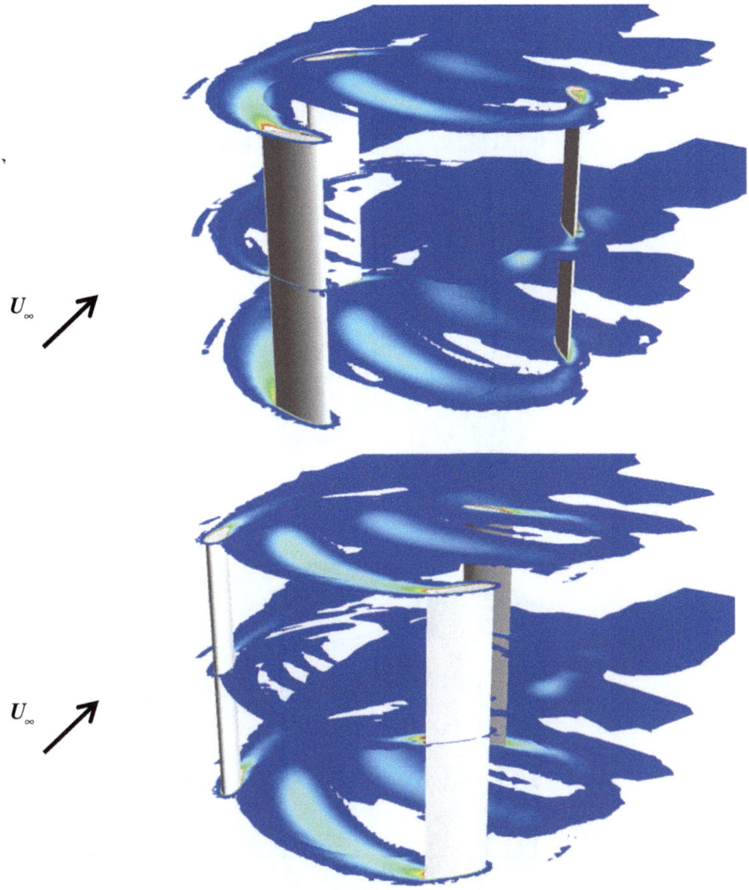

Fig. 23 Blade tip vortex visualisation in two different angles in a revolution with a phase lag of 60°

natural frequency of the structure, it causes resonance leading even to the collapse of the structure, in this case the CFWT. This phenomenon is known to have caused the accelerated destruction of guide vanes in Francis turbines [2].

From Fig. 23, it is possible to see the disposition and strength of the tip vortices as the rotor blades find themselves at different phase angles. Moreover, once detached from blades, such vortices are convected downstream for the incident flow reducing their intensity and eventually interacting with the next blade (Fig. 23 bottom). This will cause changes in both the surface pressure distribution of that rotor blade, but may also change the laminar to turbulent flow transition locations in the rotor blade. This is due to the elevated levels of turbulence in the wake and the relatively low Reynolds numbers of the flow over the rotor.

Shortly after the rotor blade experience maximum lift (in some angular location before $\theta = 90°$, following Fig. 1), the tip vortex will be at maximum strength,

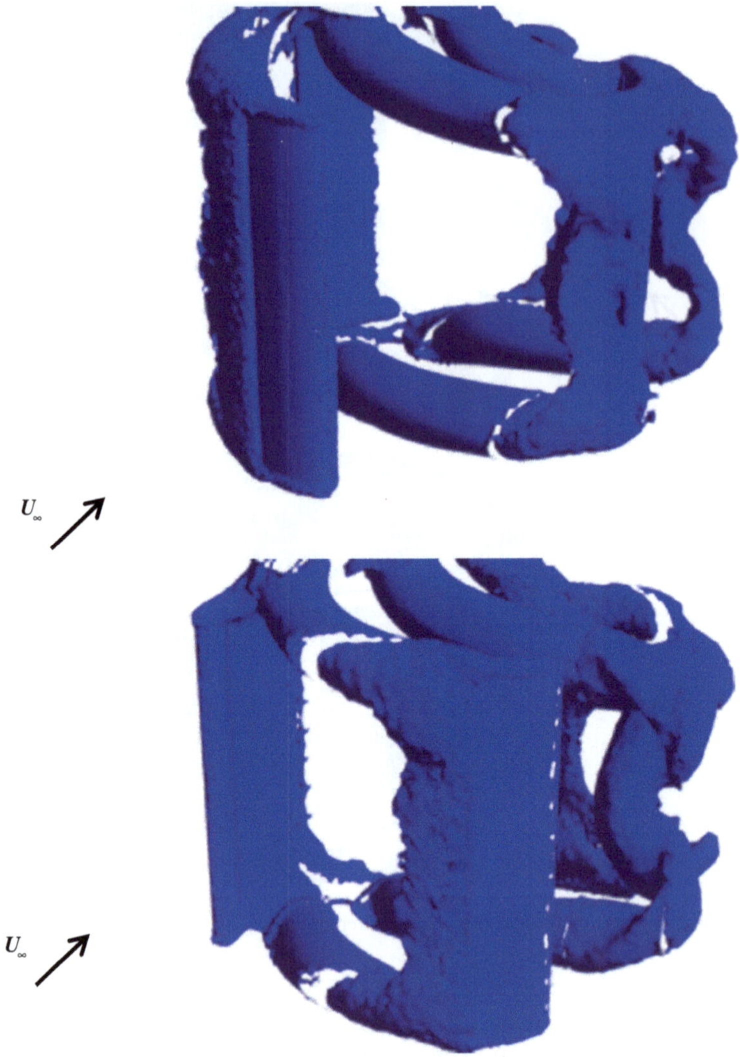

Fig. 24 Iso-vorticity levels at two different angles in a revolution with a phase lag of 60°. Free trailing vortices at blade tip and bound vortex at blade trailing edge are illustrated. Profile NACA0025

creating the maximum induced drag. The intensity of the vortex core and its extent are maximum at this position. As this rotor blade reaches phase angles beyond $\theta = 180°$, the extent of the over tip vorticity reduces, while by the time the rotor reaches locations between 270 and 360°, it appears to be at a minimum.

The changing in strength of the trailing tip vortices is due to the changing lift experienced by the blade as it rotates through different phase angles. However, there is likely a small delay between maximum lift being developed and the maximum intensity of the tip vortex occurring, because the time required for the flow to respond to the changing lift around the rotor blade.

Figure 24 show visualizations of isocontours of the vorticity shed from the rotor blades illustrating the shape of the tip trailing vortices. It is very clear from this figure not only the extension of the wake caused by the tip vortex but also the development and detachment of the blade bound vortex. Additionally, it can be appreciated from Fig. 24 that the blade bound vortex bends itself towards the inner side of the domain, with a Ω characteristic shape, in the process of detaching from the blade.

In order to illustrate better the shape of the detached tip and bound vortices, Fig. 25 presents the isosurface of vorticity intensity of 60 s^{-1} at two different rotor positions, in this case with blades based on the NACA4415 profile. In the upper

Fig. 25 Iso-vorticity levels of 60 s^{-1} at two different angles in a revolution. Free trailing vortices at blade tip and bound vortex at blade trailing edge are illustrated. The bending of the detached bound and trailing vortices can be appreciated as well as the shape of the shed tip vortices. Profile NACA4415

part, the geometry of the detached bound vortex can be clearly appreciated in the right blade. Near of the symmetry plane the vorticity isosurface is separated from the trailing edge but near the tip it is still stuck to the blade. The vortex bending towards the inner ring-like domain is evident and it is due to the non-linear self-interaction of the vortex. This fact can be explained as follows: the wake downstream the blade consists of a finite-thickness vortex sheet that under self-induction evolves into a hairpin vortex (which is a characteristic structure present in turbulent shear flows) with high values of shear Reynolds stresses [15]. The lower part of the Fig. 25 shows the maximum length reached in a revolution of the 60 s^{-1} vorticity contour in the trailing tip vortices, which corresponds with the higher induced drag on the blade. In this last figure also the bending of the tip trailing vortices along the direction of blade span can be devised.

Future work will be addressed to find a quantitative correlation between the intensity of the detached tip vortices, the increase in the effective drag experienced by the blade and the corresponding decrease in the resulting tangential force.

8 Conclusions

A study of the unsteady flow around a CFWT with the commercial code ANSYS-CFD has been carried out using a transient rotor–stator approximation with a moving mesh technique in turbulent flow. The factorial 3^2 experimental design performed in the present study evidenced the influence of the thickness parameter in the moment and tangential force coefficients. Both C_m and C_t asymptotically increase as the thickness of the airfoil increases. For airfoil profiles with thickness greater than 15 % of the chord length, the increase in C_m and C_t is not significant. The results of the parametric study also demonstrate the influence of the camber in the normal force coefficient. Even though the primary variable in vertical axis turbine is the performance, care must be taken for hydrodynamic vertical turbine in which the cyclic bending loads over the turbine shaft are higher than in vertical wind turbines. Based on the numerical results a cambered airfoil decreases the average load on the shaft and could improve the resistance of a specific turbine design to fatigue failure.

The three-dimensional simulations allowed the identification of the dynamics of the trailing tip vortices, those responsible for the lower performance of the turbine in the 3D case regarding the 2D simulations. Such simulations illustrated the periodic nature of the tip vortices caused by the changing lift generated by the rotor blades as they travel through each rotor revolution. There is a correlation between lift generation, detached tip vortices intensity and generation of induced drag which ultimately is the responsible for the reduced performance in the three-dimensional configuration. Moreover, the dynamics of vortex pairing shedding from the blade has been visualized as well as the bending of the bound vortex during its process of detachment. All together, the results obtained in this work demonstrate that the CFD model can effectively predict the hydrodynamic performance of cross flow water turbines.

References

1. Coiro, D.P., De Marco, A., Nicolosi, F., Melone, S., Montella, F.: Dynamic behaviour of the patented kobold tidal current turbine: numerical and experimental aspects. Acta Polytechnica Czech Tech. Univ. Prague **45**, 77–84 (2005)
2. Laín, S., García, M., Quintero, B., Orrego, S.: CFD numerical simulation of Francis turbines. Revista Facultad de Ingeniería Universidad de Antioquia **51**, 21–33 (2010)
3. Ferreira, C.J.S.: The near wake of the VAWT. 2D and 3D views of the VAWT aerodynamics. Ph.D. Thesis, Delt University of Technology (2009)
4. Maître, T., Achard, J.L., Guitet, L., Ploesteanu, C.: Marine turbine development: numerical and experimental investigations. Sci. Bulletion Timisoara Politechnic Univ. **50**, 59–66 (2005)
5. Laneville, A., Vittecoq, P.: Dynamic stall: the case of the vertical axis wind turbine. J. Sol. Energ. Eng. **108**, 140–145 (1986)
6. Nabavi, Y.: Numerical study of the duct shape effect on the performance of a ducted vertical axis tidal turbine. M.Sc. Thesis. British Columbia University (2008)
7. Dai, Y.M., Lam, W.: Numerical study of straight-bladed Darrieus-type tidal turbine. ICE-Energy **162**, 67–76 (2009)
8. Howell, R., Qin, N., Edwards, J., Durrani, N.: Wind tunnel and numerical study of a small vertical axis wind turbine. Renew. Energ. **35**, 412–422 (2010)
9. Brochier, G., Fraunie, P., Beguier, C., Paraschivoiu, I.: Water channel experiments of dynamic stall on Darrieus wind turbine blades. J. Propul. Power **2**(5), 445–449 (1986)
10. Larsen, J., Nielsen, S., Krenk, S.: Dynamic stall model for wind turbine airfoils. J. Fluids Struct. **23**, 959–982 (2007)
11. Nobile, R., Vahdati, M., Barlow, J., Mewburn-Crook, A.: Dynamic stall for a vertical axis wind turbine in a two-dimensional study. World Renew. Energ. Congr. 4225–4232 (2011). Sweden, 8–13 May 2011, Lindköping (Sweden)
12. Menter, F.R.: Two-equation Eddy-viscosity turbulence models for engineering applications. AIAA. J. **32**, 269–289 (1994)
13. Abbott, H., von Doenhoff, A.: Theory of wing sections: including a summary of airfoil data. Dover Publications, New York (1959)
14. Montgomery, D.: Design and Analysis of Experiments. Wiley, New York (2000)
15. Moin, P., Leonard, A., Kim, J.: Evolution of a curved vortex filament into a vortex ring. Phys. Fluids **29**, 955–963 (1986)

Hydrogen Generation

Dario Delgado, Glenn Hefter and Manickam Minakshi

Abstract The idea of an economy supported by hydrogen is still being considered by government bodies and major oil companies. The extensive use of hydrogen as a fuel has many applications in pollution-free technologies which could be of every-day use in society. Examples are in transportation and power generation; other uses include chemical technology, metallurgy, effluent processing and water production. For mass application the most appropriate process for hydrogen production is water electrolysis. To this end, a simplified description of the hydrogen and oxygen evolution mechanisms is presented. Understanding these mechanismswill provide a basis for making water electrolysis more efficient. For this reason special attention is given to the design of catalysts for water electrolysis.

1 Hydrogen Economy

The term 'hydrogen economy' was first coined by Bockris and Triner in 1970 [1], and it has been seen as a solution to the problems arising from the current unsustainable fossil fuel economy. Important contributions such as those by Meadows [2], Hubbert [3] and others [4], have supported the establishment of hydrogen as an alternative source of energy.

In recent years, the idea of the hydrogen economy becoming significant in the energy market has started to receive attention. The energy outlooks of the top oil

D. Delgado (✉) · G. Hefter
Murdoch University, 90 south street, Physical Sciences building, Murdoch 6150, Australia
e-mail: d.delgado@murdoch.edu.au

G. Hefter
e-mail: g.hefter@murdoch.edu.au

M. Minakshi
Chemistry Department, Murdoch University, Murdoch, WA 6150, Australia
e-mail: minakshi@murdoch.edu.au

G. Ferreira (ed.), *Alternative Energies*, Advanced Structured Materials 34,
DOI: 10.1007/978-3-642-40680-5_7,

companies: Exxon Mobil [5], Shell [6, 7] and BP [8], all suggest a more diversified future energy supply mix in which hydrogen participates to some extent.

Exxon Mobil and BP energy outlooks to 2030 mention emerging alternative sources of energy, for example BP's comments that 'the fastest growing fuels are renewables (including biofuels)'. On the other hand, Shell energy scenarios to 2050 shows hydrogen fuel cell vehicles (FCVs) competing with battery electric vehicles. In fact, Shell gives an even chance of only one of those two main technologies being fully developed in the near future.

It is hard to predict the future; oil faced opposition in the past, before it eventually came to dominate the energy market. Large majorenergy companies do not see hydrogen as being a major fuel in the near future; however, because of its unique advantages, there are institutions interested in the study of economies based on solar- and nuclear-generated hydrogen [4, 9]. There is still a fair probability of hydrogen being part of the energy mix in the future, which justifies further research of this technology.

'Hydrogen technology' as opposed to the 'hydrogen economy' brings several options that could be combined with systems already in place so as to achieve a synergistic effect. As a fuel, hydrogen produces energy (i.e. electrical, mechanical, heat), and water as by-product. Hydrogen energy could thus address both energy and fresh water supply issues. For example, remote areas where water is in short supply could benefit from this technology. An industrial complex, for instance, could adjust the use of its off-peak power to produce hydrogen, to be stored and sold.

A hydrogen economy could bring several benefits [1, 9]. In addition to covering all the electricity demand, it could positively influence other sectors of the economy such as chemical technology, metallurgy and refining, effluents, water, and transportation as a pollution-free fuel. It could also be used to convert organic waste into methane and ethylene. Hydrogen is an energy carrier and its transportation at practical distances between the source and power generation, is cheaper by pipelines than electrical lines [1]. However, public acceptance, high capital investment, and the high present cost of hydrogen with respect to existing fossil fuels are some of the barriers facing this technology.

2 Energy and Water Commodities

In 2003, Professor Richard Smalley, a Nobel prize winner in chemistry ranked the top ten challenges the world is facing as: (1) energy, (2) water, (3) food, (4) environment, (5) poverty, (6) terrorism and war, (7) disease, (8) education, (9) democracy, and (10) population.

The reason that energy and water are the numbers (1) and (2), respectively, is that both are interrelated with abundant clean energy, fresh water could readily be produced. One clean source of energy is solar energy; this source could in theory be used for the development of water electrolysis. Hydrogen produced by electrolysis provides energy as its main product and fresh water as a by-product.

2.1 Energy Sector

Fossil fuels currently represent a high proportion of the total world primary energy consumption, with the rest being provided by nuclear and other non-conventional or renewable sources of energy. Figure 1 shows that over three quarters of global energy production is provided by fossil fuels (i.e. oil, gas and coal). The main problems with fossil fuels are that they are finite, non-renewable and are a major source of CO_2 emissions into the atmosphere.

In an effort to decrease CO_2 emissions, most countries are developing covert and overt policies based on their natural resources. Such policies can open up many opportunities, for example, countries rich in solar energy could aim to develop renewable, which could be combine with hydrogen generation for transportation. One known case is Iceland, which has important sources of geo-thermal energy, and currently has a hydrogen generation plant which is powered by geo-thermal (supported by the Shell company). Other well known example is the Kyoto protocol which is an international environmental treaty with the goal of achieving the reduction of emissions of greenhouse gases. These few examples show how many countries are willing to increase their renewable energy share substantially and by doing it so, they recognize that there has to be a change in their energy supply to make it more sustainable.

These changes in the energy supply will give more opportunities for new sources of energy, and this is important because if technologies like solar and wind increase their share on the energy supply, it will give the foundations for a hydrogen economy to emerge.

2.2 Water Sector

Most countries cover their water demand by:

- rainwater and storm water.
- groundwater.

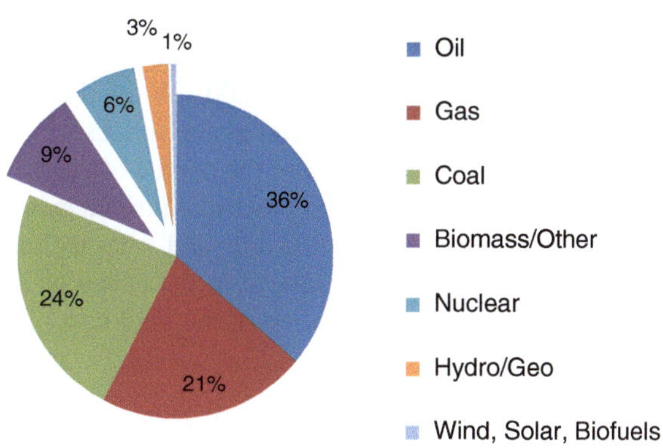

Fig. 1 Global energy mix [5]

- desalination of brines using reverse osmosis (RO), evapotranspiration, electro-dialysis, distillation andother methods.
- re-use and recycling.

Countries with fresh water supply issues use desalination, however there have been cases where they intentionally change the path of rivers to cover the water demand of certain areas. Desalination processes are energy intensive; this is the main factor that determines the water cost in such processes. For economic reasons, RO is usually the best option for desalination. Note that water produced from hydrogen usage is never mentioned; as argued before, the hydrogen economy is still distant to realize, therefore, at the moment, adding it as a supply option is not feasible.

3 Hydrogen Production

Hydrogen can be prepared [9] by thermochemical, electrochemical, and bio-chemical processes and examples are the action of steam on heated carbon, the decom-position of certain hydrocarbons with heat, thermolysis of water, water electrolysis, displacement from acids or bases by certain metals, anaerobic digestion, etc. Of all of these methods, only water electrolysis is likely to be suitable for large scale production of hydrogen [1, 4].

Energy sources that might becost-effective in the future for water electrolysis are: photobiological, photochemical and thin-film solar processes, with wind energy as the last option [4]. In addition, some government bodies forecast [10] the use of tidal and geothermal hot dry rock for hydrogen generation. It has to be mentioned that nuclear-generated hydrogen is also a valid approach but the use of such a technology is controversial. Off-peak energy from conventional power plants is seen as a more reasonable approach, especially in the hypothetical case where CO_2 sequestration were available.

Electricity from fossil fuels would be the cheapest option to drive water electrolysis, but negative externalities from this source such as CO_2 emissions have not been included because the cost of CO_2 geosequestration has not yet been assessed. Non-conventional sources of energy in general are becoming more cost competitive. In particular solar power, is forecast cost 10 US cents/kWh for a solar photovoltaic of approximately 500 kW in the year of 2020 [11] technology-promisingconsidering currentfossil fuel prices.

The appropriate selection of the source of electricity to drive water electrolysis is a combination of cost and other circumstances such as the population needs for water and energy in a specific area and what technology could be developed. For instance tidal energy is produced in the ocean or along coast lines.

A formula for the estimation of the total cost of electrolytic hydrogen has been published [1]:

$$\text{Cost} = 229Ec + 80 \tag{1}$$

where:

Cost cost of an MBtu of H_2 in US cents (1974)
E cell voltage (V)
c cost of a kWh_e in US cents (1974).

The cell voltage is defined as:

$$E = E_e - \eta_c + \eta_a + IR \qquad (2)$$

where:

E_e equilibrium cell potential (V)
η_c and η_a overpotential of the cathode or anode, respectively (V)
IR Ohmic potential drop in the electrolyte (V).

According to Eqs. 1–2, it is important to decrease the cell voltage (i.e. overpotentials and IR drop) and find a cheap source of electricity. As argued before, the cost point of view is important but other factors will determine the final technology that will drive the electrolysis. The purpose of Eq.1 is to show the main variables that determine the cost to produce hydrogen, rather than giving a good estimate of its cost. The parameters of this equation depend on technology, operational costs, government subsidies, etc. For example, it is forecast that the cost of the electrolyzer can be further lowered [4], this would change the parameters of Eq. 1.

4 Water Electrolysis

Water electrolysis involves the splitting of water into its elemental constituents, hydrogen and oxygen. This electrochemical process is not spontaneous: it needs a source of electrical energy to drive it.

Figure 2 shows a scheme of an electrolytic cell for splitting (electrolyzing) water. Here, the two electron-transfer reactions occurring at the cathode and anode simultaneously producing hydrogen and oxygen, respectively, are driven endergonically by an external source of electrical power.

Fuel cell technology uses the spontaneous direction of the water formation reaction to produce energy and water as by-product. Equation 3 shows the reaction of water in the spontaneous direction.

$$2\{H_2 + \{O_2 \rightarrow 2\{H_2\{O \qquad (3)$$

The two half-cell reactions of water electrolysis that occur at the surfaces of the electrodes and the overallreaction are shown in Table 1.

At standard conditions of temperature and pressure (T = 298 K (25 °C), P^o = 1 bar), water is liquid, while hydrogen and oxygen are gaseous. The changes in the standard Gibbs energy ΔG^o, enthalpy ΔH^o, and entropy ΔS^o, accompanying the

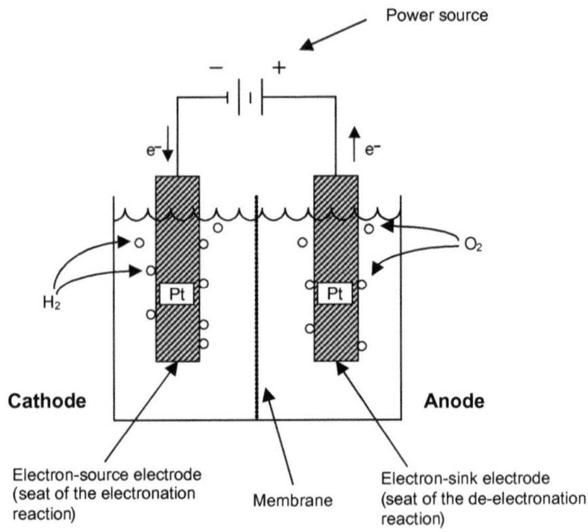

Fig. 2 A driven cell or substance producer [12]

Table 1 Overall and Half Cell Reactions of Water Electrolysis

Half cell reactions	Acid solution	Alkaline solution
Cathodic	$2H^+ + 2e^- \rightarrow H_2(g)$	$2H_2O + 2e^- \rightarrow 2OH^- + H_2(g)$
Anodic	$2H_2O \rightarrow O_2(g) + 4H^+ + 4e^-$	$4OH^- \rightarrow 2H_2O + 4e^- + O_2(g)$
Overall reaction	$2H_2O(liq) \rightarrow 2H_2(g) + O_2(g)$	

dissociation of water (i.e. the reverse of the formation of water) correspond to the thermodynamically reversible (i.e. equilibrium) electrolysis potential for water $E^o{}_e$ are:

$\Delta G^o = 237.22$ KJ mol^{-1}
$\Delta H^o = 285.84$ KJ mol^{-1}
$\Delta G^o = 163.15$ J mol^{-1} K^{-1}
$E^o{}_e = 1.229$ V

At equilibrium, the cathodic and anodic current densities at the electrode are equal in magnitude and opposite in direction; this is called the exchange current density (i^o).

The energy that the power supply must deliver for reaction on Table 1 to happen is the Gibbs energy change of the water dissociation reaction (ΔG_d) (kJ mol^{-1}):

$$\Delta G_d = -nFE_e \quad (4)$$

where:

n number of electrons exchanged during the electrochemical splitting of water (n = 2, dimensionless)

F Faraday constant 96,484 (C mol^{-1}).

E_e water equilibrium potential (V).

The value of $\Delta G^o{}_d$ includes energy provided from the environment as heat by the $T\Delta S^o{}_d$ term (*cf.* Eq. 6). In the hypothetical case that no heat were available from the environment, then the energy required to split water would increase. The heat ($T\Delta S^o{}_d$) provided from the environment at the standard condition divided by 2F represents 0.25 V, therefore without this extra source of energy the thermodynamic electrolysis voltage would increase to approximately 1.48 V. This is called the thermo-neutral potential at which heat is neither being takenfrom the surroundings to create hydrogen, nor being rejected to the surroundings. Under this condition the electrolysis is carried out entirely by electricity [[1, 13] and this potential is used as a reference to run the electrolysis. Above this potential heat is being given out as excess (i.e. there are energy losses). Equation 5 defines the thermo-neutral potential:

$$E = \frac{\Delta H}{nF} \tag{5}$$

The Gibbs energy change of the water dissociation reaction is a function of the operating temperature and pressure, as shown in the Eq. 6

$$\Delta G_d = \Delta H_d - T\Delta S_d > 0 \tag{6}$$

ΔH_d enthalpy change of the reaction (kJ mol^{-1})
ΔS_d entropy change of the reaction (J mol^{-1} K^{-1}).

Another important condition for electrolysis is the pressure; the selection of the operating pressure is based on costs. The higher the pressure the hydrogen is produced at, the better, because energy and money can be saved by removing the need for a compressor. The recommended operating pressure for water electrolysis is 2,000 psi (13.8 MPa) [13], it should be mentioned that even higher pressures have been researched, for example, ultrahigh-pressure electrolysis operating at 10,000–5,000 psi (68.9–34.5 MPa) [14]. Similarly, the current density is also determined by costs. A current density of 100 mA cm^{-2} shows a reasonable balance when considering costs, the higher the current density the higher the energy losses and operational costs, and vice versa, lower current densities would make the process economically unviable.

Another important component of the water electrolysis is the membrane which selectively permits the flow of hydrogen ions produced at the anode to the cathode where they are reduced to molecular hydrogen. Liquid water is consumed at the anode but water from the cathode migrates to the anode by a process known as electro-osmosis. The membrane material mostly used is Nafion [13], this material is extensively used for the development of proton exchange membrane (PEM) cells.

A proton exchange membrane (PEM) water electrolyzer is a reactor for hydrogen generation. Electrolyzers can be developed [1] as unipolar or bipolar cells. In the former, the electrodes are in parallel, and in the later, the electrodes are in series.

Further research required on electrolyzers is [1]:

- Low overpotential electrocatalysts.
- Reduction of the IR drop due to the effect of bubbles.
- The use of flow devices to reduce possible super-saturation hold up.
- Minimization ofgas cross over. Hydrogen and oxygen cross over through the-membrane when operating at high pressures [13].

Figure 3 is a scheme of a single stack PEM cell. Similar to Fig. 2, this simple design shows the cathodic and anodic chambers, where hydrogen and oxygen evolve, respectively. The power supply is direct current and not alternating current, the reason is that oxygen and hydrogen when combined could explode; this would pose an important health and safety hazard. From this concept design, it could be seen that the *IR* drop in the electrolyte has been minimized as the membrane only separates the cathode and anode (*cf.* Eq. 2).

Finally, the most important pieces in an electrolytic cell are the catalysts (i.e. cathode and anode surfaces), where the H_2 and O_2 are produced. The catalysts determine the overpotential of the electrolysis and its cost (*cf.* Eqs. 1–2). Overpotential in water electrolysis is the difference between the required potential and the thermodynamic potential that is needed to drive the reaction at an appropriate rate (Eq. 7).

$$\eta = a - b \log i \tag{7}$$

where:

η overpotential (mV)
a constant (mV)
b 2.303 RT/βF (Tafel's equation, 1905)
i current density (mA cm^{-2}).

Fig. 3 Concept design of a single stack PEM cell

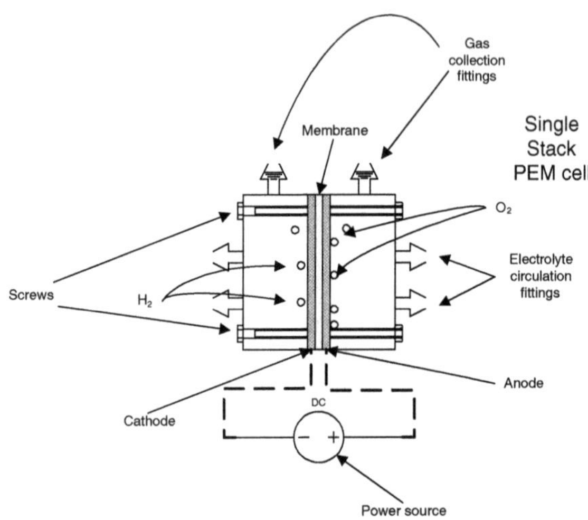

At present, most electrode catalysis research relevant to improving hydrogen technology, specifically for water electrolysis, includes [15]:

- More active at low temperature;
- Lower overpotential;
- Cheaper electrodes using less noble metals;
- Multifunctional catalysts.

However, there are fundamental issues that need clarification regarding the efficiency of catalysts in general [15]:

- Understanding catalytic activity from one system to the next;
- Understanding selectivity and trends in selectivity (i.e. preferential reactions);
- Understanding deactivation mechanisms (i.e.catalyst poisoning);
- Defining and understanding the chemical and structural state of the active siteduring catalysis (computer chemistry);
- Understanding the size effect in catalysts;
- Understanding metal–metal interaction in bi- and multimetallic catalysis; and
- Finding ways to produce novel micro- and mesoporous solids (i.e. catalyst's geometric factor enhancement).

4.1 Reaction Mechanisms of Water Electrolysis

Why bother about determining a mechanism?. The overall electrochemical reaction is a combination of two half-cell reactions. A common error found in industrial laboratories [12] is excessive focus on the overall reaction. In order to make progress one should look in detail at the individual reactions at the electrodes.

To minimize the energy and hence economic cost of the losses caused by overpotential, it is essential to know the rate-determining step (rds) in the reaction sequence. This would give a better insight on improving catalytic surfaces, otherwise the approach would be just exploring hunches. Other electrochemical approaches in combination with the reaction pathway help to determine the rds; for example, a systematic use of the Tafel equation can be used to establish the rds.

The overall reaction for water electrolysis can be separated into two half-cell reactions: the hydrogen and oxygen evolution reactions. In addition, the chlorine evolution reaction (CER) has to be considered because this reaction competes with oxygen evolution in seawater electrolysis. Saline and alkaline waters require different types of anodes (i.e. water oxidizer electrode). In saline-water electrolysis, chloride ions are oxidized at the electrode surface producing gaseous chlorine which is corrosive and harmful. Therefore, selective electrodes that produce oxygen over chlorine should be used. This is not the case for alkaline water electrolysis, where oxygen is the only product being generated, requiring no selective electrodes.

4.2 Hydrogen Evolution Reaction (HER)

Electrochemical reactions in general behave differently under acidic and basic conditions. Electrolysis in practical conditions keeps the solution in the cathode compartment (i.e. the catholyte) basic and the solution in the anode compartment (i.e. the anolyte) acidic. The hydrogen evolution reaction (HER) mechanism has three distinguishable steps [12, 16–19] as shown in Table 2.

Here "M" stands for the participating active-site atom (i.e. not only a metal atom) from the electro-catalyst. The desorption and catalytic paths are two alternatives for the evolution of gaseous hydrogen, which may occur along either, depending on the electrode. The Tafel slope indicates the rds, for example a slope of *2RT/F* indicates that the rds is proton discharge. The Tafel approach is appropriate but not sufficient, other tests are required to determine the mechanism [20].

4.3 Oxygen Evolution Reaction (OER)

The model for the oxygen evolution reaction (OER) is still being developed; there are several paths which fit all the constraints. For instance different ideas [21, 22] have been proposed for reactions happening in the double layer interface (as opposed to the catalyst's surface).

Certainly the OER mechanism appears to be more complicated than the HER process. Table 3 contains the various pathways for oxygen evolution on metal oxide surfaces that are consistent with electrochemical theory. Note that "M" stands for the participating active site on the electro-catalyst, the other abbreviationsare standard chemical symbols.

In addition, recent publications [23, 24] show different OER mechanisms than the ones shown in Table 3. A mechanism can be developed from kinetic or

Table 2 HER mechanism in acid and alkaline solutions

Step name	Acid solution	Alkaline solution
Electrochemical/charge transfer steps Volmer/hydrogen adsorption/proton discharge	$M + H_3O^+ + e^- \rightleftarrows M - H + H_2O$	$M + H_2O + e^- \rightleftarrows M - H + OH^-$
Heyrovsky/electrochemical desorption by recombination path	$M - H + H_3O^+ + e^- \rightleftarrows H_2 + H_2O + M$	$M - H + H_2O + e^- \rightleftarrows H_2 + OH^- + M$
Chemical step: Tafel/catalytic path	$M - H + M - H \rightleftarrows H_2 + 2M$	$M - H + M - H \rightleftarrows H_2 + 2M$

Table 3 Mechanisms of oxygen evolution on metal oxides [25]

Acid solution	
Electrochemical oxide path [26]	Oxide path [26]
$M + H_2O \rightarrow M - OH + H^+ + e^-$	$M + H_2O \rightarrow M - OH + H^+ + e^-$
$M - OH \rightarrow M - O + H^+ + e^-$	$2M - OH \rightarrow M - O + M + H_2O$
$2M - O \rightarrow 2M + O_2$	$2M - O \rightarrow 2M + O_2$
Krasil'shchkov path [27]	Wade and Hackerman's path [28]
$M + H_2O \rightarrow M - OH + H^+ + e^-$	$2M + 2H_2O \rightarrow MO + MH_2O + 2H^+ + 2e^-$
$M - OH \rightarrow M - O^- + H^+$	$MO + 2MOH^- \rightarrow 2M + MH_2O + O_2 + 2e^-$
$M - O^- \rightarrow M - O + e^-$	
$2M - O \rightarrow 2M + O_2$	
Alkaline solution	
Electrochemical oxide path [26]	Oxide path [26]
$M + OH^- \rightarrow M - OH + e^-$	$M + OH^- \rightarrow M - OH + e^-$
$M - OH + OH^- \rightarrow M - O + H_2O + e^-$	$2M - OH \rightarrow M - O + M + H_2O$
$2M - O \rightarrow 2M + O_2$	$2M - O \rightarrow 2M + O_2$
Krasil'shchkov path [27]	Yeager's path [29, 30]
$M + OH^- \rightarrow M - OH + e^-$	$M + OH^- \rightarrow M - OH + e^-$
$M - OH + OH^- \rightarrow M - O^- + H_2O$	$M^z - OH \rightarrow M^{z+1} - OH + e^-$
$M - O^- \rightarrow M - O + e^-$	$2M^{z+1} - OH + 2OH^- \rightarrow 2M + 2H_2O + O_2$
$2M - O \rightarrow 2M + O_2$	
Bockris path [31]	
$M + OH^- \rightarrow M - OH + e^-$	
$MOH + OH^- \rightarrow M - H_2O_2 + e^-$	
$M - H_2O_2 + OH^- \rightarrow M - O_2H^- + H_2O$	
$M - H_2O_2 + M - O_2H^- \rightarrow H_2O + OH^- + O_2$	

thermodynamics tools, the former approach in general gives more accurate results, however narrowing down possible paths is done with the aid of the latter.

The models in Table 3 are developed by an electrochemical approach (i.e. kinetic). It is well known that even if a species is thermodynamically favored this does not mean it is kinetically dominant, and what it is important is whether the species is kinetically dominant or not. Which mechanism or approach is the correct one?. There are several proposed OER mechanisms butthere must be only one of those dominant. It should be considered that a mechanism primarily depends on catalyst material, electrostatic potential, and electrolyte.

4.3.1 Chlorine Evolution Reaction (CER)

Several mechanisms have been proposed for the CER based on electrochemical analyses [32, 33]

Volmer-Tafel reaction:

$$2M + 2Cl^- \rightarrow 2Cl - M + 2e^- \rightarrow 2M + Cl_2 + 2e^- \quad (8)$$

Volmer-Heyvrosky reaction [34]:

$$M + 2Cl^- \rightarrow Cl - M + e^- + Cl^- \rightarrow M + Cl_2 + 2e^- \quad (9)$$

Khrishtalik reaction [35, 36]:

$$\begin{aligned} &M + 2Cl^- \rightarrow Cl - M + e^- + Cl^- \\ &\rightarrow (Cl - M)^+ + 2e^- + Cl^- \rightarrow M + Cl_2 + 2e^- \end{aligned} \quad (10)$$

The reaction mechanisms for chlorine evolution are still being formulated. This is especially in regard to what happens at the atomic scale, which is known to depend strongly on the catalyst material, the electrostatic potential, and the electrolyte. Under standard conditions, the equilibrium potentials for Cl_2 and O_2 evolution are 1.36 and 1.23 V, respectively. However, depending on the catalyst, the overpotential [23] for the OER could be either higher or lower than that of the CER. In this respect metal oxides always show a potential for the OER larger than that of the CER [23] as shown on Fig. 4 (i.e. a so called volcano plot for the CER).

A volcano plot is an useful tool for the analysis of different materials towards the evolution of a gas. A volcano plot is created by plotting on the abscissa a function involving some property of the catalyst, e.g. heat of sublimation, interatomic distance among the surface atoms or the bond strength of the intermediate species with the catalyst.

Figure 4 is a theoretical CER and OER volcano develop by density functional theory DFT (see also Fig. 7). This Fig. illustrates the activity of some selected rutile oxides, although TiO_2 does not appear note that because its activity was

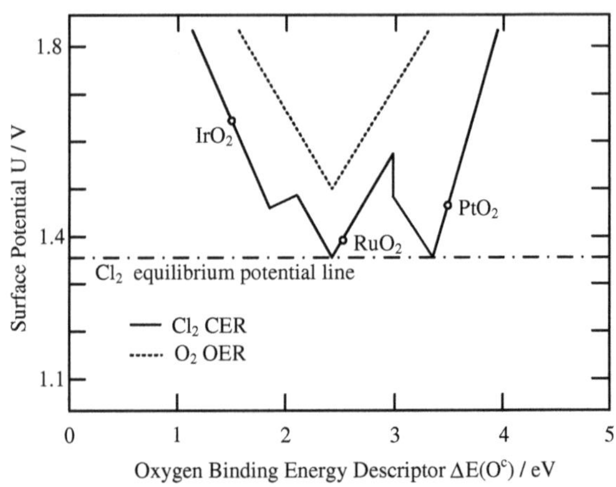

Fig. 4 Theoretical volcano plot for the CER/OER of a few selected metal oxides [23]

found to be out of the scale (i.e. low activity). Figures 4 and 7 cannot be directly compared as they are determined at different electrolyte conditions.

Note that Fig. 4 does not have the regular volcano shape, because at the studied conditions, there are different CER mechanisms competing with each other, giving a different than usual volcano shape. However, the OER in Fig. 4 does have an usual volcano shape. This Fig. 4 shows that most materials will produce chlorine preferentially over oxygen from a thermodynamic point of view.

Recent studies [23] show reaction mechanisms involving reactive oxygen species which are OER intermediates being involved in the CER. Intermediates such as ClOM, $Cl(M)_2$, ClM, have been proposed where 'M' for this particular case is surface oxygen from one of the OER intermediates. The same publication [23] which had a quantum-mechanistic approach mentions that Pt/MnO_2 catalyst in acid has energy of 0.3 eV higher for oxygen evolution than the potential of chlorine evolution.

This information is important for the development of oxygen selective electrodes for saline waters.

This resulted [23] in the potential for chlorine evolution to be 0.4 V lower than the potential for oxygen evolution on Pt/MnO_2. First of all, this is a thermo-mechanistic approach; thermodynamic unknowns are filled in with the aid of quantum mechanics. This thermodynamic approach does not explain why oxygen evolution is the preferred product from MnO_2 based catalysts, if fact, it suggests the opposite. However, a thermodynamic approach is important but not sufficient, what can be extracted is that definitely this material has some sort of selectivity towards either the OER or CER.

Is the selected reaction mechanism for the OER and CER on MnO_2 electrodes during calculations, the right one?. Different mechanisms have different activation energies, the conclusions would be biased if such a selection has not been properly taken care of. Computer chemistry is a field that is gaining more popularity as computing power increases, however this approach requires several approximations and they could sometimes lead to wrong conclusions.

If thermodynamics fails to explain the experimental observation of MnO_2 favoring the production of O_2, what is left is a kinetic approach. It has been explained [37] for iridium based electrodes that:

a. Chlorine evolution is a potential-controlled reaction limited by the number of active sites on the catalyst surface.
b. Oxygen evolution is surface transport controlled.

This kinetic approach might help to explain the selectivity for the OER on MnO_2-based electrodes. For instance, operating at higher temperatures will improve the transport of oxygen intermediates along the catalyst surface, leaving less active sites for the CER. There have been publications involving manganese oxide based electrodes where temperature improves the OER selectivity [38]. In addition, it has been found [39] that divalent ion doping helps stabilize Mn^{IV}/Mn^{III} oxygen intermediate species, which could be involved as mediators in oxygen reaction mechanisms (*cf.* Table 3). Combining all this information gives an insight about the factors involved in the selectivity of MnO_2-based electrodes for the OER over the CER. Another example [40] is found in photosynthesis II of green plants and algae, where the manganese oxidation state plays a significant role in explaining the water oxidation mechanism on the water-oxidizing complex (WOC).

4.4 Catalysts for Water Electrolysis

In the present context, a catalyst electrode is a material that lowers the energy threshold for both, the hydrogen and oxygen evolution reactions.

As such, it is a material that speeds up the electrochemical processes by allowing them to take place more easily, thereby lowering energy consumption, producing less waste and influencing product selectivity.

In general, catalysts operate by lowering the activation energy ($\Delta G^{\ddagger}$) of a reaction (Fig. 5). This is achieved by stabilization of the transition state, destabilization of the reactants or by creating a new reaction path.

Figure 5 is a reaction coordinate plot. A simple analogy is looking at it as if it were a hill to be climbed. Going from oneside of the hill tothe other could be done by two ways, the catalyzed and the un-catalyzed. The activation energy is the energy required to climb the uphill side and it can be seen that this value is different for the two paths. The downhill side does not have any constraints to stop the reaction for product formation.

There are twosimilar equationsthat describe the energy of activation, the Arrhenius and the Eyring–Polanyi equation. Eq. 11 and 12, respectively.

$$k = Ae^{-\frac{E_a}{RT}} \tag{11}$$

$$k = \frac{k_B T}{h} e^{-\frac{\Delta G^{\ddagger}}{RT}} \tag{12}$$

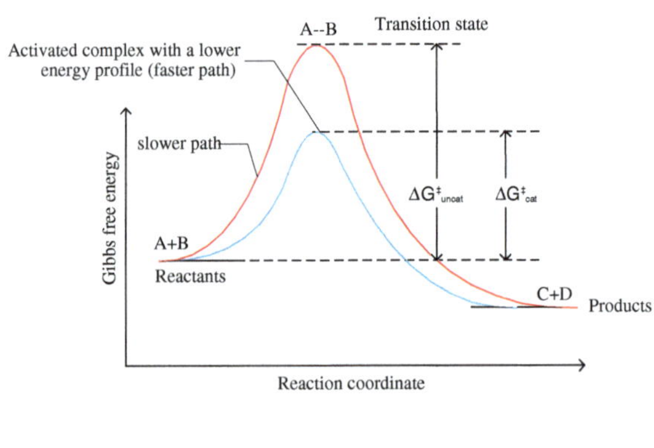

Fig. 5 The reaction coordinate and the transition state

where:

k	Rate constant of a chemical reaction (units depends on the reaction order).
A	Pre-exponential constant or frequency factor (units depends on the reaction order).
E_a	Activation energy (J mol^{-1}).
R	Universal gas constant (8.31 J mol^{-1} K^{-1}).
T	Thermodynamic temperature (K).
k_B	Boltzmann constant (1.38×10^{-23} J K^{-1}).
h	Planck's constant (6.62×10^{-34} J s).

These Eqs. 11, 12 also explain why it is beneficial to run reactions at higher temperatures. The rate constant k is a measure of the speed of a chemical reaction. Its dependence on temperature means that the higher the temperature, the faster the reaction. In addition, increase temperature lowers the equilibrium potential of a chemical reaction (Eqs. 4, 6). This lowers the electrical power needed to drive the electro chemical reaction.

The Arrhenius equation is another approach to determine the rds. As shown on the Fig. 5, different reaction paths have different activation energies. This Fig. shows two mechanisms with different speeds and in canbe seen that the path with transition state with the smaller Gibbs energy is more energy efficient.

There are two recognized ways to improve the catalytic properties of an electrode [12]. The first is to alter the structure of the electrode surface to increase its roughness (geometric factor). The second is to enhance the 'electronic factor' of the electrode; this refers to the catalyst-intermediate bonding strength which is also known as the Sabatier principle. The Sabatier principle states that the interactions between the catalyst and the intermediate species should be neither too strong nor too weak, it should be intermediate. In this respect, changes at the molecular level on the electrode are important in the development of more efficient electrodes or catalysts.

There are sound approaches for improving the geometric and electronic factors of catalysts (also known as size and intrinsic effects). Several articles have focused on the enhancement of the electronic factor, for example by using mixed metal oxides. While others have worked on the geometric factor, for instance with Raney nickel based electrodes.

The influence of the geometric factor on an electrochemical reaction is straightforward: the greater the available area the more molecules can react for the same period of time. On the other hand, understanding the electronic factor requires knowledge of how the energy profile of the activated complex can be altered to make the electrochemical reaction happen at lower activation energies. The electronic factor is analyzed by computer chemistry, one example is the recent [40] deciphering of the mechanism of the oxygen evolving complex (OEC) for oxygen evolution on green plants (i.e. photosynthesis), this is an important find towards the artificial production of hydrogen and oxygen. This process is known to be highly efficient and this is a step towards mimicking it.

There are several methods of catalyst fabrication [41]. One simple method is electrodeposition, however, this technique [41] is unsuited to producing controlled porous structures. This shortcoming could be fixed by using porous catalyst carriers or using specialized techniques like pulse electrodeposition [42], low current density electrodeposition [43] or using organic additives to modify crystal growth during electrodeposition [12]. Techniques for the preparation of thin-film catalysts are: electrolytic codeposition, spin or dip coating, high energy ball milling of powder, sol–gel, flame- and plasma-spraying of thermal decomposition with subsequent reduction to metal alloys or sputtering, arc ion plating, cold-rolling and sintering [41, 44–48], etc.

4.4.1 Catalyst Materials

HER Catalysts

Materials such as the platinum group metals; platinum, rhodium, iridium, rhenium, ruthenium and to some extent gold are the best elements for the HER as shown in "volcano-type" plots [12].

Figure 6 shows the best pure metals (i.e. not alloys) for the HER. These are metals that have both a high exchange current density, which is a measure of catalyst activity, and intermediate metal-hydrogen bond strength, which is a measure of binding strength descriptor. Nonetheless, the platinum group metals are extremely expensive, particularly for industrial scale application.

Platinum is the most widely used material as a catalyst because of its stability, availability and its good catalytic power. Other noble metals (except for gold) are more difficult to find in wire or plate forms. These are expensive materials, however a cheap metal could be substantially enhanced, e.g., aslead when used as lead dioxide PbO_2, also known as 'the poor man's platinum'. However, lead is carcinogenic and the technology concept that it should be after is about being environmentally friendly in addition to being economic and socially acceptable.

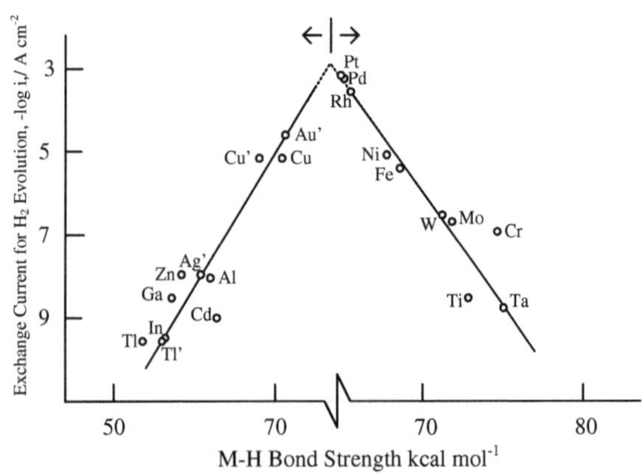

Fig. 6 Experimental volcano plot for the hydrogen evolution reaction [12]

OER Catalysts

Non-platinum and platinum electrocatalytic systems have been widely studied [48–50] and have been grouped into what has been called the hypo-hyper *d*-intermetallic electrocatalysts. The main types are:

1. Intermetallic catalysts.

   Mo–Co, $MoPt_3$, $MoNi_3$, $MoFe_3$, $LaCo_5$, etc.

2. Sulfides, phosphides.

   Ni–W–S, Ni-W–P, Fe–Ni–P, etc.

3. Catalysts based on Raney-Ni.

   Ni–Al–Ti, Ni–Al–Mo, Ni–Al–Cr, Ni–Zn-Ti, Ni–Zn–Mo, etc.

4. Mixed Oxides.

   TiO_2–RuO_2, TiO_2–IrO_2, $Ru_xTi_{1\text{-}x}O$, RuO_2–IrO_2, etc.

5. Hypo oxide-hyper metal.

   Pt–TiO_2, Pt–Ru–WO_3, Ni–V_2O_4, etc.

It is difficult to determine a volcano plot for the oxygen evolution reaction: textbooks [12] in general do not show it. The OER is a more complicated reaction than the HER; there are four electrons involved to produce one oxygen molecule as opposed to the two electrons in the HER. There are several proposed mechanisms for the OER (*cf.* Table 3) and they involve intermediates interacting with the participating active site atoms from the electrode.

For the HER case, with a unique intermediate being atomic hydrogen, is simple when it comes to plotting. However, there has been identified at least two intermediates for the OER: the hydroxyl ion and atomic oxygen. In addition, there could be other intermediates (i.e. reactive oxygen species) not yet been properly identified [22, 45]. It is difficult to make a plot for the OER under these circumstances.

Nevertheless, recent publications [23, 24] show what is called a single descriptor, which groups the binding energies of all the oxygen intermediates into a single value (i.e. ΔE_O (eV)). This makes it possible to create a volcano plot for the OER. Figure 7 shows the theoretical catalytic activity versus the single descriptor (i.e. pseudo-oxygen binding strength).

Figure 7 illustrates how metal oxides have different overpotentials for the OER. There are areas labeled electrolytic cell and fuel cell, respectively. The former being the one of interest, the latter has been included, as well, only to show how these two chemical processes are related even though they operate at different directions of the water electrolysis reaction. It can be seen that for both labeled areas, their respective triangles are like mirror images around the O_2 equilibrium

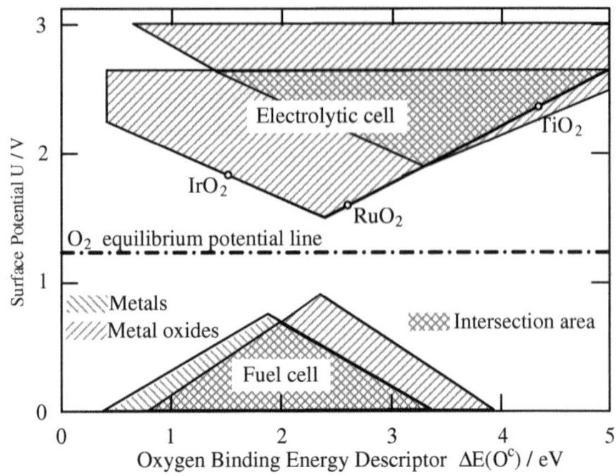

Fig. 7 Theoretical volcano plot for the OER at metals and metal oxides [24]

line. Therefore, a catalyst that is good for fuel cells is also likely to be good for electrolytic cells.

From the previous discussion a material can be both an oxidizing and reducing catalyst. For example, transition metal doping of MnO_2-catalysts has been found to improve their catalytic activity for oxygen reduction. It is valid to say that this could be the case for oxygen oxidation, as well. Another observation from Fig. 7 is when comparing triangles of a specific labeled area. The triangles for metal and metal oxides for electrolytic cells do not completely intersect, the area for the metal oxide is shifted towards an intermediate bonding strength. This is the bonding strength where the catalyst are more efficient. As a consequence, metal oxides have a lower overpotential than metals for the OER.

In Fig. 7, RuO_2 is at the top of the Volcano plot, very close to the 1.23 V thermodynamic minimum threshold limit for the OER to happen, RuO_2 has the lowest reported [51] overpotential for OER in acid. These materials could be further improved by enhancing their geometrical factor, however in practice this is cumbersome. For instance, using carbon blacks (e.g.Vulcan XC-72) [50,48] as catalyst support do improve the activity. Unfortunately, at high current densities, 300–200 mA cm^{-2}, as used in commercial chlor-alkali cells [52] such materi-alsclog, translating into a substantial drop in the effective area due to gas bubbles not being removed properly.

In addition to improving the geometric and electronic factors of these catalysts, the selection of the base material for the catalyst is also important. One approach could be toconsider only non-precious materials, therefore all the combinations involving any of the platinum group metals would be discarded without any further consideration.

On this sense, one could focus on improving the catalytic activity of running up materials. A running up material has a catalytic power lower than any of the platinum group based electrodes but could be abundant or cheap. For example, manganese dioxide which is an important choice for the development of selective oxygen evolution electrodes, requires manganese as raw material. Manganese is abundant, cheap and environmentally friendly. If its catalytic activity could be substantially increased then it would be a better option than platinum group based electrodes.

Acknowledgments The authors would like to thank AINSE Ltd for providing financial assistance (Award No ALNGRA12020/10366 and AINSE Post Graduate Research Award 10595) to enable work on the catalyst surfaces. We would also like to acknowledge the technological support from ANSTO (Australian Nuclear Science and Technology Organization) and the grant provided by Australian Synchrotron (grant AS123/HRIR 5428A) for use of far infrared beamline.

The global R&D Centers Program of NRF (National Research Foundation of Korea), funded by MSIP (Ministry of Science, ICT & Future Planning) at KIGAM (Korean Institute of Geoscience and Mineral Resources) for instrument time. Special thanks to Dr. Danielle Meyrick, Dr. Justin McGinnity, Dr. Trevor Pryor, Dr. Kim D. J., Dr. Chung K. W., Dr. Mihail Ionescu, Dr. Gamini Senanayake and Sue Farr for their suggestions and support.

References

1. Bockris, J.O.M.: The Solar-hydrogen Alternative. Architectural Press, London (1975)
2. Meadows, D.H., Meadows, D.L., Randers, J., Behrens, W.W.: The Limits to Growth. Universe Books, New York (1972)
3. Hubbert, M.K.: Resources and Man. W.H Freeman, San Francisco (1969)
4. The Hydrogen Economy.: The National Academies Press, Washington, D.C (2004)
5. Colton, W.M.: The Outlook for Energy a view to 2030. Exxon Mobil Corporation (2011)
6. Voser, P.: Shell Energy Scenarios to 2050—Signals and Signposts. Royal Dutch Shell plc (2011)
7. Grundy, P.: Shell Energy Scenarios to 2050—Energy. Royal Dutch Shell plc (2011)
8. Dudley, B.: BP Energy Outlook 2030. BP plc (2011)
9. Lide, D.R.: The Elements. CRC Handbook of Chemistry and Physics. 86th edn. CRC Press, New York (2005)
10. Needham, S.: The Potential for Renewable Energy to Provide Baseload Power in Australia. Parliament of Australia, Commonwealth of Australia (2011)
11. El-Moneim, A.A., Bhattarai, J., Kato, Z., Izumiya, K., Kumagai, N., Hashimoto, K.: Mn-Mo-Sn oxide anodes for oxygen evolution in seawater electrolysis for hydrogen production. ECS Trans. **25**, 127–137 (2009)
12. Bockris, J.O.M., Reddy, A.K.N., Gamboa-Aldeco, M.: Modern Electrochemistry. vol. 2A. 2nd edn. Kluwer Academic/Plenum Publishers, New York (2000)
13. Millet, P., Ngameni, R., Grigoriev, S.A., Mbemba, N., Brisset, F., Ranjbari, A., Etievant, C.: PEM water electrolyzers: from electrocatalysis to stack development. Int. J. Hydrogen Energy **35**, 5043–5052 (2010)
14. Jackson, T.: XI.13 High-efficiency, ultra-high pressure electrolysis with direct linkage to photovoltaic arrays (Phase II Project) (2005)
15. Dresselhaus, M.: Basic research needs for the hydrogen economy (2003)
16. Brown, D.E., Mahmood, M.N., Man, M.C.M., Turner, A.K.: Preparation and characterization of low overvoltage transition metal alloy electrocatalysts for hydrogen evolution in alkaline solutions. Electrochim. Acta **29**, 1551–1556 (1984)
17. Hashimoto, K., Sasaki, T., Meguro, S., Asami, K.: Nanocrystalline electrodeposited Ni-Mo-C cathodes for hydrogen production. Mat. Sci. Eng. A-Struct. **375–377**, 942–945 (2004)
18. Chen, L., Lasia, A.: Study of the kinetics of hydrogen evolution reaction on nickel-zinc alloy electrodes. J. Electrochem. Soc. **138**, 3321–3328 (1991)

19. Vandenborre, H., Vermeiren, P., Leysen, R.: Hydrogen evolution at nickel sulphide cathodes in alkaline medium. Electrochim. Acta **29**, 297–301 (1984)
20. Yamashita, H., Yamamura, T., Yoshimoto, K.: The relation between catalytic ability for hydrogen evolution reaction and characteristics of nickel-tin alloys. J. Electrochem. Soc. **140**, 2238–2243 (1993)
21. Conway, B.E., Liu, T.-C.: Characterization of electrocatalysis in the oxygen evolution reaction at platinum by evaluation of behavior of surface intermediates states at the oxide film. Langmuir **6**, 268–276 (1990)
22. Birss, V.I.: Oxygen evolution at platinum electrodes in alkaline solutions. J. Electrochem. Soc. **133**, 1621–1625 (1986)
23. Hansen, H.A., Man, I.C., Studt, F., Abild-Pedersen, F., Rossmeisl, J.: Electrochemical chlorine evolution at rutile oxide (110) surfaces. Phys. Chem. Chem. Phys. **12**, 283–290 (2010)
24. Rossmeisl, J., Qu, Z.-W., Zhu, H., Kroes, G.-J., Norskov, J.K.: Electrolysis of water on oxides surfaces. J. Electroanal. Chem. **607**, 83–89 (2007)
25. Matsumoto, Y., Sato, E.: Electrocatalytic properties of transition metal oxides for oxygen evolution reaction. Mater. Chem. Phys. **14**, 397–426 (1986)
26. Bockris, J.O.M.: Kinetics of activation-controlled consecutive electrochemical reactions: anodic evolution of oxygen. J. Chem. Phys. **1956**, 817–827 (1956)
27. Krasilshchikov, A.I.: Intermediate stages of anodic oxygen evolution. Russ. J. Phys. Chem. **37**, 273 (1963)
28. Wade, W.H., Hackerman, N.: Anodic phenomena at an iron electrode. Trans. Faraday Soc. **53**, 1636–1647 (1957)
29. O'Grady, W., Iwakura, C., Huang, J., Yeager, E.: Ruthenium oxide catalysts for the oxygen electrode. In: Breiter, M.W. (ed.) Proceedings of Symposium Electrocatalyst, pp. 286–302 (1974)
30. O'Grady, W.E., Iwakura, C., Yeager, E.: Oxygen electrocatalysts for life support systems. Am. Soc. Mech. Eng. (76-ENAs-37), 11 (1976)
31. Bockris, J.O.M., Otagawa, T.: The electrocatalysis of oxygen evolution on perovskites. J. Electrochem. Soc. **131**, 290–302 (1984)
32. Kelly, E.J., Heatherly, D.E., Vallet, C.E., White, C.W.: Application of ion implantation to the study of electrocatalysis. I. Chlorine evolution at ruthenium-implanted titanium electrodes. J. Electrochem. Soc. **134**, 1667–1675 (1987)
33. Trasatti, S., O'Gradey, W.E.: Properties and applications of ruthenium dioxide-based electrodes. Adv. Electroch. El. Eng. **1981**, 177–261 (1981)
34. Trasatti, S.: Studies in physical and theoretical chemistry. vol. 11. Electrodes of conductive metallic oxides, Pt. A (1980)
35. Krishtalik, L.I.: Kinetics and mechanism of anodic chlorine and oxygen evolution reactions on transition metal oxide electrodes. Electrochim. Acta **26**, 329–337 (1981)
36. Hepel, T., Pollak, F.H., O'Grady, W.E.: Chlorine evolution and reduction processes at oriented single-crystal ruthenium dioxide electrodes. J. Electrochem. Soc. **133**, 69–75 (1986)
37. Kuznetsova, E., Krtil, P., Heyrovsky, J.: Parallel oxygen and chlorine evolution on iridium based oxides. Paper presented at the electrocatalysis: Present and future and ELCAT meeting, Alicante, Spain, 14–17 (2011)
38. Ghany, N.A.A., Kumagai, N., Meguro, S., Asami, K., Hashimoto, K.: Oxygen evolution anodes composed of anodically deposited Mn-Mo-Fe oxides for seawater electrolysis. Electrochim. Acta **48**, 21–28 (2002)
39. Vondrak, J., Klapste, B., Velicka, J., Sedlarikova, M., Reiter, J., Roche, I., Chainet, E., Fauvarque, J.F., Chatenet, M.: Electrochemical activity of manganese oxide/carbon-based electrocatalysts. J. New Mat. Elect. Syst. **8**, 209–212 (2005)
40. Gatt, P., Petrie, S., Stranger, R., Pace, R.J.: Rationalizing the 1.9 A crystal structure of photosystem II—a remarkable Jahn-Teller balancing act induced by a single proton transfer. Angew. Chem. Int. Ed. **51**, 1–5 (2012)

41. Guizard, C., Princivalle, A.: Preparation and characterization of catalyst thin films. Catal. Today **146**, 367–377 (2009)
42. Djokic, S.S.: Modern Aspects of Electrochemistry, vol. 48. Springer, New York (2010)
43. Chen, Y.-C., Hsu, Y.-K., Lin, Y.-G., Lin, Y.-K., Horng, Y–.Y., Chen, L.-C., Chen, K.-H.: Highly flexible supercapacitors with manganese oxide nanoshhet/carbon cloth electrode. Electrochim. Acta **56**, 7124–7130 (2011)
44. Kawashima, A., Sakaki, T., Habazaki, H., Hashimoto, K.: Ni-Mo-O alloy cathodes for hydrogen evolution in hot concentrated NaOH solution. Mat. Chem. Eng. A **267**, 246–253 (1999)
45. Choquette, Y., Menard, H., Brossard, L.: Hydrogen discharge on a Raney nickel composite-coated electrode. Int. J. Hydrogen Energy **14**, 637–642 (1989)
46. Raush, S., Wendt, H.: Morphology and utilization of smooth hydrogen-evolving Raney nickel cathode coatings and porous sintered-nickel cathodes. J. Electrochem. Soc. **143**, 2852–2862 (1996)
47. Endoh, E., Otouma, H., Morimoto, T., Oda, Y.: New Raney nickel composite-coated electrode for hydrogen evolution. Int. J. Hydrogen Energy **12**, 473–479 (1987)
48. Paunovic, P.: Toward sustainable hydrogen economy: electrode materials-non-platinum electrocatalytic systems. J. Univ. Chem. Tech. Metall. **45**, 149–160 (2010)
49. Jaksic, M.M., Lacnjevac, C.M., Grgur, B.N., Krstajic, N.V.: Volcano plots along intermetallic hypo-hyper-*d*-electronic phase diagrams and electrocatalysis for hydrogen electrode reactions. J. New Mat. Elect. Syst. **3**, 169–182 (2000)
50. Paunovic, P., Popovski, O., Dimitrov, A., Slavkov, D., Lefterova, E., Jordanov, Sh: Improvement of performances of complex non-platinum electrode materials for hydrogen evolution. Electrochim. Acta **52**, 1810–1817 (2006)
51. Trasatti, S., Buzzanca, G.: Ruthenium dioxide: a new interesting electrode material. Solid state structure and electrochemical behavior. J. Electroanal. Chem. **29**, 1–2 (1971)
52. Thompson, R.: The Modern Inorganic Chemicals Industry, vol. 31. London (1977)

Structure and Transport Properties of Polymer Electrolyte Membranes Probed at Microscopic Scales

Sandrine Lyonnard

Abstract The synthesis and manufacturing of polymer electrolyte membranes with improved functional properties such as high proton conductivity and chemical stability is an actual challenge to increase the performances of Proton Exchange Membrane Fuel Cells. To achieve this goal, a microscopic understanding of the relation between the primary chemical nature of the electrolyte, the morphology, the proton transfer and water diffusion mechanisms, and the effective properties is essential. Multi-scale experimental strategies need to be developed for studying the structure/transport interplay in these complex charged polymers. In this chapter we focus on complementary spectroscopic techniques that operate at molecular, nanoscopic or mesoscopic scales. Both structural and dynamical characterizations of two representative polymer electrolytes are detailed: the benchmark perfluorinated Nafion membrane and an alternative polyaromatic material, the Sulfonated Polyimide. A review of state-of-the art numerical simulations is also provided to complement the experimental findings. The first section is dedicated to small angle scattering studies of polymer microstructure. The second section is devoted to the water and proton dynamics studied by quasi-elastic neutron scattering and NMR relaxometry. Finally the last section is dedicated to model self-assembled surfactant systems where the effect of confinement on proton mobility is explored in a systematic way.

S. Lyonnard (✉)
Structure et Propriétés d'Architectures Moléculaires, UMR 5819 (CEA-CNRS-UJF),
Laboratoire des Polymères Conducteurs Ioniques, INAC/SPrAM, CEA-Grenoble,
38054 Grenoble Cedex 9, France
e-mail: sandrine.lyonnard@cea.fr

G. Ferreira (ed.), *Alternative Energies*, Advanced Structured Materials 34,
DOI: 10.1007/978-3-642-40680-5_8,

1 Introduction

Fuel Cells are promising clean and efficient electrochemical systems developed as an alternative to combustion engines. A fuel cell is composed of an ionic conducting electrolyte sandwiched in between two electrodes. The anode is supplied with the fuel, which is most generally hydrogen, and the cathode by an oxydizing gas, usually oxygen. The energy of oxidation/reduction catalytic reactions is converted into electric energy and heat. The ions transferred through the electrolyte are generally protons that are produced at the anode by the fuel oxidation. Several fuel cells technologies have been developed to target different applications. Among them, the Proton Exchange Membrane Fuel Cells (PEMFC) [1] employs a hydrated ionomer membrane for low temperatures applications ($T < 120$ °C), mainly light-duty vehicles and on-site power generation. Operation of a PEMFC requires to control and optimize the mass transfer of all species (gas, electrons, protons and water molecules) through the functional layers of a fuel cell: gas diffusion layers, catalyst layers and electrolyte.

The key component of the PEMFC is the polymer electrolyte membrane, which mostly determines the operating range of the fuel cell, i.e. accessible temperatures as well as gas humidification and pressure. In addition to its essential properties of proton conductivity and chemical stability, it has to fulfil a number of demanding requirements: good mechanical properties under constraints, high glass transition temperature (to increase the mechanical stability at high temperature), barrier properties to oxygen and hydrogen, low electronic conductivity, reasonable synthesis and manufacturing costs. The main factors currently limiting an industrial development are the materials durability and performances, and cost reduction. Modern approaches to design new electrolytes include the control of membrane morphology [2], which imposes its connectivity and the confinement of the ionic phase thus controlling water and protons transport, and the preparation of hybrid/composite materials [3].

An actual challenge is thus to understand the relation between the primary chemical structures of the materials, the composition of the heterogeneous media, the effective properties and the performances [4]. However a detailed understanding of the interplay between structure and transport properties at all relevant scales, from molecular to macroscopic level, is still lacking in most membranes, including the reference perfluorinated materials such as Nafion. We have developed a coupling approach that combines spectroscopic techniques for determining both the structural and the dynamical features on an extended time- and length-scale from molecular level to mesoscopic scales (Fig. 1). In this chapter we focus on the microscopic properties of two types of representative ionomer membranes: the benchmark Nafion on one hand, and one of its polyaromatic challenger, a Sulfonated Polyimide (sPI). The first section is dedicated to the study of polymer membrane nanostructuration as a function of its hydration state, mainly by small angle scattering techniques (SANS/SAXS) in complement to Molecular Dynamics (MD) simulations. Next, the following section is devoted to the study of protons

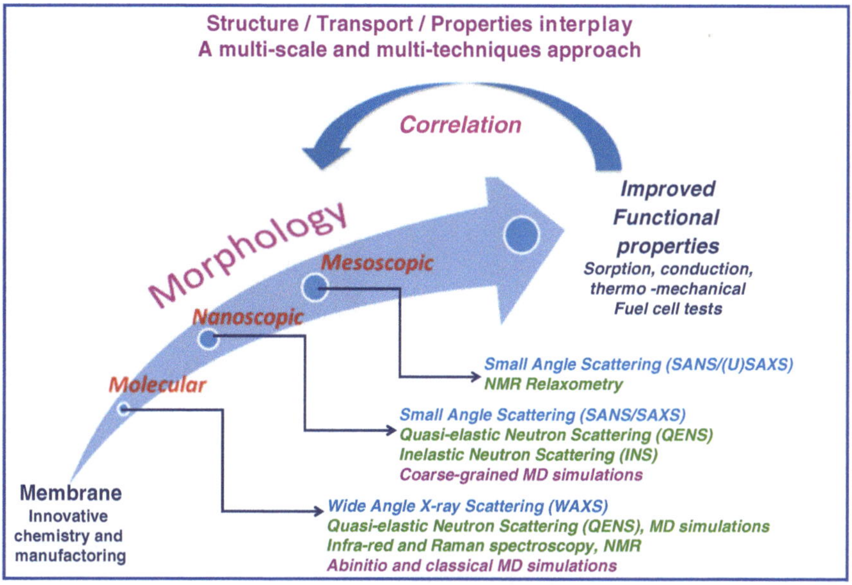

Fig. 1 Schematic multi-scale strategy to correlate functional properties to the chemical nature of new electrolytes, their morphology and the transport properties. Experimental spectroscopic techniques for structural (*blue*) and dynamical (*green*) studies complement molecular dynamics simulations (*purple*)

and water dynamics in hydrated membranes, by quasi-elastic neutron scattering (QENS) and NMR relaxometry. In the last section, the use of model systems to quantify the effect of confinement on local proton mobility is explored. Overall we show how sophisticated experimental tools can be interestingly used to bridge the different scales and compare competitive materials, and we make a portrayal of an ideal membrane.

2 Morphology of Hydrated Membranes

Ionomer membranes for PEMFC are hydrated acidic polymers, generally sulfonic [5]. The coexistence of ionic functionalities and hydrophobic polymer backbone results in an hydrophilic/hydrophobic phase separation, as schematically drawn in Fig. 2. Mechanical properties are insured by the polymer skeleton that forms an hydrophobic host matrix, while the hydrophilic acidic functions serve as proton donors and confer to the material its sorption and swelling properties. Such nanometric structuration is a common feature to all ionomer membranes and provides a multiply connected network of ionic channels/domains accessible to hydration water. Proton transfer and migration takes place in the hydrophilic phase and is mediated by water molecules through a dense H-bond network. The overall

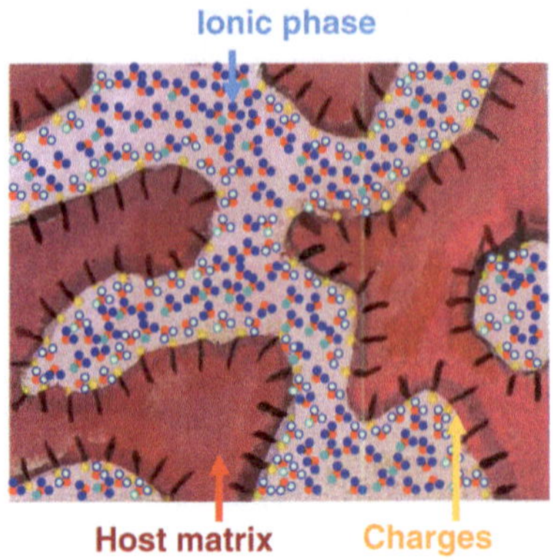

Fig. 2 Schematic representation of the nanophase separation in a ionomer membrane. The ionic phase contains the sulfonic acid groups (*yellow*), water molecules and hydronium ions, confined inside the hydrophobic host polymer matrix

proton conductivity highly depends on the water content and the local water-polymer interactions. The hydration state of a membrane is quantified by a local variable, λ, defined as the number of water molecules per ionic site. This variable is widely employed to compare sorption and transport properties of membranes with different charge densities and chemical nature.

Today's benchmark materials are Perfluorinated Sulfonic Acid Membranes (PFSA) that show unchallenged proton conductivities and stabilities at temperatures below 80 °C. Among them the Nafion [6] produced by Dupont de Nemours is the reference polytetrafluoroethylene (PTFE)-based PFSA membrane (Fig. 3a). It is actually the only one with sufficiently high ionic conductivity ($\sim$0.1 S/cm) to allow PEMFC operation in standard conditions: T = 80 °C and Relative Humidity = 80 %. These conditions however do not permit an industrial development of the fuel cell technology. Operation at higher temperatures is highly desirable in order to enhance the electrochemical reactions kinetics at the electrodes and moderate CO poisoning of the catalysts. Lower hydration is also targeted to facilitate water management and gas humidification and simplify the cooling and compression circuits. Therefore an optimal operation at higher temperature (>120 °C) and lower hydration (20 % RH) is recommended. To achieve this goal the efficiency of proton transport mechanisms must be clearly improved in the low water content regime and the temperature range must be extended. The synthesis and developments of new materials has thus become a high stake in the last years. Moreover, cost reduction, degradation problems and poor resistance to swelling cycles have triggered extensive efforts to develop alternative membranes to Nafion. Improved perfluorinated membranes with shortened or longer side-chains length have been continuously developed as commercially available alternatives to Nafion: Flemion (Asahi Glass Co.), Aciplex (Asahi Glass Co), Dow (Dow Chemical Companies), Gore-tex (Gore and Associates), Fumion (Fumatech), Aquivion (Solvay Solexis). Since 15 years non-fluorinated aromatic polymers have attracted a growing interest, their main advantages being a reduced synthesis cost and high chemical versatility. A large variety of chemical composition has been explored: polystyrene-divinylbenzene acid [7], poly(arylene ether sulfone) [8], polyether

ether ketone (PEEK) [9], sulfonated polyimides (sPI, Fig. 3b) [10, 11], poly(phenylene) sulfones [12, 13], polyphosphazenes [14] etc. More recently, composite or hybrid membranes are also developed [3]. A review of recent progresses in materials synthesis can be found in references [2, 4, 15–19].

2.1 Microstructure Probed by Small-Angle Scattering

Polymer electrolyte membranes are hardly observable by microscopy techniques because of a lack of self-organisation and high sensitivity to electron beam [20]. Recent technical improvements and developments of AFM [21] and tunnelling microscopy [22] have permitted new observations but membrane structures were mostly studied by spectroscopic techniques [23] and molecular dynamics simulations.

Experimentally, the polymer microstructure has been mainly investigated by X-ray Scattering ((U)SAXS/WAXS [24–26]) and Neutron Scattering (SANS [27–32]). A broad correlation peak—so-called ionomer peak—is observed in all ionomer membranes and interpreted as the signature of phase separation at nanometric scale (Fig. 3). The understanding of polymer morphology has been mainly based on the interpretation of this peak and its variation with relevant parameters: water content, thermal history, temperature, nature of solvent and

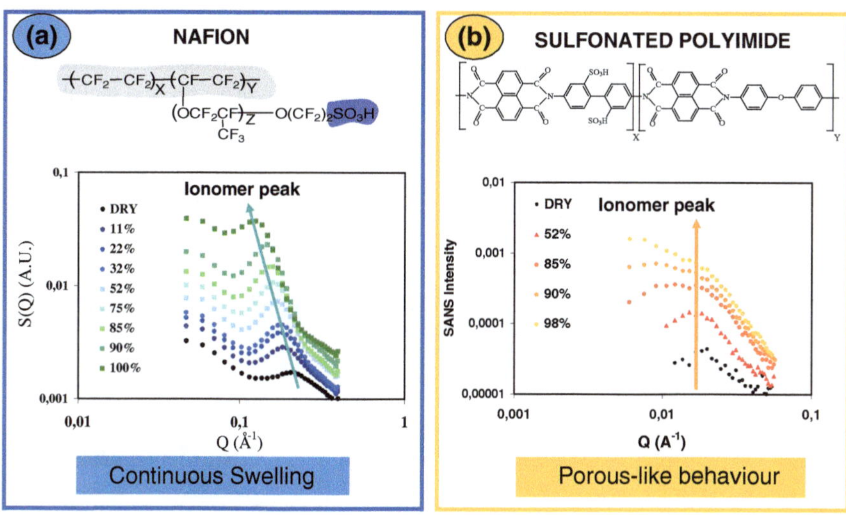

Fig. 3 Small angle scattering spectra of **a** Nafion and **b** Sulfonated Polyimide membranes as a function of relative humidities, from dry state to fully saturated. Arrows indicate the ionomer peak position, that shifts upon hydration in Nafion and remains unshifted in Polyimide, thus evidencing in the former the nanometric swelling and in the latter a porous-like behaviour

counter-ions, manufacturing process. Figure 3 shows the spectra recorded on Nafion (a) and Sulfonated Polyimide (b) as a function of increasing water content.

In the Nafion, the ionomer peak shifts to small wavevectors Q with increasing hydration, which is the signature of a continuous swelling process at nanometric scale with characteristic ionic domains size growing from 1 to 3 nm. In contrast, a different behaviour is observed in the sPI: not only the ionomer peak position is unchanged upon hydration, but the typical nanophase separation distance is an order of magnitude higher than in Nafion. Sulfonated Polyimide membranes indeed possess a residual porosity due to solvent evaporation during film preparation, and thus behaves as a rigid porous matrix with large pores being filled by water.

While attempts to elucidate polyaromatic or composite structures on the basis of scattering spectra remain scarce, many models have been proposed since the 80′s to describe the Nafion: ionic spherical clusters [24], bilayer structures [33], lamellar structure [34], sandwich-like structures [26], connected cylindrical structure [35]. More recently, structure factors S(Q) have been recorded on a much extended wavevector range thanks to technical improvement of scattering techniques. Multi-scale models have then been proposed to account for all features in addition to ionomer peak: intense small-angle upturn, lower-q correlation bump referred to as matrix knee, diffraction peaks including crystalline and amorphous components. Among these modern models, the fibrillar model [31, 32] describes ribbon-like hydrophobic aggregates embedded in a continuous ionic medium and organized in large bundles (size > 500 Å), while in the water rod-like channels model [36], nanometric parallel water channels are hexagonally arranged inside the perfluorinated matrix. Both models are depicted on Fig. 4a, together with the structure proposed for Sulfonated Polyimide [37] shown in Fig. 4b. sPI polymers form inter-connected platelets of ~20 nm organized at larger scale in a 2D layered structure. The morphology of PFSA membranes thus appears to be very different with that of sPI, and more generally aromatic hydrocarboned membranes.

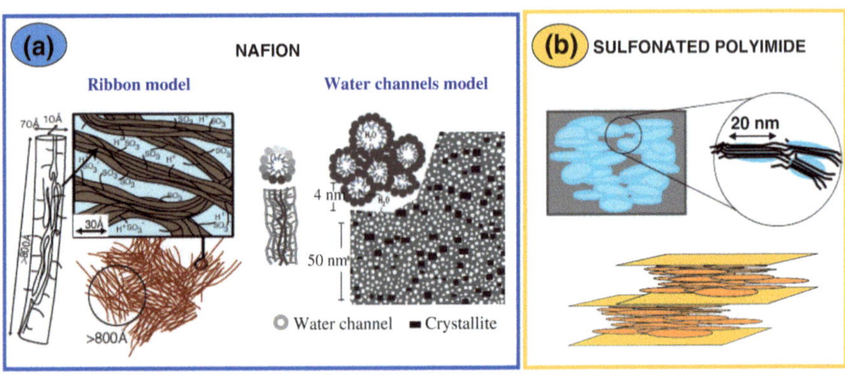

Fig. 4 Schematic representation of the fibrillar model [31] and the water rod-like channels [36] proposed for the microstructure of Nafion, and a Sulfonated Polyimide model [37] based on the staking of anisotropic particles. Ionic domains in Nafion are typically few nanometers in size, whatever the model

2.2 Numerical Simulations

Since ten years, a tremendous effort has been made from the simulation side at all relevant scales: atomic ab initio calculations [38–41], classical MD simulations [42–46], and lately mesoscopic [47, 48] and coarse-grained methods [49, 50]. The simulations aim at determining the shape, the size and the organisation of ionic domains in the ionomer membranes as a function of certain chemical or structural parameters such as charge density, chain length, block sequence, hydration, etc. They have however to face a harsh dilemma: details of the chemical architecture and local interactions must be included, but large length and time scales are necessary to describe the aggregation phenomena (typically 50 nm and 100 ns). Hardly are those antagonist requirements met in a single simulation. It is worth noting also that most of studies have been performed on PFSA membranes.

Mesoscopic techniques have brought important insights into the microstructure of Nafion, as they can describe the nanophase separation. The first trials have been reported by Khalatur [51]. Each CF_2/CF_3 group has been modelled by a unit bead, and a uniform pendant chain distribution has been considered. A three layer segregation has been observed: a water-rich central region, two external layers containing acid groups highly associated to the hydration water molecules. The linear dependence of the swelling with λ, attributed to the inter-fibrillar swelling, is in agreement with the experimental dilution laws [52]. In 2006, Wescott and Galperin [48, 53] have used mean-field simulations to evaluate the morphological modifications as a function of temperature and hydration. They observe the existence of isolated ionic clusters of spherical shape at low hydration ($\lambda < 6$). At high water content ($\lambda > 8$) the clusters adopt an elliptic shape. The Dissipative Particle Dynamics (DPD) method has also been used by several groups to study the mesoscopic structure of Nafion at different water content. In 2003 Yamamoto and Hyodo [54] have shown the existence of a cluster network at low hydration, and a sponge-like structure at high swelling. The cluster size and inter-cluster distance linearly increase with λ. More recent DPD simulations performed by Visnyakov [55] and Malek [49] show the existence of a percolation transition between isolated clusters at low λ and a three-dimensional network of randomly connected channels at high hydration. Wu, Paddison and Elliott [47, 56, 57] have performed DPD simulations since 2008 to evaluate the impact of pendant chain length and charge density on the transport and structure properties. The isolated clusters formed at low hydration increase in size and form at high hydration continuous regions resembling to channels or pores. The size and the dispersion of the aqueous regions increase with increasing charge density and chain length. However, both DPD and mean-field methods are not capable to describe the properties related to correlation functions (for instance diffusion), which limit their relevance for studying structure/transport interplay. An alternative method is the Coarse Grained Molecular Dynamics (CGMD). In 2008, Malek and Eikerling [49] have reported CGMD simulations with the longest Nafion chain ever used (Fig. 5).

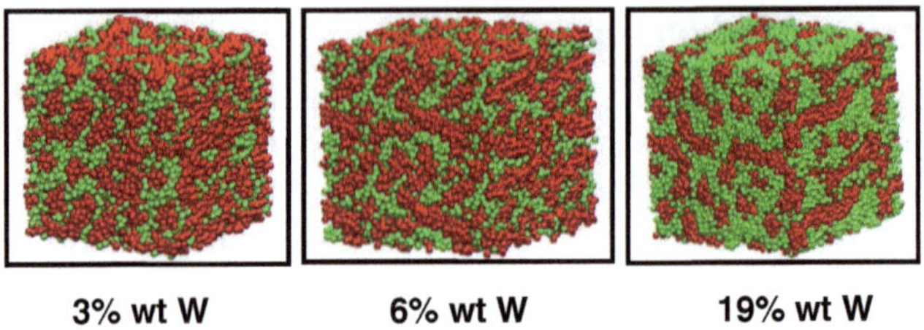

Fig. 5 Microstructure of Nafion predicted by CGMD simulations at three water content [49]. Hydrophilic domains (protons, water, pendant chains) are shown in green, hydrophobic domains in red

Analysis of density maps and radial distribution functions indicated the existence of poorly connected water clusters of spherical shape at low hydration, that grow (from 1 to 3 nm typically) and evolve into elongated aggregates of cylindrical shape, highly connected at high hydration. The percolation was observed at $\lambda = 4$ (10 % wt water). Above this threshold, protons, water molecules and anionic pendant chains form a random three-dimensional network immersed in the hydrophobic matrix.

Overall the size and the distribution of ionic domains remain a controversial issue. Knox and Voth [50] have performed recently a multi-scale MD simulations to evaluate inter-clusters phenomena at atomic resolution (2 millions of atoms, simulation box 30 nm). They have shown that it is impossible to discriminate the main structural models on the basis of the Nafion experimental scattering spectra S(Q). However, hydrophobic/hydrophilic segregation appears to be an essential property, with a well-marked interface due to the high interfacial energy between water molecules and fluorinated matrix. All studies reveal a complex structure organized at different length scales, from the local chemical architecture of polymer up to the random and heterogeneous morphology of the membrane at macroscopic scale. Ionic clusters are relatively anisotropic in shape and heterogeneously spatially distributed. The polymer matrix self-organizes into interconnected fibrillar aggregates. Dissociation of the hydrophilic acid groups $-SO_3H$ rejects the mobile protons in the aqueous phase that fills the inter-aggregates space. Pendant chains that carry the hydrated anions SO_3^- are fixed at the surface of hydrophobic aggregates, and they form a charged flexible interfacial layer along which protons and water molecules move. The structure of this interface strongly influences the state and dynamical properties of water molecules, the ion mobility, the pendant chain vibrational modes, the strength of polymer/water/proton interactions and the membrane stability. The relative performances of membranes of different chemical nature might largely originate from differences in this complex interfacial region.

2.3 Sorption Properties

Water management is a critical issue for fuel cells operation, because the distribution, the state and properties of water molecules in the membrane, inside the electrodes, and in the gas diffusion layers, largely determine the performances. The water distribution within the gas distribution channel, the gas diffusion layers and the membrane has been analysed in situ using sophisticated techniques such as neutronography, magnetic resonance imaging, confocal Raman microscopy, small-angle neutron scattering or x-ray diffraction using a microbeam. However, the basic processes of water transport and sorption are not correctly described especially within and through the membrane, which still limits the development of realistic water management models.

All the properties of proton conducting polymers—conductivity, water and proton diffusion coefficients, electro-osmotic drag coefficient, mechanical properties, temperature stability, life-time, degradation processes—strongly depend on their hydration state [58]. Moreover, the role of water is also preponderant in the structural formation of the electrolyte membranes and in elementary proton conduction mechanisms [59–61]. Water sorption and transport in Nafion membranes have been mainly studied at macroscopic scale by gravimetric and permeation methods. Pulse-field Gradient Nuclear Magnetic Resonance has been also extensively used to measure the water self-diffusion coefficient. References can be found in a recent review published on the subject [62]. Interestingly, the diffusion coefficients obtained by the three methods vary by several orders of magnitude: values of $D_s \approx 10^{-6}$ cm^2 s^{-1} are found by NMR, $\approx 10^{-7}$ cm^2 s^{-1} by permeation measurements and $\approx 10^{-8}$ cm^2 s^{-1} by standard gravimetric analysis. Such discrepancy suggests competition between interfacial and bulk transport mechanisms. Interfacial vaporization-exchange kinetics might indeed limit the water sorption in vapour phase, due to the presence of a thin skin on the Nafion surface which hydrophilicity depends on the environment [63]. Recently, Small Angle Neutron Scattering experiments have been performed to study the kinetics of vapor and liquid water sorption at a microscopic level [64]. In vapor phase, the kinetics are very slow since the swelling equilibrium is not attained after two days. Water concentration profiles through the membrane were found to be completely flat during all the sorption process (Fig. 6).

This result suggests that the vapor sorption process is limited by the transfer at the membrane interface and confirms that water diffusion within the membrane is extremely fast. Moreover, the integration of the water concentration profiles suggests that the sorption kinetics can be analyzed as a purely Fickian process at least for short times. Noteworthy also is the dependence of the water sorption with the logarithm of time. Contrary to this, a swelling kinetic experiment performed in the presence of liquid water is a very fast process (Fig. 7), as confirmed by recent SAXS experiments with sub-second resolution [65]. The diffusion of liquid water inside the membrane is finished in less than a minute. Another mechanism involving long-term polymer reorganization upon swelling has been also

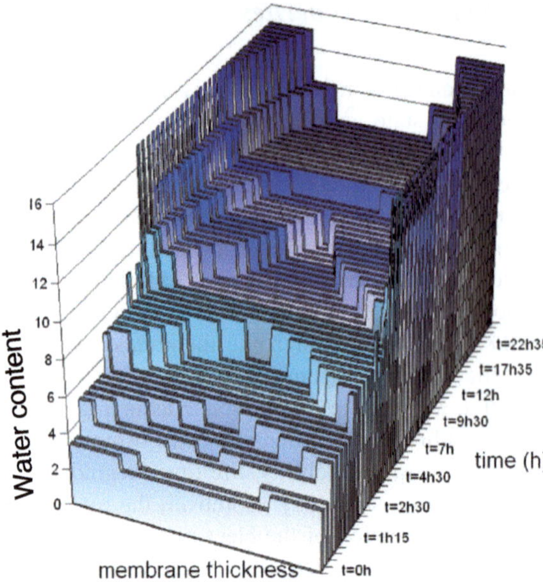

Fig. 6 Water concentration evolution along the Nafion membrane thickness as a function of time obtained from SANS spectra analysis [64]

evidenced by SANS performed on Nafion membranes immersed in water since few years [66]. The characteristic ionic domain size slightly increases with a logarithmic time dependence. This results demonstrates that Nafion is never at equilibrium.

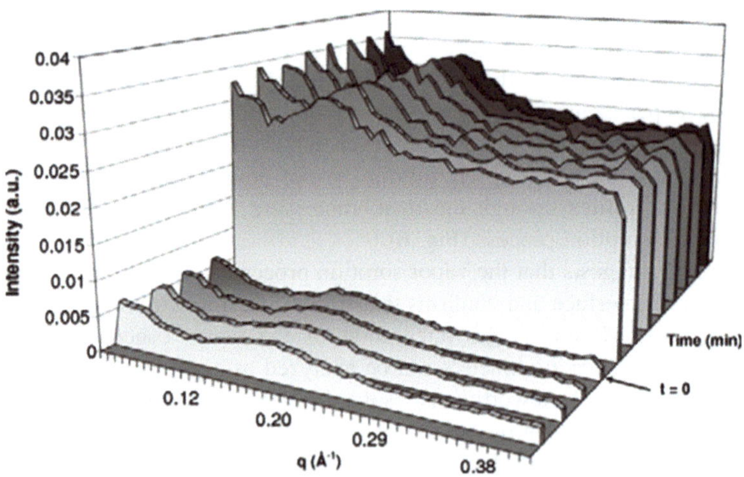

Fig. 7 SANS spectra recorded on Nafion [64]. Time t = 0 corresponds to liquid water injection. After 1 min the membrane is almost fully swelled

3 Dynamics of Water and Protons in Membranes

Proton conduction mechanisms in hydrated ionomer membranes are intimately connected to the existence of a denser H-bond network as hydration increases, to the state of water confined at nanometric scale, its structural and dynamical properties, notably at the hydrophilic/hydrophobic interface. In addition to the geometrical confinement, interactions with the charges at the matrix/solvent interface profoundly affect the water molecule mobility. The existence of several types of water molecules has been observed in the Nafion by various experimental techniques and molecular dynamics simulations. In 1999, Elliott et al. [67] have studied the dynamics of small molecules in a model Nafion membrane. They observed the segregation of "bounded" water associated to sulfonate groups and "free" water molecules. Recent time-resolved infra-red experiments [68–72] have shown that the vibrational relaxation is bi-exponential at all hydrations, thus indicating the existence of at least two different local environments. By dielectric spectroscopy, Lu et al. [73] have observed in 2008 the existence of three types of water molecules : free, highly bounded (observed through the rotation of the sulfonate groups) and more loosely bound, three times slower than free water. Differential Scanning Calorimetry studies have also reported the existence of freezing and non-freezing water molecules [74]. The distinction between bounded and unbounded water populations is also consistent with water adsorption heat measurements [75]. Infra-red experiments on Nafion [76–79] and other PFSA membranes [80] have also evidenced different local environment of water and hydrated complexes (hydronium ions, Eigen and Zundel entities). The relative proportion of the water populations evolve upon hydration, with an increasing number of "bulk-like" water molecules.

3.1 Numerical Simulations

The existence of "interfacial" water and more volumic water related to the strength of water-polar groups interactions via hydrogen bonding, is discussed in most numerical simulations dedicated to proton transport mechanisms. Paddison has proposed in 2002 [81] the existence of a surface conduction process occurring at the water-matrix interface in the first hydration layer, when no "bulk" phase is available for mass diffusion. This mechanism operates at very low hydrations ($\lambda < 2$). Eikerling et al. [82, 83] employed a dielectric continuum approach to study proton conductivity in model pores with well defined geometries. They concluded that two different conduction mechanisms coexist: the first one is similar to that of bulk water and dominates at the center of the pores, while the second dominates at the interface along the charge network. Surface mobility is generally reduced with respect to volume diffusion, however, this can be compensated by the excess of protons at the interface. In small pores, the surface mechanism might even be predominant. The region of surface conduction is limited to a monolayer of water near the interface, where the electrostatic potential is highly modulated by polar groups. Proton mobility in the vicinity of sulfonate groups is reduced because of high Coulomb barriers. Increasing charge densities, thermal fluctuations as well as charge delocalisation reduce the Coulomb barrier and thus facilitate proton mobility. The transition between surface conduction and "bulk" regime occurs at low hydration. The balance between the two mechanisms is determined by the density of charges and the size of the pores. Choi et al. [84] also developed in 2005 a pore model where three conduction mechanisms were taken into account : surface diffusion, Grotthuss mechanism and mass diffusion. Surface diffusion involves proton jumps from one sulfonate to a neighbouring one via the bounded water molecules. However, Grothuss mechanisms involving complex H-bond reorganisations in hydrated clusters would be predominant at all hydrations.

Classical MD simulations neglect H-bonds and only provide information on hydronium ions diffusion. Alternatively, Empirical Valence Bond (EVB) models can include the details of the structural diffusion due to H-bond breaking and forming, in addition to the vehicular component. In 2004 Spohr [85] has incorporated the details of non-classical proton dynamics in the MD simulations of a single pore, using a two-states EVB model. This study showed that structure diffusion contributes to a large extent to proton diffusion, and that it diminishes at low hydration due to the formation of ion pairing $H_5O_2^+$–SO_3^-. More complete multi-state EVB models have been used since by Petersen et al. [86, 87] and Kornyshev et al. [88–90] to study the effect of confinement in nanometric pores and the role of charged walls on solvatation and proton transport. They have shown the inhibiting effect of sulfonate groups on proton mobility : SO_3^- act as proton traps and limit the hydronium ion diffusion. It was also suggested that vehicular and structural mechanisms are equally important but negatively correlated. Also, the effect of charge delocalisation, sulfonate groups mobility and pendant chain conformational motions has been evidenced. Lately, in 2006 protonic transport has been evaluated in model "planar" geometries [91]. The authors concluded that there is no relevant distinction between surface and volume mechanisms: a single structure diffusion mechanism is modulated by the topology of the sulfonate groups distribution.

Another approach has been conducted simultaneously by a number of groups, focusing on the interaction effects rather than the competition between transport mechanisms due to confinement effect. Urata et al. [42] used an all-atom description of fully dissociated pendant chains together with a coarse-grained polymer skeleton. They observed aggregation of sulfonate groups that share some water molecules at low hydration. Strong water–SO_3^- interactions suppress water mobility. At high λ, free and bounded water molecules frequently exchange. In the years 2005–2006, Paddison and Elliott [92, 93] have performed electronic structure calculations on fragments of short-side chains. They found that the separation distance between sulfonic acid groups disposed on the backbone affects the

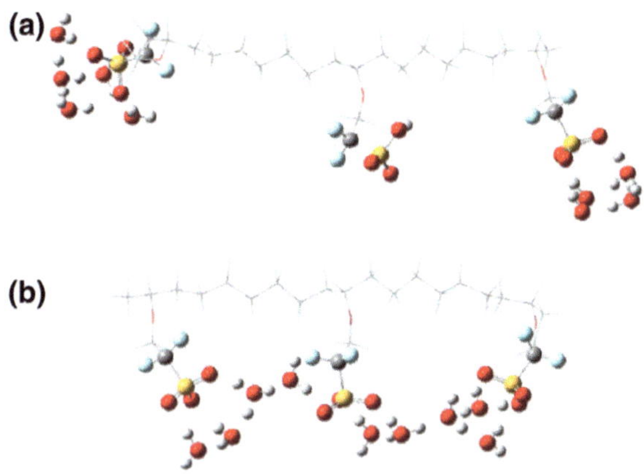

Fig. 8 Optimized structures of Short-Side Chains fragments carrying 3 pendant chains with 9 water molecules ($\lambda = 3$) [94]. **a** Clusters of 4–5 water molecules hydrate only the external sulfonate groups. **b** Water molecules hydrate each ionic group, forming a continuous H-bond network

amount of water molecules necessary for proton transfer to the first hydration layer, and the nature of the hydrated proton. Less than three water molecules are needed when the charges are moved closer. Hydrated protons bridge the acid groups at low hydration. Charge dissociation and Zundel ion formation are favoured by skeleton distortions [38, 94] (Fig. 8). In 2008, Hristov and Paddison have completed this work by studying the impact of the density of charge. The hydrated morphology reveals the formation of H-bond bridges between distant sulfonate groups. At small hydration ($\lambda < 3$), ionic cages made of hydronium ions linked to three sulfonate functions have been evidenced. Other more recent studies from the same group [45] have focused on the effect of chain length, charge density and molecular weight. The most striking result is that water diffusion is governed by a competition between the effects of confinement and connectivity. The balance between these two phenomena is subtle. Long pendant chains for instance reduce connectivity but also reduce confinement, thus producing two antagonist effects with little impact on proton diffusivity.

Dupuis and coworkers [43, 95, 96] have recently performed all-atoms classical MD simulations with a huge number of atoms (40 pendant chains of Nafion spaced by 7 monomers, up to 700 water molecules) and large time scales (2 ns). They obtained the microstructure of Nafion, radial distribution functions, relaxation times and diffusion coefficients of hydronium and water molecules. They show that sulfonate groups move apart upon hydration, as well as hydronium ions. The number of bridged configurations SO_3^-–H_3O^+-SO_3^- decreases with water content but never vanishes, even at full hydration.

To summarize, these abundant numerical studies underline the importance of the interactions between hydronium ions and sulfonate groups, which are modulated by water molecules. Bridges play a predominant role and prevent hydronium ions mass diffusion. The sulfonate groups sterically hinder the hydronium hydration and limit structural diffusion at low hydration. Both vehicular and structural diffusion are slowed down at low hydration.

3.2 Motions at Molecular Level Probed by QENS

Incoherent QENS is usually a very efficient technique to investigate individual proton motions, especially in the situation of confinement [97–99]. The energy transfer (few meV) is comparable to that of relaxation processes and translational/rotational molecular motions. The total scattering function S(Q, ω), where Q is the momentum transfer and hω the energy transfer, is proportional to the contrast between scattering atoms which is determined by the coherent and incoherent scattering cross sections of nuclei. Coherent (incoherent) scattering allows to characterize collective (individual) motions. The hydrogen atom is a natural incoherent scatterer: its incoherent scattering cross-section is an order of magnitude higher than that of any other atom. This property makes the neutrons well suited to investigate individual dynamics of protons or water confined in polymer electrolyte membranes.

The difficulty of the QENS technique however is to operate in Fourier space: appropriate diffusion models in real space are needed to analyse the inelastic spectra in a wide range of scattering vectors [100]. Very few QENS studies have been reported on ionomer membranes despite the obvious interest to obtain microscopic dynamical data that can be compared to MD simulations and serve as input parameters in mass transfer models. In the 80s, a saturated Nafion membrane has been studied on a time-of-flight (TOF) spectrometer [101]. In the years 2005–2006, this work was followed by QENS investigation of hydration [102] and temperature [103] effects on water dynamics in Nafion. In 2007 we have proposed a more sophisticated approach relying on the combination of spectrometers (time-of-flight at different incoming neutron wavelengths and backscattering) to extend the time-scale from pico- to nanoseconds [104]. A specially designed Gaussian model for localized translational diffusion [105] has been developed to analyse the numerous and correlated data. We show in the next subsections that the QENS technique can be used in a qualitative way to obtain raw valuable information on molecular processes in competing membranes, while more quantitative analysis provide deep insights into mechanisms and give a unique access to important dynamical parameters.

3.2.1 QENS: A Qualitative Tool for Comparing Membranes

QENS measurements have been performed on Nafion [104] and sPI [106] to obtain molecular level information on proton mobility, with the objective to understand the discrepancy of their functional properties: sPI membranes actually fail to challenge the PFSA membranes, their macroscopic conductivities remaining much lower even at much higher charge densities. Raw comparison of typical sPI and Nafion scattering function S(Q, ω) obtained on the Time-Of-Flight spectrometer Mibemol (LLB, Saclay) is shown on Fig. 9 as a function of the energy transfer ω. The membranes have been equilibrated at 98 % RH, leading to values of

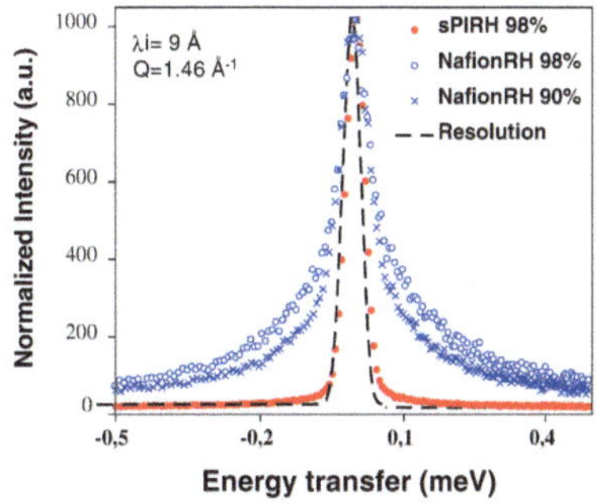

Fig. 9 Quasi-elastic spectra measured by time-of-flight technique on sPI and Nafion membranes at high hydration (RH 98 % and RH 90 %). The resolution is shown together with normalized experimental S(Q, ω) scattering functions plotted at Q = 1.46 A^{-1} in the range of energy transfer [−0.5, 0.5] meV

$\lambda = [H_2O]/[SO_3^-] = 15.3$ for Nafion and 9.5 for sPI. Another reference Nafion sample prepared at 90 % RH corresponds to a comparable local hydration ($\lambda = 10$). The spectra have been normalized for visualization and are reported at a representative value Q = 0.46 $Å^{-1}$. The resolution of the instrument is also shown.

The quasi-elastic broadening due to proton motions is observed out of the zero-energy transfer elastic contribution for both membranes. But a striking difference comes from the quasi-elastic/elastic intensity ratio, which is considerably higher in Nafion. As elastic scattering arises from all non-moving atoms at the time-scale of the experiments, i.e. polymer backbone but also a fraction of immobilized protons (if any), Fig. 9 shows unambiguously the existence of a large amount of immobile hydration proton atoms in the sPI structure. Such result is in agreement with the partial wetting of the aromatic backbone with water showed by Infra-Red measurements [107]. It is clear that these immobile protons are lost for conduction. In contrast, all protons are mobile in the Nafion [104]. Such great discrepancy at molecular level between a completely hydrophobic polymer backbone (Nafion) and a "partially hydrated" backbone (sPI) is further reflected in significantly reduced sPI macroscopic conductivity.

Additional qualitative information can be obtained on series of membranes prepared at various hydrations. Figure 10 shows the Nafion and sPI S(Q, ω) spectra measured on Mibemol spectrometer in dry state and increasing hydration up to full saturation. In Nafion, the quasielastic signal is absent in the dried polymer and appears at the lowest hydration (11 % RH, $\lambda \sim 2$). Then a continuous intensity increase and broadening of the quasielastic line is observed, evidencing the increased number of mobile protons. We have shown that the signal arises from all protons (H^+ and H_2O) in the whole hydration range. The typical correlation time of the observed motions, τ, can be obtained qualitatively from the Half-Width at Half Maximum Γ of the quasielastic Lorentzian-shaped signal. The apparent broadening (or relaxation time decrease) is the signature of the acceleration of molecular motions upon hydration. In contrast, the sPI spectra reveal a two-step dynamical behaviour with a threshold at 32 % RH ($\lambda \sim 3$). Spectra recorded in dry state, 11, 22 and 32 % RH superimpose and there is no significant quasielastic component. All protons are strongly immobilized at the time-scale of

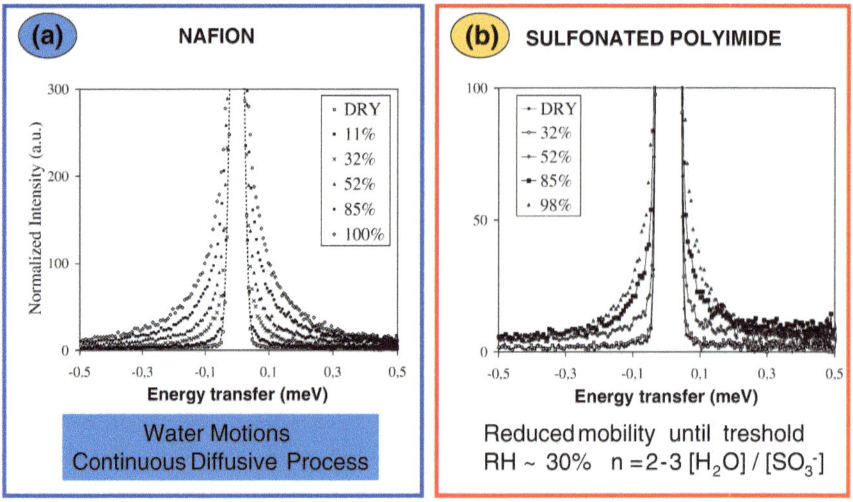

Fig. 10 Raw S(Q, w) spectra of Nafion (**a**) and sPI (**b**) membranes at various hydration level, from dry to saturated. Data are shown at Q = 1.46 A^{-1} in the range of energy transfer [−0.5, 0.5] meV

the experiment (1–50 ps). At higher hydration, protons become mobile and a continuous evolution of the quasielastic intensity and width reflects the increased molecular mobility upon hydration. Such behaviour is typical of porous-like materials. At first, water molecules are strongly interacting with the polymer and bounded to the porous interface. When a water monolayer has been adsorbed onto the pore surface, additional water molecules are used to fill the pore and form a "bulk" phase with increased proton mobility. These findings are clearly correlated to polymer nanostructuration as discussed in the first section.

Raw comparison of QENS spectra thus allows a first diagnosis of membrane's dynamical properties at molecular level, giving important clues on the different mechanisms at work and the existence or not of threshold for enhanced proton mobility. It allows to directly compare the propensity of a given polymer to free (Nafion-like) or trap (sPI-like) a large number of protons, which in turn impacts dramatically the local transport properties and have a knock-on effect on conductivity.

3.2.2 QENS: A Quantitative Tool for Molecular Motions and Mechanisms

In addition to the qualitative description of raw data, a more quantitative analysis of S(Q, ω) spectra can be achieved provided a well-chosen diffusive model is used to generate theoretical S(Q, ω) functions to be compared to the experimental ones in the whole Q-region. In the following, we first describe the sophisticated model used to analyse confined motions. Then we focus on the proton transport mechanisms observed in partially hydrated Nafion membrane. At last, we discuss the variation of the obtained confinement sizes, relaxation times and diffusion coefficients with water content.

Gaussian Model for Localized Translational Diffusion

The self-diffusion of protons and the reorientations of water molecules inside the membrane are "hindered" and generally slowed down by the confinement and the existence of a charged interface. As the water molecule can only diffuse in restricted volumes, the $S(Q, \omega)$ function is affected and classical models (such as diffusion in an infinite medium) are not applicable. A number of models have been developed to describe the confinement situation : diffusion on a segment, on a circle, a cylinder or a sphere. They have been used for describing diffusion of molecules in porous matrixes, supercooled liquids or polymers. Canonical jump diffusion models have also been successfully employed to take into account the granularity that produces a deviation to Fick's law at high Q. However, none of the existing model can reproduce in a satisfying way the complex set of data recorded in Nafion on several QENS instruments on a time-scale ranging from the picosecond—lifetime of a H-bond—to the nanoseconds—typical time for exploring ionic nanochannels. Thus we have developed the so-called Gaussian model [105] which has three main advantages: (1) there is no discontinuity inherent to a boundary condition (2) the mathematical expression of $S(Q, \omega)$ is rather simple, (3) the model takes into account localized translational proton diffusion in a confining domain, the microscopic details of the elementary jump process, and the possible existence of long-range diffusion.

Figure 11 shows the principle of the model, which is based on the use of a random Gaussian variable to describe the proton displacement. Two types of motions are considered: protons are confined in water droplets of size 2σ, they move with a local diffusion coefficient D_{local}. We define τ_{mi} as the jump relaxation time. The second mechanism correspond to a possible proton diffusion from a water confining droplet to another, described by a standard Fickian process and characterized by a long-range diffusion coefficient D_{LR}.

Proton and Water Diffusion Mechanisms

The hydrated Nafion $S(Q, \omega)$ spectra, recorded on the Time-of-Flight spectrometer IN5 and backscattering IN16 of the Institut Laue Langevin, showed the existence of two different quasielastic contributions, revealing the existence of two populations of protons that are not exchangeable at the nanosecond timescale: "slow" and "fast" protons [104]. The slow protons are non-diffusive, their motions are well described by a two-sites jump model, with typical jump time $\tau_L \sim$ few hundreds of picoseconds. On the contrary, fast protons are diffusive with

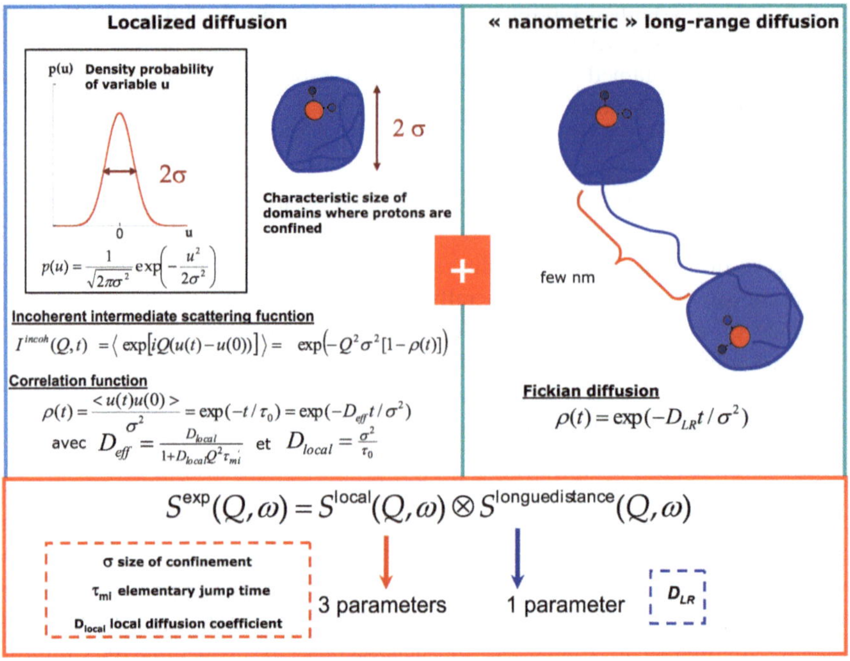

Fig. 11 Principle of Gaussian model applied to incoherent neutron scattering and related dynamical parameters [105]

characteristic times $\tau_{\mathrm{mi}} < 10$ picoseconds. Figure 12 shows that both characteristic times decrease upon hydration, as the signature of enhanced local dynamics. A striking result comes from the number of slow protons which is found constant over the whole hydration range and equal to 3, i.e. the number of protons in the hydronium entity. These protons are the only moving atoms at low hydration ($\lambda < 3$). Moreover, the associated slow jump motions are still observed even at full hydration. In contrast, the number of fast protons is proportional to the number of water molecules as soon as $\lambda = 3$. Fast mobile protons are thus mostly hydration protons forming the H-bond network in the ionic channels. Such results are consistent with MD simulations [95], that showed the presence of a small amount of H_3O^+ in the vicinity of SO_3- whatever the swelling, together with the existence of surface conduction process in the very low water content regime.

Analysis of Quantitative Diffusion Parameters

The quasielastic contribution corresponding to fast diffusive proton motions has been analysed by the Gaussian model in the whole hydration range. On Fig. 13 the size of the confining domains, named dynamical correlation distance d_{dynamics}, is plotted against the structural size of the ionic channels $d_{\mathrm{structure}}$ obtained from

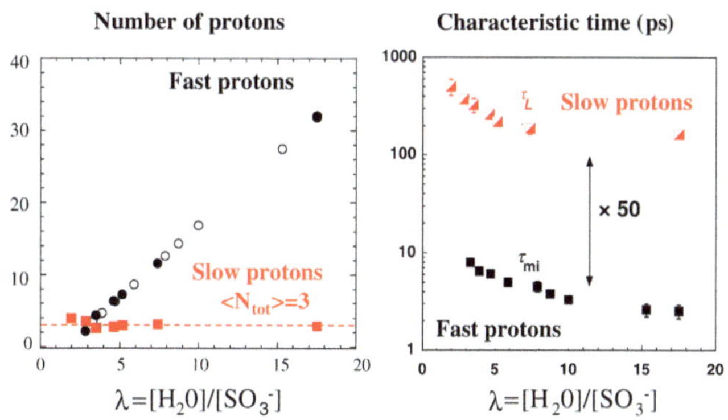

Fig. 12 Number of fast and slow protons as a function of the Nafion hydration λ, and characteristic times of the associated motions

SANS data. The evolution as a function of the local hydration parameter λ shows two regimes. At low hydrations ($\lambda < 10$) both distances are strongly correlated: geometrical restrictions are piloting the confinement of protons and water molecules. In contrast, the correlation is lost at high hydrations ($\lambda > 10$): structural swelling does not impact any more the dynamical confinement size. Interestingly, the threshold size is ~ 1 nm, which is the size of the hydrated cluster formed by the first and second hydration shells of a hydronium ion. This entity is at the heart of the structure diffusion process that has been proposed to describe the proton transfer in acidic solutions [108] and in polymer electrolytes. Continuous and correlated breaking and forming of H-bonds in the outer part of the clusters allow an extremely fast transfer of the excess charge.

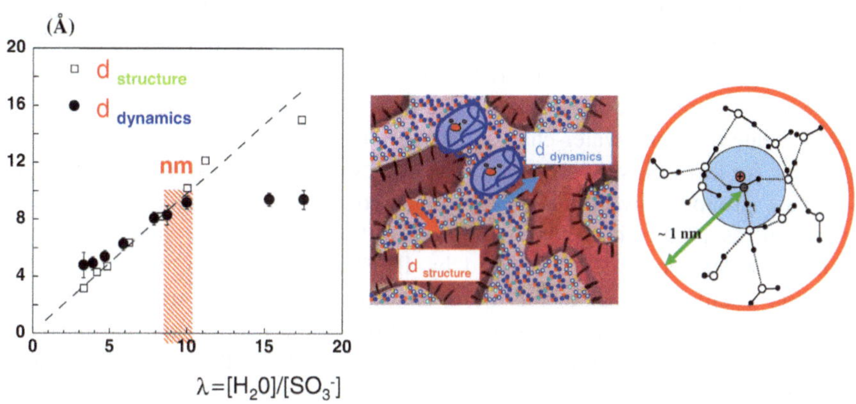

Fig. 13 Size of confining water droplets $d_{dynamics}$ obtained from the QENS analysis, compared to the mean ionic channel size $d_{structure}$ obtained from SANS data, as a function of Nafion local hydration. The pictures represent the nanophase separation where the structural and dynamical distances have been reported, and the hydration cluster of H_3O^+

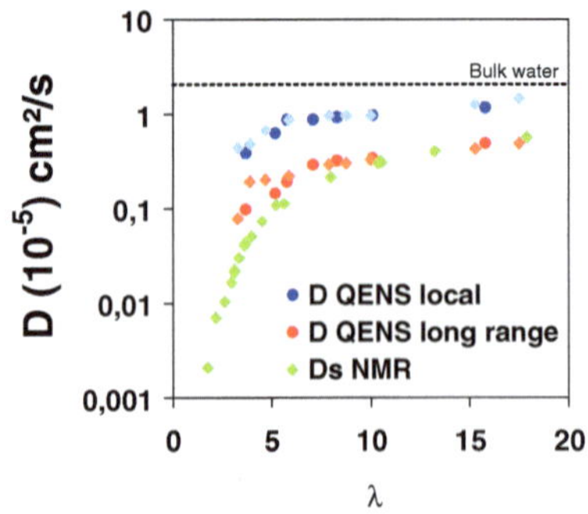

Fig. 14 Water diffusion coefficients measured at molecular and nanometric scale by QENS compared to the self-diffusion coefficient obtained at micronic scale by Pulsed-Field Gradient NMR

The diffusion coefficients associated to localized motions inside the water droplets (D_{local}) and the long-range Fickian diffusion at nanometer scale, from one droplet to a neighbouring one, D_{LR}, are shown on Fig. 14. Clearly, both diffusion coefficients increase when the membrane hydration is increased, and reach a plateau-like behaviour at high water content, typically for $\lambda > 10$. The local diffusion is then extremely fast, almost bulk-like, while the long-range diffusion is reduced by a factor ~ 4 because of the tortuosity of the polymer matrix. The local bulk-like regime corresponds to the building of hydrated clusters forming a well-connected H-bond phase (Fig. 12). These QENS diffusion coefficients are interestingly compared to the self-diffusion water coefficient D_s measured at a completely different length scale (μm) by Pulsed-Field Gradient NMR. Surprisingly, the values of D_s and D_{LR} are very similar at high hydration, showing that proton motions are limited at ionic cluster level and no supplementary obstacle to diffusion takes place above the nanometer scale. Thus, sub-nanometric phenomena controlled by local interactions in the proton-water-polymer system determine the mobility as soon as the dense hydrogen bonding network has been formed. In contrast, at low hydrations, the confinement is predominant and proton motions are drastically reduced at all scales. Further insights to understand this finding have been brought by the mesoscopic studies performed by NMR relaxometry, detailed in the next section. The spectacular reduction of local proton mobility observed at $\lambda < 5$ is due to strong water-charges interaction in the absence of a "bulk" phase available for fast diffusion processes. The QENS results confirm the importance of sulfonate groups - hydronium ions strong correlations in the early stages of hydration, as predicted by MD studies [43, 95, 96], and the high number of bounded water molecules evidenced by IR. The evolution of H_2O and H_3O^+ diffusion coefficients calculated from all atom simulations are also in good agreement with the QENS values. At last, the molecular restrictions evidenced and quantitatively characterized in this work are responsible for the major drop in total conductivity also observed for $\lambda < 5$.

3.3 Interactions at Mesoscopic Level Probed by NMR Relaxometry

To cover the gap between molecular and macroscopic studies, the field-cycling nuclear magnetic relaxation (also called NMR relaxometry) is one of the few techniques available to study the motion at the mesoscopic scale (100 ns $< t <$ 0.1 ms) [109]. The nuclear spin–lattice relaxation rate $R_1 = 1/T_1$, where T_1 accounts for the longitudinal relaxation time, is an NMR parameter sensitive to the rate and the nature of the motion of fluid molecules. The fast-field cycling NMR is a technique based on the very fast commutation (with respect to T_1) of the strength of the magnetic field B. The field can be tuned to record the longitudinal relaxation rate over a large range of Larmor frequencies ν (typically 10 kHz to 10 MHz).

The NMR relaxometry is a powerful tool to investigate fluid dynamics in confinement [110]. The dispersion law $R_1(\omega = 2\pi\nu)$ is directly related to the fluid dynamics. The shape of the $R_1(\omega)$ laws has been predicted in some cases, for instance water embedded in porous materials or at the surface of colloidal particles of various geometry [111, 112]. The NMR relaxometry shares with the QENS technique two aspects: valuable qualitative information can be obtained from the inspection of raw data, but more sophisticated quantitative analysis requires the use of ad hoc diffusion models.

To carry on the comparison of Nafion and sPI properties, we have measured the relaxation rate dispersions $R_1(\omega)$ of fully hydrated membranes at room temperature (Fig. 15) [113, 114]. In the Nafion, the magnetization decays at all relaxation fields are well described by a mono-exponential behavior characterized by a single T_1. The situation looks somewhat more complex in the case of the polyimide, where at least two relaxation times are observed. The polyimide dispersion law of Fig. 15 depicts the average relaxation rates.

There is a clear difference in the dispersion shapes. The Nafion exhibits an almost flat frequency dependence, characteristic of a non-wetting behavior. The polyimide on the contrary shows a strongly frequency dependent dispersion in the whole spectral range, indicating a strong interaction of water with "interfacial"

Fig. 15 Dispersion of the relaxation rate R_1 as a function of the NMR frequency at 298 K, for fully hydrated Nafion (*black dots*) and sulfonated polyimide (*empty dots*)

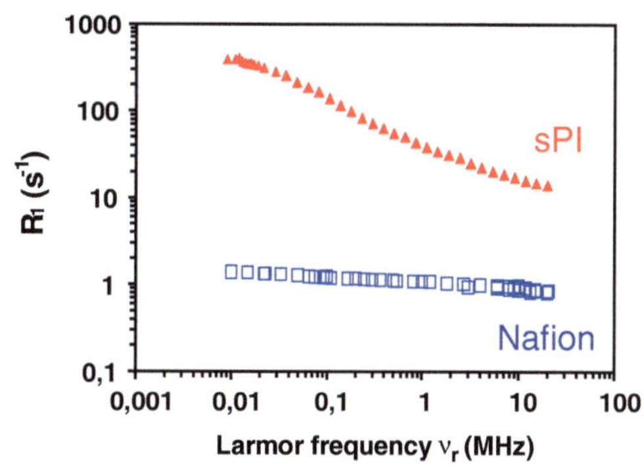

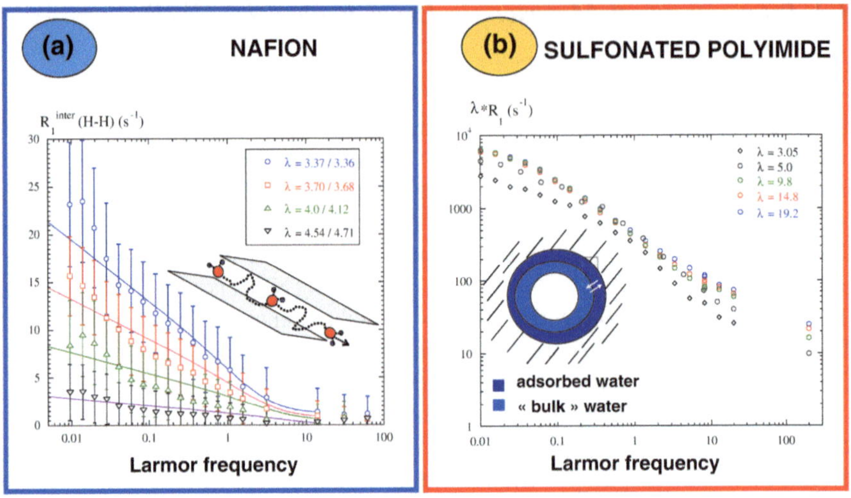

Fig. 16 **a** Relaxation rate of protons in Nafion membrane as a function of hydration, fitted by the 2D diffusion model schematically represented on the inserted picture. **b** Relaxation rate $R_1(\omega)$ multiplied by the hydration variable λ in the sPI membrane, for different water contents. The behavior is typical of a porous material with surface pore coverage at $\lambda < 5$, and then pore filling (existence of a master curve)

hydrophilic groups of the polymeric matrix. At intermediate values (0.02 to 2 MHz), $R_1(\omega)$ follows a $\omega^{-1/2}$ trend, which is the dispersion law characteristic of fluid diffusion in a 3-D multiconnected porous network [112]. This spectacular difference thus teaches at first glance that the proton motion inside the fully hydrated Nafion and sPI is radically different. The confinement effect is enhanced in the polyimide: the proton dynamics as probed at the mesoscopic scale is much slower, or more anisotropic, than in the Nafion. This result is consistent with the QENS studies and the nanostructuration of the membranes. It confirms that sPI membranes behave as wetting rigid porous-like materials where the local water-host interactions play a decisive role in reducing proton mobility.

A molecular model is required at this stage to further interpret both the shape and the magnitudes of the relaxation rates dispersions and get more insights on the mechanisms. Variations of the relaxation rates with water uptake have been measured in both Nafion and sPI (Fig. 16) [115]. The two-step hydration process already observed by QENS in the sPI was evidenced again: solvation and formation of disconnected aqueous clusters near polar groups, followed by the formation of a continuous H-bond network. This is evidenced on the $\lambda \times R_1(\omega)$ plot, showing the existence of a master curve above a threshold at $\lambda \sim 5$. In the Nafion, the values of $R_1(\omega)$ remain much lower than in sPI, and weak variations are observed upon hydration. At low water content, $R_1(\omega)$ evolves logarithmically suggesting a confined two-dimensional diffusion of protons in the microsecond time range. The theoretical dispersion curves calculated with a diffusion in between two planes model [116] is shown on the figure: a good agreement is found with the experimental data. Such a situation is lost at higher swelling ($\lambda > 5$) where a plateau related to 3D diffusion is observed (Fig. 15).

Thus, a severe geometrical confinement of protons in semi-lamellar local environments is responsible for the low mobility in the early stages of hydration. This corresponds to the drastic drop of diffusion coefficients as measured by QENS and PFG-NMR in this region. In addition, the transition to a 3D connected structure occurs around $\lambda \sim 5$, where the local and nanometric dynamics revealed by QENS are significantly enhanced. The threshold value is comparable to that found in CGMD simulations between isolated clusters and 3D interconnected ionic network ($\lambda \sim 4$).

To summarize, the driving property for improved proton transport in polymer electrolyte membranes seems to be a balance between interactions and confinement. In the struggle to rationalize structure/transport interplay in competing fuel cells materials, NMR relaxometry should grow on as it is a very powerful tool as a decisive complement to QENS and other techniques.

4 Model Systems for Controlled Confining Structures

Correlating the transport properties to the microstructure of a polymer electrolyte membrane is often limited by the high level of structural complexity of the material. To overcome such difficulty, model systems with tunable and adjustable nanoporous geometries could help to decouple purely geometrical confinement effects and water-host interactions. An interesting lyotropic model system is formed by the self-assembling of perfluorinated sulfonic surfactant molecules in water. The perfluoroalkylsulfonic acids are particularly interesting as their chemical structure is close to that of Nafion: strong acidity of the terminal ionic group and high hydrophobicity of the main chain. Their topology can be continuously modulated by varying the water content and/or the temperature. The phase diagram of the PerFluoroOctaneSulfonic-H^+ (PFOS) shows the existence of hexagonal and lamellar phases (Fig. 17). These systems are also excellent samples for both neutron scattering techniques and NMR spectroscopy: as a consequence of the well-marked phase separation between highly hydrophobic aggregates and nanopores, the protons are unambiguously located in the ionic phase.

At the local scale, the dynamical behaviour of water in an hexagonal and a lamellar PFOS phase have been investigated by QENS experiments [117]. A qualitative analysis of the neutron scattering spectra with a single Lorentzian used to describe the translational motions has shown three important features (Fig. 17). First, an effect of confinement has been observed in these systems. The Q-dependence of the Half Width at Half Maximum of the quasielastic line shows a large low-Q flat regime that is ascribed to spatially confined motions. Secondly, the geometry of the confinement affects the diffusive motions of the water molecules: in the hexagonal phase, all the parameters indicate an almost bulk-like behaviour at the molecular level. In contrast, in the lamellar phase, the effect of

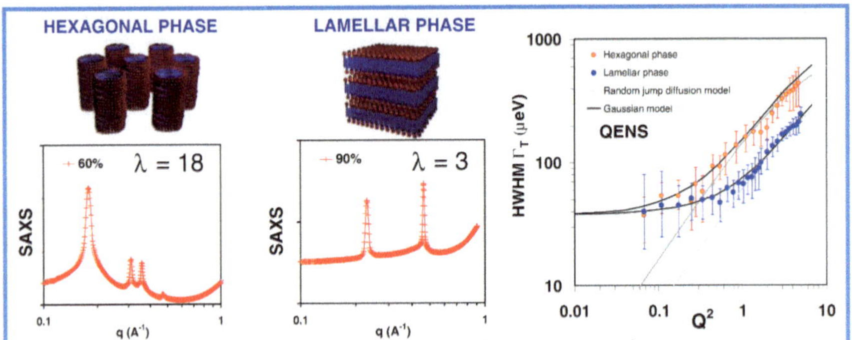

	$d_{structure}$ (Å)	$d_{dynamics}$ (Å)	τ_{mi} (ps)	D_{loc} (10^{-5} cm²/s)	D_{LD} (10^{-5} cm²/s)
Hexa 60	7.3	4.2±0.4	2±1	2.1±0.2	0.7±0.3
Nafion λ=17	15	4.2±0.3	2.5±0.4	2±0.1	0.5±0.05
Lamellar 90	3.2	2.5±0.4	2±1	0.8±0.2	0.2±0.05
Nafion λ=3	3	2.1±0.4	8±0.45	0.45±0.03	0.08±0.03

Fig. 17 Representation of a lamellar phase and an hexagonal phase formed by PFOS surfactant in water (90 and 60 % wt in surfactant, $\lambda = 3$ and 18, respectively). The X-ray scattering spectra are shown with the typical Bragg peaks allowing to determine the structure and the confinement size $d_{structure}$. The right hand side graph shows the width of the quasielastic signal as a function of Q, together with a Gaussian model fit [117]. The dynamical parameters extracted from the fit (dynamical confinement size, characteristic time and diffusion coefficients at local and nanometric scales) are given in the table together with the values obtained on Nafion membrane at comparable λ

confinement is much stronger and results in significantly decreased diffusion coefficients. The third important observation is the similarity of proton and water dynamics in PFOS and Nafion. Indeed, fast confined motions in water droplets of few Å have been observed in PFOS, accompanied by slower long-range translational diffusion at the nanometric scale, as in Nafion. The confinement sizes, diffusion coefficients and residence time obtained by a Gaussian model analysis are given in the table included in Fig. 17. They are very similar to those of Nafion membranes at comparable λ: the lamellar (hexagonal) PFOS geometry is a good model for low (high) hydration regime. A systematic study of the correlation between the geometrical conformation and the confinement size is thus very interesting, and parameters such as chain length, chain flexibility, presence of ether as in Nafion pendant chains, branched molecules, oligomers and copolymers of Nafion, are currently investigated in our group to rationalize the transport/structure interplay in perfluorinated sulfonic compounds.

5 Conclusions

The design of new polymer membranes for fuel cells is an astounding challenge that requires a range of strategies, including a detailed understanding of the structure/transport interplay. Polymer self-aggregation phenomena lead to hierarchical and complex morphologies that determine confinement and connectivity, therefore impacting functional properties. In particular, the proton conductivity measured through membranes of different chemical nature may vary by orders of magnitudes. In standard fuel cells conditions, the benchmark Nafion conductivity is 0.1 S/cm, a value that is hardly challenged despite tremendous efforts to develop alternative membranes, including composite systems or polyaromatic polymers. In the latter family, the Sulfonated Polyimide are high Tg polymers with interesting chemical modularity: their properties can be tuned by selecting the hydrophilic/hydrophobic blocks and their sequence. However, even at high density of charges they fail to reach Nafion's conductivity at full hydration, and it is even worst in the low water content regime. A microscopic study of these differences is essential. We have developed a unique approach for investigating both structural and dynamical properties of Nafion and sPI simultaneously, by combining Small Angle Scattering techniques, QuasiElastic Neutron Scattering and NMR relaxometry. We have shown that the functional discrepancies originate from an intricate body of processes that take place at different time and length scales. Thus the multi-scale method is relevant to tackle the effective comparison of different electrolytes.

The sPI possess a hydrocarbonated backbone that carry ionic functions but also hydrophilic carboxyls and carbonyls polar groups. Strong water-skeleton interactions result in a wetting behaviour that is not favourable to conductivity, because a large number of adsorbed water molecules are bounded to the polar groups and thus immobilized. This situation was evidenced by NMR relaxometry performed at mesoscopic scales in a series of hydrated membranes, and confirmed by QENS and structure measurements. An additional finding is that sPI are rigid porous-like membranes characterized by a two-step dynamical regime: first, coverage of pore surface with polar groups solvatation and formation of disconnected clusters centered on ionic functions, followed by the filling of porosities and continuous H-bond network building. The specific surface available for the first layer adsorption is important in this material where the pores are neither spherical nor regularly shaped. No proton mobility is observed in the first phase at a picosecond timescale. Almost all water molecules are thus used to hydrate the pore surface instead of mediating elementary proton transfer processes. On the contrary, the Nafion membrane is a non-wetting polymer with a well-defined interface between the polymer matrix and the ionic phase. The structure continuously rearranges upon hydration or dehydration, accommodating the water molecules in semi-lamellar local environments that evolve into three-dimensional interconnected network. The dynamical parameters reach 50 % of their maximal values for a local hydration of 5 water molecules per ionic site. This explains why a Nafion membrane is efficient in operation even if the fuel cell is only partially hydrated. At full hydration, water and proton mobilities in the confining domains are almost bulk-like and controlled by sub-nanometric constraints. A common finding to both QENS and NMR measurements in Nafion is the prime importance of poor water-matrix interactions.

Together all these results give clues on the essential effect of nanostructuration and water-polymer interactions in polymer membranes for fuel cells. Pre-existing porosities generated by solvent evaporation during processing are not suited. Poor water-matrix interactions is a fundamental property that should be obtained together with an abrupt, well-defined and preferably chemically inert interface. Combination of a high T_g skeleton with grafted perfluorinated pendant chains could be a means to separate water molecules from the backbone and avoid partial wetting while taking advantage of the chemical potentialities of aromatic polymer architectures.

An important output of the multi-scale study is also a set of quantitative experimental parameters on phase separation, proton and water mobilities at molecular and mesoscopic scales. The dynamical data were lacking, although they must be used against the numerous numerical simulations to discuss contradictory conjectures such as percolation threshold, existence of transient bridges between ionic clusters, correlation and competition between Grotthuss-like proton conduction mechanisms, specific surface diffusion processes and vehicular hydronium ions transport. Progress in numerical simulations could be achieved by optimizing the force fields and compare relevant parameters such as diffusion coefficients to the experimental values measured at the same scales. Controlled chemical modifications would also be very useful to tune some selected properties and shed light on the controversial debate on Nafion microstructure, but such strategy is not available with commercial products. The use of model surfactant systems is a promising route to correlate the ion and water mobility to self-assembled tunable nanostructures with high chain hydrophobicity and strong acid functionalities. Both experimental and numerical investigations in the different mesomorphous phases formed by a variety of surfactants could help providing a clear correlation between geometrical confinement, interactions and transport properties, which is a fundamental step for the optimization of real electrolytes and membranes.

References

1. Fuel Cells I. In: Scherer G.G. (ed.) Advances in Polymer Science, vol. 215. Springer, Berlin (2008)
2. Yang, Y., Holdcroft, S.: Synthetic strategies for controlling the morphology of proton conducting polymer membranes. Fuel Cells **5**(2), 171–186 (2005)
3. Jones DJ, Rozières J: Advances in the development of inorganic-organic membranes for fuel cell application. In: Scherer, G.G. (ed.) Advances in Polymer Science, Fuel Cell I, pp. 219. Springer, Berlin (2008)
4. Peckham, T.J., Holdcroft, S.: Structure-morphology-property relationships of non-perfluorinated proton-conducting membranes. Adv. Mater. **22**(42), 4667–4690 (2010)

5. Peighambardoust, S.J., Rowshanzamir, S., Amjadi, M.: Review of the proton exchange membranes for fuel cell applications. Int. J. Hydrogen Energy **35**(17), 9349–9384 (2010)
6. Mauritz, K.A., Moore, R.B.: State of understanding of Nafion. Chem. Rev. **104**, 4535–4585 (2004)
7. Byun, H.S., Burford, R.P., Fane, A.G.: Sulfonation of cross-linked asymmetric membranes based on polystyrene and divinylbenzene. J. Appl. Polym. Sci. **52**(6), 825–835 (1994)
8. Zschocke, P., Quellmalz, D.: Novel ion-exchange membranes based on an aromatic polyethersulfone. J. Membr. Sci. **22**(2–3), 325–332 (1985)
9. Alberti, G., et al.: Polymeric proton conducting membranes for medium temperature fuel cells (110–160 degrees C). J. Membr. Sci. **185**(1), 73–81 (2001)
10. Genies, C., et al.: Soluble sulfonated naphthalenic polyimides as materials for proton exchange membranes. Polymer **42**(2), 359–373 (2001)
11. Marestin C, et al.: Sulfonated polyimides. In: Scherer, G.G. (ed.) Advances in polymer Science, Fuel Cells II, pp. 185. Springer, Berlin (2008)
12. de Araujo, C.C., et al.: Poly(p-phenylene sulfone)s with high ion exchange capacity: ionomers with unique microstructural and transport features. Phys. Chem. Chem. Phys. **11**(17), 3305–3312 (2009)
13. Schuster, M., et al.: Highly sulfonated poly(phenylene sulfone): preparation and stability issues. Macromolecules **42**(8), 3129–3137 (2009)
14. Wycisk, R., Pintauro, P.N.: Polyphosphazene membranes for fuel cells. Fuel Cells II **216**, 157–183 (2008)
15. Maier, G., Meier-Haack, J.: Sulfonated aromatic polymers for fuel cell membranes. In: Scherer, G.G. (ed.) Advances in Polymer Science, Fuel Cells II, pp. 1–62. Springer, Berlin (2008)
16. Smitha, B., Sridhar, S., Khan, A.A.: Solid polymer electrolyte membranes for fuel cell applications—a review. J. Membr. Sci. **259**(1–2), 10–26 (2005)
17. Savadogo, O.: Emerging membranes for electrochemical systems—Part II. High temperature composite membranes for polymer electrolyte fuel cell (PEFC) applications. J. Power Sources **127**(1–2), 135–161 (2004)
18. Hickner, M.A., et al.: Alternative polymer systems for proton exchange membranes (PEMs). Chem. Rev. **104**(10), 4587–4611 (2004)
19. Hickner, M.A., Pivovar, B.S.: The chemical and structural nature of proton exchange membrane fuel cell properties. Fuel Cells **5**(2), 213–229 (2005)
20. Rieberer, S., Norian, K.H.: Analytical electron-microscopy of Nafion ion-exchange membranes. Ultramicroscopy **41**(1–3), 225–233 (1992)
21. Lehmani, A., Durand-Vidal, S., Turq, P.: Surface morphology of Nafion 117 membrane by tapping mode atomic force microscope. J. Appl. Polym. Sci. **68**(3), 503–508 (1998)
22. Aleksandrova, E., et al.: Electrochemical atomic force microscopy study of proton conductivity in a Nafion membrane. Phys. Chem. Chem. Phys. **9**(21), 2735–2743 (2007)
23. Gebel, G., Diat, O.: Neutron and x-ray scattering: Suitable tools for studying ionomer membranes. Fuel Cells **5**(2), 261–276 (2005)
24. Gierke, T.D., Munn, G.E., Wilson, F.C.: The morphology in Nafion perfluorinated membrane products, as determined by wide-and small-angle x-ray studies. J. Poly. Sci. **19**, 1687–1704 (1981)
25. Halim, J., et al.: Characterization of perfluorosulfonic acid membranes by conductivity measurements and small-angle x-ray-scattering. Electrochim. Acta **39**(8–9), 1303–1307 (1994)
26. Haubold, H.G., et al.: Nano structure of Nafion: a SAXS study. Electrochim. Acta **46**(10–11), 1559–1563 (2001)
27. Gebel, G., Lambard, J.: Small-angle scattering study of water-swollen perfluorinated ionomer membranes. Macromolecules **30**(25), 7914–7920 (1997)
28. Gebel, G., Moore, R.B.: Small-angle scattering study of short pendant chain perfuorosulfonated ionomer membranes. Macromolecules **33**(13), 4850–4855 (2000)

29. Rollet, A.-L., et al.: A SANS determination of the influence of external conditions on the nanostructure of Nafion membrane. J. Polym. Sci., Part B: Polym. Phys. **39**, 548–558 (2001)
30. Rollet, A.-L., Diat, O., Gebel, G.: A new insight into Nafion structure. J. Phys. Chem B **106**(12), 3033–3036 (2002)
31. Rubatat, L., Gebel, G., Diat, O.: Fibrillar structure of Nafion: Matching fourier and real space studies of corresponding films and solutions. Macromolecules **37**(20), 7772–7783 (2004)
32. Rubatat, L., et al.: Evidence of elongated polymeric aggregates in Nafion. Macromolecules **35**, 4050–4055 (2002)
33. Tovbin, Y.K., Dyakov, Y.A., Vasyutkin, N.F.: Study of water molecule diffusion in naphione membranes by the molecular-dynamics technique. Zh. Fiz. Khim. **67**(10), 2122–2125 (1993)
34. Litt, M.H.: A reevaluation of Nafion morphology. Abstr. Pap. Am. Chem. Soc. **213**(2), 33 (1997)
35. Kreuer, K.D.: On the development of proton conducting polymer membranes for hydrogen and methanol fuel cells. J. Memb. Sci. **185**, 29–39 (2001)
36. Schmidt-Rohr, K., Chen, Q.: Parallel cylindrical water nanochannels in Nafion fuel-cell membranes. Nat. Mater. **7**(1), 75–83 (2008)
37. Blachot, J.-F., et al.: Anisotropy of structure and transport properties in sulfonated polyimide membranes. J. Memb. Sci. **214**, 31–42 (2003)
38. Elliott, J.A., Paddison, S.J.: Modelling of morphology and proton transport in PFSA membranes. Phys. Chem. Chem. Phys. **9**(21), 2602–2618 (2007)
39. Paddison, S.J., Elliott, J.A.: On the consequences of side chain flexibility and backbone conformation on hydration and proton dissociation in perfluorosulfonic acid membranes. Phys. Chem. Chem. Phys. **8**(18), 2193–2203 (2006)
40. Zhou, X.Y., et al.: Atomistic simulation of conduction and diffusion processes in Nafion polymer electrolyte and experimental validation. J. Electrochem. Soc. **154**(1), B82–B87 (2007)
41. Idupulapati, N., Devanathan, R., Dupuis, M.: Ab initio study of hydration and proton dissociation in ionomer membranes. J. Phys. Chem. A **114**(25), 6904–6912 (2010)
42. Urata, S., Irisawa, J., Takada, A.: Molecular dynamics simulation of swollen membrane of perfluorinated ionomer. J. Phys. Chem. B **109**, 4269–4278 (2005)
43. Devanathan, R., Venkatnathan, A., Dupuis, M.: Atomistic simulation of Nafion membrane: I. Effect of hydration on membrane nanostructure. J Phys Chem B **111**(28), 8069–8079 (2007)
44. Cui, S.T., et al.: Comparison of the hydration and diffusion of protons in perfluorosulfonic acid membranes with molecular dynamics simulations. J Phys Chem B **112**(42), 13273–13284 (2008)
45. Liu, J.W., et al.: On the relationship between polymer electrolyte structure and hydrated morphology of perfluorosulfonic acid membranes. J. Phys. Chem. C **114**(25), 11279–11292 (2010)
46. Brandell, D., et al.: Molecular dynamics studies of the Nafion (R), Dow (R) and Aciplex (R) fuel-cell polymer membrane systems. J. Mol. Model. **13**(10), 1039–1046 (2007)
47. Wu, D.S., et al.: Mesoscale modelling of hydrated morphologies of 3 M perfluorosulfonic acid-based fuel cell electrolytes. Langmuir **26**(17), 14308–14315 (2010)
48. Wescott, J.T., et al.: Mesoscale simulation of morphology in hydrated perfluorosulfonic acid membranes. J Chem Phys **124**(13), 14 (2006)
49. Malek, K. et al.: Nanophase segregation and water dynamics in hydrated Nafion: molecular modeling and experimental validation. J. Chem. Phys. **129**(20) (2008)
50. Knox, C.K., Voth, G.A.: Probing selected morphological models of hydrated nation using large-scale molecular dynamics simulations. J. Phys. Chem. B **114**(9), 3205–3218 (2010)
51. Khalatur, P.G., Talitskikh, S.K., Khokhlov, A.R.: Structural organization of water-containing Nafion: the integral equation theory. Macromol. Theory Simul. **11**(5), 566–586 (2002)

52. Hsu, W.Y., Gierke, T.D.: Elastic theory for ionic clustering in perfluorinated ionomers. Macromolecules **15**(1), 101–105 (1982)
53. Galperin, D.Y., Khokhlov, A.R.: Mesoscopic morphology of proton-conducting polyelectrolyte membranes of Nafion((R)) type: a self-consistent mean field simulation. Macromol. Theory Simul. **15**(2), 137–146 (2006)
54. Yamamoto, S., Hyodo, S.A.: A computer simulation study of the mesoscopic structure of the polyelectrolyte membrane Nafion. Polym. J. **35**(6), 519–527 (2003)
55. Vishnyakov, A.: Final report for US Army Research Office, DAAD190110545, Editor. (2005)
56. Wu, D.S., Paddison, S.J., Elliott, J.A.: A comparative study of the hydrated morphologies of perfluorosulfonic acid fuel cell membranes with mesoscopic simulations. Energy Environ. Sci. **1**(2), 284–293 (2008)
57. Wu, D.S., Paddison, S.J., Elliott, J.A.: Effect of molecular weight on hydrated morphologies of the short-side-chain perfluorosulfonic acid membrane. Macromolecules **42**(9), 3358–3367 (2009)
58. Takimoto, N., et al.: Hydration behaviour of perfluorinated and hydrocarbon-type proton exchange membranes: relationship between morphology and proton conduction. Polymer **50**(2), 534–540 (2009)
59. Kreuer, K.D., et al.: Transport in proton conductors for fuel-cell applications: simulations, elementary reactions, and phenomenology. Chem. Rev. **104**, 4637–4678 (2004)
60. Eikerling, M., Kornyshev, A.A., Kucernak, A.R.: Water in polymer electrolyte fuel cells: Friend or foe? Phys. Today **59**(10), 38–44 (2006)
61. Saito, M., et al.: Mechanisms of ion and water transport in perfluorosulfonated ionomer membranes for fuel cells. J Phys Chem B **108**(41), 16064–16070 (2004)
62. Majsztrik, P.W., et al.: Water sorption, desorption and transport in Nafion membranes. J. Membr. Sci. **301**(1–2), 93–106 (2007)
63. Bass, M., et al.: Surface structure of Nafion in vapor and liquid. J. Phys. Chem. B **114**(11), 3784–3790 (2010)
64. Gebel, G. et al.: The kinetics of water sorption in Nafion membranes: a small-angle neutron scattering study. J. Phys.-Condens. Matter 23(23) (2011)
65. Kusoglu, A., et al.: Subsecond morphological changes in Nafion during water uptake detected by small-angle x-ray scattering. Acs Macro Lett. **1**(1), 33–36 (2012)
66. Lyonnard, S.: Unpublished data
67. Elliott, J.A., et al.: Atomistic simulation and molecular dynamics of model systems for perfluorinated ionomer membranes. Phys. Chem. Chem. Phys. **1**, 4855–4863 (1999)
68. Moilanen, D.E., Piletic, I.R., Fayer, M.D.: Tracking water's response to structural changes in Nafion membranes. J. Phys. Chem. A **110**(29), 9084–9088 (2006)
69. Moilanen, D.E., Piletic, I.R., Fayer, M.D.: Water dynamics in Nafion fuel cell membranes: The effects of confinement and structural changes on the hydrogen bond network. J. Phys. Chem. C **111**(25), 8884–8891 (2007)
70. Moilanen, D.E., et al.: Confinement or the nature of the interface? Dynamics of nanoscopic water. J. Am. Chem. Soc. **129**(46), 14311–14318 (2007)
71. Spry, D.B., et al.: Proton transport and the water environment in Nafion fuel cell membranes and AOT reverse micelles. J. Am. Chem. Soc. **129**(26), 8122–8130 (2007)
72. Moilanen, D.E., Spry, D.B., Fayer, M.D.: Water dynamics and proton transfer in Nafion fuel cell membranes. Langmuir **24**(8), 3690–3698 (2008)
73. Lu, Z.J., et al.: State of water in perfluorosulfonic ionomer (Nafion 117) proton exchange membranes. J. Electrochem. Soc. **155**(2), B163–B171 (2008)
74. Kalapos, T.L., et al.: Thermal studies of the state of water in proton conducting fuel cell membranes. J. Power Sources **172**(1), 14–19 (2007)
75. Duplessix, R. et al.: Water absorption in acid Nafion membranes. Water in polymers (1980)
76. Falk, M.: An infrared study of water in perfluorosulfonate (Nafion) membranes. Can. J. Chem. **58**, 1495–1501 (1980)

77. Laporta, M., Pegoraro, M., Zanderighi, L.: Perfluorosulfonated membrane (Nafion): FT-IR study of the state of water with increasing humidity. Phys. Chem. Chem. Phys. **1**, 4619–4628 (1999)
78. Gruger, A., et al.: Nanostructure of Nafion membranes at different states of hydration. An IR and Raman study. Vib. Spectrosc. **26**, 215–225 (2001)
79. Iwamoto, R., et al.: Water in perfluorinated sulfonic acid Nafion membranes. J. Phys. Chem. B **106**(28), 6973–6978 (2002)
80. Ferrari, M.C., et al.: FTIR-ATR study of water distribution in a short-side-chain PFSI membrane. Macromolecules **45**(4), 1901–1912 (2012)
81. Paddison, S.J., Paul, R.: The nature of proton transport in fully hydrated Nafion (R). Phys. Chem. Chem. Phys. **4**(7), 1158–1163 (2002)
82. Eikerling, M., et al.: Mechanisms of proton conductance in polymer electrolyte membranes. J. Phys. Chem. B **105**(17), 3646–3662 (2001)
83. Eikerling, M., Kornyshev, A.A.: Proton transfer in a single pore of a polymer electrolyte membrane. J. Electroanal. Chem. **502**(1–2), 1–14 (2001)
84. Choi, P., Jalani, N.H., Datta, R.: Thermodynamics and proton transport in Nafion—II. Proton diffusion mechanisms and conductivity. J. Electrochem. Soc. **152**(3), E123–E130 (2005)
85. Spohr, E.: Molecular dynamics simulations of proton transfer in a model Nafion pore. Mol. Simul. **30**(2–3), 107–115 (2004)
86. Petersen, M.K., et al.: Excess proton solvation and delocalization in a hydrophilic pocket of the proton conducting polymer membrane narion. J. Phys. Chem. B **109**(9), 3727–3730 (2005)
87. Petersen, M.K., Voth, G.A.: Characterization of the solvation and transport of the hydrated proton in the perfluorosulfonic acid membrane Nafion. J. Phys. Chem. B **110**(37), 18594–18600 (2006)
88. Spohr, E., Commer, P., Kornyshev, A.A.: Enhancing proton mobility in polymer electrolyte membranes: Lessons from molecular dynamics simulations. J. Phys. Chem. B **106**(41), 10560–10569 (2002)
89. Eikerling, M., Kornyshev, A.A., Spohr, E.: Proton-conducting polymer electrolyte membranes: water and structure in charge. In: Scherer, G.G. (ed.) Advances in Polymer Science, Fuel Cells I, pp. 15–54. Springer, Berlin (2008)
90. Kornyshev, A.A., et al.: Kinetics of proton transport in water. J. Phys. Chem. B **107**(15), 3351–3366 (2003)
91. Dokmaisrijan, S., Spohr, E.: MD simulations of proton transport along a model Nafion surface decorated with sulfonate groups. J. Mol. Liq. **129**(1–2), 92–100 (2006)
92. Paddison, S.J., Elliott, J.A.: Molecular modelling of the short-side-chain perfluorosulfonic acid membrane. J. Phys. Chem. A **109**(33), 7583–7593 (2005)
93. Paddison, S.J., Elliott, J.A.: The effects of backbone conformation on hydration and proton transfer in the 'short-side-chain' perfluorosulfonic acid membrane. Solid State Ionics **177**(26–32), 2385–2390 (2006)
94. Paddison, S.J., Elliott, J.A.: Selective hydration of the 'short-side-chain' perfluorosulfonic acid membrane. An ONIOM study. Solid State Ionics **178**(7–10), 561–567 (2007)
95. Devanathan, R., Venkatnathan, A., Dupuis, M.: Atomistic simulation of Nafion membrane. 2. Dynamics of water molecules and hydronium ions. J. Phys. Chem. B **111**(45), 13006–13013 (2007)
96. Devanathan, R., et al.: Atomistic simulation of water percolation and proton hopping in nation fuel cell membrane. J. Phys. Chem. B **114**(43), 13681–13690 (2010)
97. Bée, M.: Quasielastic neutron scattering. Adam Hilger, Bristol and Philadelphia (1988)
98. Bellissent-Funel, M.C.: Status of experiments probing the dynamics of water in confinement. Eur. Phys. J. E **12**(1), 83–92 (2003)
99. Proceedings of the International Conference on Neutron Scattering. In: Physica B. (1998)

100. Dianoux, A.J., Pineri, M., Volino, F.: Neutron incoherent scattering law for restricted diffusion inside a volume with an anisotropic shape—Application to the problem of water in Nafion membranes. Mol. Phys. **46**(1), 129–137 (1982)
101. Volino, F., et al.: Water mobility in a water-soaked Nafion membrane: A high-resolution neutron quasielastic study. J. Polym. Sci. **20**(3), 481–496 (1982)
102. Pivovar, A.M., Pivovar, B.S.: Dynamics behaviour of water within a polymer electrolyte fuel cell membrane at low hydration levels. J. Phys. Chem. B **109**, 785–793 (2005)
103. Paciaroni, A.: Temperature-dependent dynamics of water confined in nafion membranes. J. Phys. Chem. B **110**(28), 13769–13776 (2006)
104. Perrin, J.C., Lyonnard, S., Volino, F.: Quasielastic neutron scattering study of water dynamics in hydrated nafion membranes. J. Phys. Chem. C **111**(8), 3393–3404 (2007)
105. Volino, F., Perrin, J.C., Lyonnard, S.: Gaussian model for localized translational motion: Application to incoherent neutron scattering. J. Phys. Chem. B **110**(23), 11217–11223 (2006)
106. Lyonnard, S., Gebel, G.: Neutrons for fuel cell membranes: structure, sorption and transport properties. Eur. Phys. J. Special Topics **213**, 195–211 (2012)
107. Jamróz, D., Maréchal, Y.: Hydration of sulfonated polyimide membranes. II. water uptake and hydration mechanisms of protonated homopolymer and block copolymers. J. Phys. Chem. B **109**, 19664–19675 (2005)
108. Tuckerman, M., et al.: Ab-initio molecular-dynamics simulation of the solvation and transport of hydronium and hydroxyl ions in water. J. Chem. Phys. **103**(1), 150–161 (1995)
109. Noack, F.: NMR field-cycling spectroscopy: principles and applications. Progr. NMR Spectrosc. **18**, 171–276 (1986)
110. Kimmich, R.: NMR tomography diffusometry relaxometry, p. 526. Springer, Berlin (1997)
111. Levitz, P., Korb, J.-P., Petit, D.: Slow dynamics of embedded fluid in mesoscopic confining systems as probed by NMR relaxometry. Eur. Phys. J. E **12**, 29–33 (2003)
112. Levitz, P.E.: Slow dynamics in colloidal glasses and porous media as probed by NMR relaxometry: assessment of solvent Levy statistics in the strong adsorption regime. Magn. Res. Imag. **21**, 177–184 (2003)
113. Perrin, J.C., et al.: Water dynamics in Ionomer membranes by field-cycling NMR relaxometry. Fuel Cells **6**(1), 5–9 (2006)
114. Perrin, J.C., et al.: Water dynamics in ionomer membranes by field-cycling NMR relaxometry. J. Phys. Chem. B **110**(11), 5439–5444 (2006)
115. Perrin, J.C., et al.: Water dynamics in ionomer membranes by field-cycling NMR relaxometry. Magn. Reson. Imaging **25**(4), 501–504 (2007)
116. Korb, J.-P., Xu, S., Jonas, J.: Confinement effects on dipolar relaxation by translational dynamics of liquids in porous silica glasses. J. Chem. Phys. **98**(3), 2411–2422 (1993)
117. Lyonnard S. et al.: Perfluorinated surfactants as model charged systems for understanding the effect of confinement on proton transport and water mobility in fuel cell membranes. A study by QENS. Eur. Phys. J.-Spec. Topics **189**(1), 205–216 (2010)
118. Cornet, N., Beaudoing, G., Gebel, G.: Influence of the structure of sulfonated polyimide membranes on transport properties. Sep. Purif. Technol. **22–23**, 681–687 (2001)
119. Diat, O. et al.: Anisotropie structurale et des propriétés de transport dans les membranes en polyimide sulfoné. J. Phys. IV France **12**, 2-63—2-71 (2002)

Exergy Analysis as a Tool to Analyze the Performance of Water Depuration Processes

A. Lozano, J. Uche and A. Martínez

Abstract In this chapter, water and energy interaction was analyzed from the point of view of a detailed exergy analysis of the two operating wastewater plants (WWTP) of a Spanish city in Spain: Zaragoza. The analysis was focused on the energy efficiency of the diverse wastewater stages of both WWTPs, and especially in the sludge valorization. The two plants present different capacity and alternative systems for sludge processing: dry incineration or biogas cogeneration. The unit exergy costs of water technologies (k*) were used as the indicators to evaluate the WWTPs exergy performance along a year. A better energy efficiency was found for the La Almozara WWTP with respect to La Cartuja WWTP, since biogas reduced the energy consumption of the plant by means of decreasing the exergy content of that plant residue.

Nomenclature

b	Specific exergy (J/l)
B	Total exergy (kJ)
BOD	Biological oxygen demand
COD	Chemical oxygen demand
η	Exergy efficiency
F	Fuel
I	Ionic strength
ICE	Internal combustion engine
k*	Unit exergy cost
LCA	Life cycle assessment
LHV	Light heating value
n	Mass weight

A. Lozano · J. Uche (✉) · A. Martínez
CIRCE Research Institute, University of Zaragoza, María de Luna s/n,,
50005 Zaragoza, Spain
e-mail: javiuche@unizar.es

A. Lozano
e-mail: javiuche@unizar.es

G. Ferreira (ed.), *Alternative Energies*, Advanced Structured Materials 34,
DOI: 10.1007/978-3-642-40680-5_9,

P	Product
R	Residue
RE	Reference environment
TOC	Total organic carbon
WWTP	Wastewater treatment plant
Y	Molar weight

Subscripts

a	Air
c	Cold
ca	Combustion air
cg	Combustion gases
cond	Condenser
cool	Cooling
d	Dry
EC	Electricity consumption
El	Electricity
go	Gas oil
i	Input, component
in	Inorganic
h	Hot
o	Output
poly	Polymer
proc	Process
prod	Product
Q	Heat
R	Residues
rm	Removed
s	Sand
sl	Sludge
tf	Thermal fluid
w	Wastewater
0	Reference environment

Superscripts

o	Standard

1 Introduction

Energy involved in the water cycle is a very important issue to bear in mind in our efforts toward a sustainable future. As a matter of fact, nowadays it represents around the 7 % of the consumed electricity in Spain [1]. Through the water cycle, water is evaporated and consumed and its quality gets degraded as well. Thus, some additional energy is required to restore the former quality. When no large infrastructures are required to collect water or desalination is required to provide drinking water energy required to depurate wastewaters is usually the highest consumer in the water cycle of a city. A recent study applied to the same city [2] using Life Cycle Assessment (LCA) corroborated that WWTPs were the high impacting processes in the water cycle. That LCA study evaluated global environmental impacts related to the construction of the WWTPs, their land use, and energy and chemical dosing consumed in the operation of both WWTPs.

Energy consumption in a WWTP is mainly focused to reduce the exergy content of wastewaters due to its organic matter concentration, by activating the corresponding physic-chemical processes. Two main separated output streams are obtained, but only the sludge has really the exergy content, since depurated wastewaters are usually very low exergy streams. Consequently, a low energy efficiency is expected in a WWTP process unless sludge is properly valorized. This former analysis is widely explained in this chapter: the analysis of the energy efficiency of two real WWTPs in the same city with diverse sludge valorization schemes.

A detailed exergy analysis of a WWTP is a complex task to carry out. First, a detailed knowledge of the water and sludge processes is compulsory. Second, exergy assessment of very different flow types (waste and clean waters, sludge, sands, fats, chemical reagents, chlorine) is required. In this point, adequate selection of reference environment (RE) to calculate exergies is crucial. Thirdly, some internal WWTP flows, which are usually not measured or whose data are not available have also to be known. Consequently, the application of the exergy analysis to WWTPs is not very common. Hellström [3, 4], firstly evaluated the exergy content of chemical dosing and organic matter as well and nitrogen and phosphor in wastewater and then applied the exergy analysis to diverse nutrient removal methods. Then, Mora and Oliveira [5, 6] introduced the exergy and pollution ratios to analyze the energy efficiency of WWTPs. Gallegos and Zaleta [7] introduced the exergy ratio of consumed resources with respect to raw water. A generic exergy efficiency definition can be found for diverse water treatment plants in [8], which will be partly used in this chapter. A first approach was also given by the authors regarding exergy analysis and the sludge valorization in WWTPs [9], for a small WWTP nearby Zaragoza.

2 WWTPs Description

Zaragoza is placed on the northeast of Spain (see Fig. 1). It is in the middle of the Ebro River main course (910 km). Ebro Valley (85,000 km^2) is characterized by a wet period (spring and autumn) and a dry one (winter and summer), being natural resources about 14,000 Mm^3 per year [10]. Beside being the biggest urban area of the Ebro Valley (700,000 inhabitants), urban demand is totally guaranteed (Ebro, Yesa and Itoiz dams). After domestic, industrial and gardening uses, polluted water is collected and transported to the two existing WWTPs: La Almozara and La Cartuja.

La Almozara WWTP has a capacity of 100,000 equivalent inhabitants, and operates at full capacity since 1991. It collects wastewater from the west side of the right bank of the city (Delicias suburb) as well as the nearby industrial area (Logroño road). The plant consist of a conventional primary treatment (solids, sand and fat removal with a primary decanter), and a biologic aerobic secondary treatment helped by fan blowers. Net power consumption of the WWTP is reduced thanks to the biogas generated in the anaerobic sludge treatment coming from the sludge purges of primary and secondary decanters. Biogas is stored in a gasometer and then burned in a 250 kW internal combustion engine. A building-blocks scheme of the plant is presented next in Fig. 2.

La Cartuja WWTP is in operation since 1993. It collects the remaining majority of the Zaragoza wastewaters, having a design capacity of 1,000,000 equivalent inhabitants, therefore it does not work at full capacity. Wastewaters coming from Malpica industrial area (left bank of the Ebro) slightly increased the wastewater pollution and salinity of urban wastewaters. Both primary and secondary treatments are inside of a building, thus a complete deodorization system is required. Sludge purged in secondary decanters is thickened, dried and further combusted in a fluidized bed at 850 °C. The produced vapor is consumed in a steam turbine of 550 kW. Figure 3 shows the simplified scheme of La Cartuja WWTP. Main wastewater characteristics coming to the WWTPS are shown in Table 1.

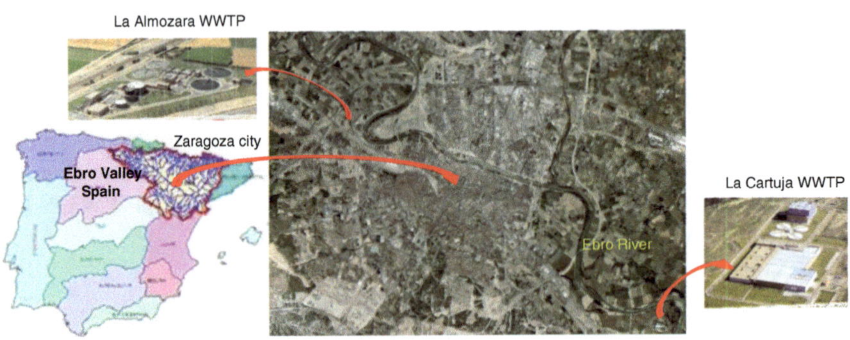

Fig. 1 Zaragoza city and La Cartuja and La Almozara WWTP locations. *Source* Google maps

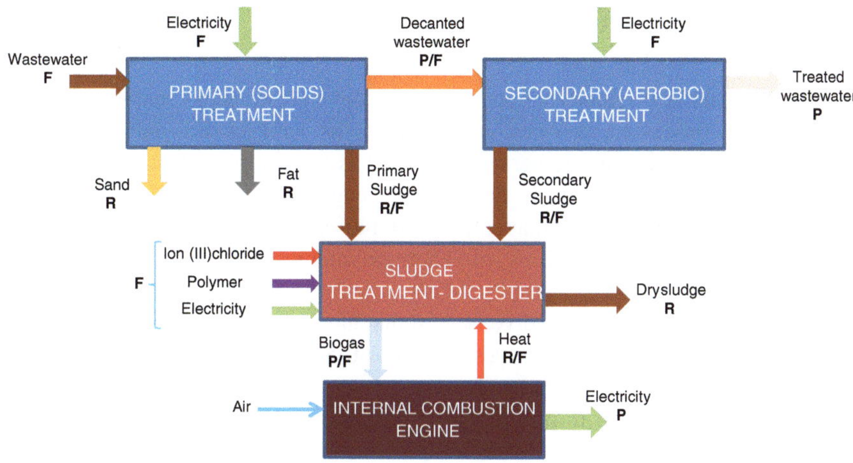

Fig. 2 Blocks scheme of La Almozara WWTP to apply the exergy analysis, including the F–P-R definition of each block

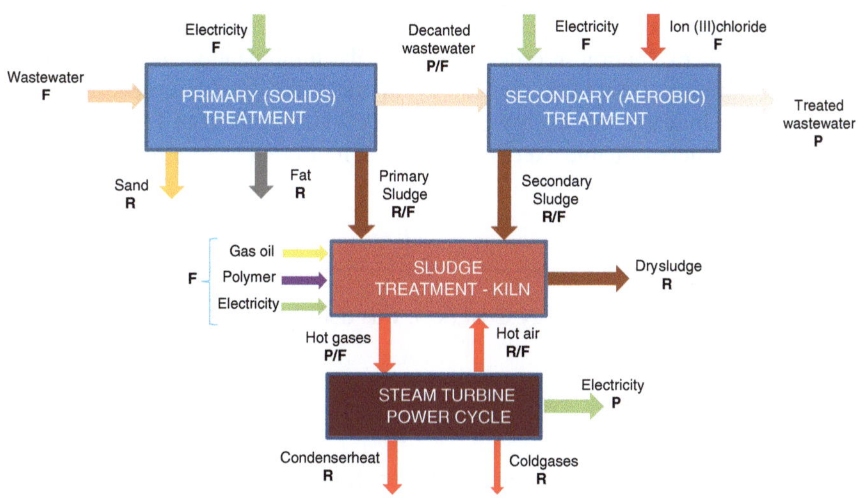

Fig. 3 Conceptual scheme of La Cartuja WWTP prior to the exergy analysis

Table 1 Key technical data of the two operating WWTPs in the city of Zaragoza (Spain)

Inflow characteristics (2010)	La Almozara	La Cartuja
Capacity (equivalent inhabitants)	100,000	1,000,000
Treated flow (m^3/day)	30,249	152,767
Conductivity (s/cm)	1,457.2	2,121
Chemical oxygen demand (mgO_2/L)	511.2	691
Biological oxygen demand (mgO_2/L)	220.1	324

3 Methodology: Exergy Analysis

Exergy analysis is thermodynamic-based tool devoted to assess and improve the efficiency of processes, devices and systems, as well as for enhancing their environmental and economic performance. It has been applied in different fields, mainly focused on mechanical and chemical engineering. However, for the last decades, also economics, management and biology have developed their relationships with the Second Law of Thermodynamics by applying the exergy concept.

As previously mentioned, diverse authors have applied the exergy analysis methodology to water-related issues, with positive outcomes. One of the advantages of exergy analysis is that it permits to discover energy consumed in producing losses consuming in producing by-products or wastes in a process.

Therefore, Exergy analysis gives the picture of the energy efficiency of water cleaning treatment processes and it could suggest new guidelines to reduce energy consumption in present water technologies, which are relevant to support human life needs in urban areas (once water demand strategies have been fully applied, of course).

A wide variety of ratios, parameters and indicators have been developed for energy systems evaluation in order to analyze the minimization of energy losses. Since the difference between the exergy of fuels and products determines the energy losses in a system, exergy efficiency (k*) was defined in [11] as the proportion of products valuable exergy (P) in relation of input energy in fuel flows (F) (see Eq. 1):

$$\eta = \frac{P}{F} \tag{1}$$

In this sense, the Unit Exergy Cost (k*), was defined in [11] as the inverse of η, as follows in Eq. (2):

$$k^* = \frac{F}{P} \tag{2}$$

As the exergy balance is obviously fulfilled, R is the exergy content of the flows which were considered as residues in a productive process. As a consequence, different exergy ratios could be analyzed depending on the productive purpose of each WWTP stage:

Exergy unit cost of the product (Eq. 3): product P is the exergy of depurated wastewater, and fuels F are all the inputs coming to the WWTP (wastewater, electricity, fuels, chemicals).

$$k^*_{prod} = \frac{F}{P} \tag{3}$$

Exergy unit cost of the process (Eq. 4): product P is the exergy of depurated wastewater and energy recoverable residues. Fuel F is the same as the previous case.

$$k^*_{proc} = \frac{F}{P + R} \tag{4}$$

The arising question is which of these two costs should be applied for the performance calculation searched in this work. Attending to the final target of any WWTP, it has sense to apply the cost of the product, where the product is the water that needs to be cleaned or replaced.

However, as the diverse outflows of the process are getting valuable for any other process or final user, they can be considered as products (co-products of the desired clean water) and the cost of the process could be reconsidered. That is the case when sludge valorization occurs.

At this stage, it can be now easily understood that, upon this premises, the key point of the methodology is to accurately calculate the main exergy of the WWTP streams. In the following section those exergy flows calculations are explained

3.1 Exergy Flows Calculation

In order to perform the exergy analysis of a wastewater plant, diverse types of flowstreams are considered: water flows, chemical dosing, energy (heat and electricity), residues (sand, sludge and fat), and those related to sludge valorization (biogas, gasoil, vapor, hot water, combustion air).

The adequate selection of the Reference Environment (RE) is then a relevant decision to take before those calculations. Several authors dealt with the appropriate RE to evaluate the exergy quality of natural resources [12–17]. In case of wastewaters, and as a part of the hydrologic water cycle, an in depth study [18, 19] was made to consider that the best option is to select the average seawater composition (36,700 ppm of salinity), in which organic matter as well as nitrogen and phosphorous are not included in that RE. The selected RE regarding the atmospheric and seawater environment was presented in Tables 2 and 3 respectively.

Chemical exergy of *wastewater* is calculated as follows (Eq. 6):

$$b_q = b_{IM} + b_{OM} + b_{NP} + b_s \tag{6}$$

where b_{IM} represents the exergy content in inorganic matter, that is, the one of pure water (X_{H_2o}) plus the inorganic salts (b_{salts}), b_{OM} is the organic matter exergy content, b_{NP} includes the exergy content of nitrogen and phosphor compounds, and finally b_S includes the water turbidity measured as silica concentration. Exergy of both pure water and inorganic salts could be estimated with the following formula (Eq. 7)

Table 2 Air composition (molar weight) of the RE. *Source* [26]

Component	Molar fraction
N_2	$X_{N2} = 0.7898*(1-X_{H_2O})$
O_2	$X_{O2} = 0.2099*(1-X_{H_2O})$
CO_2	$X_{CO2} = 0.0003*(1-X_{H_2O})$
H_2O	$X_{H_2O} = P_{VH_2O}(T_0)/P_0$

*P_{VH_2O} water vapor presure in air at T_0

Table 3 Seawater composition (at weight, in ppm) of the RE

Component	ppm, at weight
Hydrogen, H_2O	110,000
Oxygen, H_2O	883,000
Sodium, NaCl	10,800
Chlorine, NaCl	19,400
Magnesium, Mg	1,290
Sulphur, S	904
Potassium, K	392
Calcium, Ca	411
Bromine, Br	67.3

$$b_{in,i} = RT_0 \sum_i x_i Ln \frac{a_i}{a_0} \tag{7}$$

where x_i is the molar concentration of component i in wastewater, and a_i and a_o are respectively the activity coefficients in the solution and in RE. Those coefficients are estimated taking into account that weak solutions are dealt with:

$$a_i = \gamma_i m_i ln\gamma_i = \frac{-Az_i^2\sqrt{I}}{1 + B\phi_i\sqrt{I}} \quad I = \frac{1}{2}\sum_i m_i z_i^2 \tag{8}$$

In Eq. 8, γ_i is the activity coefficient of component i, m_i its molality, z_i its ionic charge, I the ionic strength of the ion i in diluted wastewater, and A and B are two constants whose values were 0.51 $kg^{1/2}$ $mol^{1/2}$ and 3.287×10^9 $kg^{1/2}$ m^{-1} $mol^{-1/2}$ for water at 25 °C [20]. Molality of each component m_i in wastewater could be estimated with Eq. 9 [21], as a function of wastewater salinity c,

$$m_i = w_i \left(\frac{1}{M_i}\right)\left(\frac{1000}{\frac{1-c}{1000}}\right) \tag{9}$$

being w_i the mass fraction of component i in diluted water, and M_i its molecular weight. On the other hand, pure water activity could be estimated as a function of the molality of the rest of inorganic salts m_{tot}:

$$a_{H_2O} = 1 - 0,017m_{tot} \tag{10}$$

Exergy content of organic matter in wastewaters b_{OM} was computed following the guidelines proposed by [22, 23]. Thus, different measurable parameters could be used (TOC, COD) to evaluate the exergy content of organic matter in waters:

$$b_{OM}\left(\frac{J}{l}\right) = 13,6\,TOD\left(\frac{mg}{l}\right) \approx 45\,TOC\left(\frac{mg}{l}\right) \tag{11}$$

As nitrogen as phosphorous compounds do not take part of the RE, the corresponding chemical reaction has to be performed, starting from the standard chemical exergy of individual components (Eq. 12). In the case of wastewaters, it is assumed that all the nitrogen is in form of ammonia (NH_4^+) and phosphor is in form of phosphates (PO_4^{-3}) [3].

$$\mathrm{b_{ch,f}} = \sum_{\mathrm{i}} \mathrm{y}_i\left(\Delta \mathrm{G_{fi}} + \sum_{\mathrm{e}} \mathrm{n_e b_{ch,ne}}\right) \tag{12}$$

According to Szargut [20], chemical exergy of N ($b_{ch,N}$) and P ($b_{ch,P}$) were respectively 322.1 and 204,4 kJ/mol.

Sand removal in WWTP was characterized as silica in order to assess its chemical exergy content. As in the case of N and P, it does not take part of the RE. Depending on the natural status of SiO_2 (cristobalite, amorphous), diverse standards chemical exergies could be found. Here, the value obtained in [24] was used to estimate that exergy value ($b_{S,rm}$ = 0,82 kJ/kmol).

Fat present in wastewaters are mainly composed by animal fat, domestic oil and wax. The molecule that represents the majority of fat acids in nature is $C_{18}H_{32}O_2$, with a molecular weight of 282 g/mol and a heating value of 38,874 kJ/kg [19]. Anyway, they could be considered as a solid fuel since it has C, H and O. Thus, the expression suggested by Kotas [25] for the estimation of the exergy of solid and liquid fuels as a function of its low heating value (LHV) and the mass fraction of hydrogen, carbon, oxygen and nitrogen (Eq. 13) could be used:

$$\varphi = \frac{b^0_{ch}}{LHV}\quad \varphi = 1.0437 + 0.1882\frac{h}{c} + 0.0610\frac{o}{c} + 0.0404\frac{n}{c} \tag{13}$$

Values for h, c, n and o were respectively 0,113, 0,766, 0 and 0,113 for fat acids. The same expression could be taken to evaluate the exergy content of wastewater *sludge*. However, the sludge composition of primary, secondary, purged or dried sludge sometimes is not analyzed by the WWTP managers, and bibliography or mass balances were then applied.

Regarding *chemical dosing* applied to the WWTPs, standard chemical exergy of pure chemicals (chorine, alumina sulphate and iron (III) chloride) were taken starting from the chemical reaction of each compound, as we can see in Eq. 14. As the detailed composition of the consumed polymer was not known, a conventional product was taken from [4] to evaluate its exergy content. Table 4 shows the main standard chemical exergy of chemical reagents consumed in Zaragoza WWTPs.

Table 4 Chemical standard exergy of chemical reagents consumed in Zaragoza WWTPs

Chemical reagent	b^0_{ch} (kJ/mol)
Cl_3Fe	228.1 [20]
$Al_2(SO_4)_3$	344.3 [20]
NaClO	170.17
Polymer (kJ/kg)	15.2 [4]

$$b^0_{ch} = \Delta G^0_f + \sum\nolimits_{el} n_{el} b^0_{chel} \tag{14}$$

In order to calculate the exergy content in the *biogas* produced in La Almozara WWTP, the method proposed in [26] was follow in which composition (as an ideal gas mixture, b_{JQ}), mechanical (b_{JP}) and thermal components (b_{JT}) were considered (Eq. 15):

$$b_{biogas} = \sum x_j \left(b_{jT} + b_{jP} + b_{jQ}\right) \tag{15}$$

Thermal exergy is calculated as a function of the specific heat of its gaseous components, as follows in Eq. 16. A, B, C and D coefficients are given in Table 5:

$$\begin{aligned} b_{jT}\left(\frac{cal}{mol}\right) &= (A - T_0B)(T - T_0) + \frac{1}{2}(B - T_0C)\left(T^2 - T_0^2\right) \\ &\quad + \frac{1}{3}(C - T_0D)\left(T^3 - T_0^3\right) + \frac{1}{4}D\left(T^4 - T_0^4\right) - AT_0 ln\frac{T}{T_0} \end{aligned} \tag{16}$$

Mechanical component is calculated in relation with the environmental pressure as is presented in Eq. 17:

$$b_{jP}\left(\frac{cal}{mol}\right) = RT_0 ln\frac{P}{P_0} \tag{17}$$

Chemical exergy is calculated taking into account if the gaseous compound is or not included in the gaseous RE adopted. If the gas compound is included in the aforementioned RE, a concentration rule is used, which is quite similar to Eq. 7:

Table 5 Coefficients to calculate thermal exergy of the component i of an ideal gas mixture. Source: Adapted from [26]

i	A_i	B_i	D_i	C_i
CH_4	4.75	1.20E−02	−2.63E−09	3.03E−06
H_2S	7.07	3.13E−03	−7.87E−10	1.36E−06
H_2	6.952	−4.58E−04	−2.08E−10	9.56E−07
CO	6.726	4.00E−04	−5.31E−10	1.28E−06
N_2	6.903	−3.75E−04	−6.86E−10	1.93E−06
O_2	6.085	3.63E−03	3.13E−10	−1.71E−06
CO_2	5.316	1.43E−02	1.78E−09	−8.36E−06

*C_p (cal/mol K) = $A + BT + CT^2 + DT^3$; T(K)

$$b_{jQ} = RT_0\ ln\frac{x_i}{x_i^0} \tag{18}$$

where x_i and x_i^0 are respectively the molar fraction of gaseous component in the biogas mixture and the RE. If the compound is not found in the RE, the alternative solution was taken (Eq. 19) from its constituents, in which constants included were supplied in Table 6:

$$\begin{aligned} b_{jQ} = & -\left[\Delta H_R^0 - T_0\Delta S_R^0 + \Delta C_{PR}^0(T_0 - T^0) - \Delta C_{PR}^0 T_0 ln\frac{T_0}{T^0}\right] \\ & + RT_0 ln\frac{(x_{O2,0}P_0)^{r_{O_2}}}{(x_{O2,0}P_0)^{p_{CO2}}(x_{N2,0}P_0)^{p_{N2}}} + RT_0\ln(x_jP_0). \end{aligned} \tag{19}$$

On the other hand, *combustion gases* coming from the internal combustion engine (La Almozara WWTP) or fluidized bed (La Cartuja WWTP) were also valorized from the point of view of exergy. Here, again three terms were considered: thermal, chemical and mechanical. The unique different formula with respect to biogas exergy content is that one related to thermal component (Eq. 20).

$$b_{cg,T}\left(\frac{kJ}{kg}\right) = c_p\left(T - T_0 - T_0 ln\frac{T}{T_0}\right) \tag{20}$$

Exergy of *thermal fluids* in Rankine cycle or cooling devices were calculated by the standard expression:

$$b_{tf} = (h - h_0) - T_0(s - s_0) \tag{21}$$

being T_0, s_0, and h_0 the temperature, entropy and enthalpy values in the RE. For *combustion air*, only thermal and pressure components at T_{air} and p_{air} were compiled since it was present in the RE and therefore it does not have chemical exergy (Eq. 22):

$$b_{ca} = c_{p,air}\cdot\left(T_{air} - T_o - T_o\ln\left(\frac{T_{air}}{T_o}\right)\right) + R_aT_o\ln\left(\frac{p_{air}}{p_o}\right) \tag{22}$$

Finally, a similar expression than in Eq. (13) was used from Kotas [25] to evaluate the chemical exergy content of gas oil consumed to put into operation the fluidized bed of La Cartuja WWTP, as it is included in Eq. 23:

Table 6 Main reaction parameters in the RE. *Source*: adapted from [26]

i	$\Delta C_{PR,i}$ (cal/ mol K)	$\Delta H_{R,i}$ (l/mol)	$\Delta S_{R,i}$ (cal/ mol K)
CH_4	22.32	−212.797	−58
H_2S	−6.42	−215.530	−88.65
H_2	7.6	−68.317	−38.99
CO	−1.6	−67.636	−20.74

$$\varphi = 1,04010 + 0,1728\frac{h}{c} + 0,0432\frac{o}{c} + 0,2169\frac{s}{c}\left(1 - 2,0628\frac{h}{c}\right) \quad (23)$$

Of course, heat was characterized by the Carnot factor in order to estimate its exergy content, and electricity was directly translated into useful energy (exergy).

4 Case Studies

4.1 Case Study 1: La Almozara WWTP

La Almozara WWTP was divided into four main stages: primary (1), secondary (2) and sludge treatments (3) and further energy valorization (4). In this manner, at least stages (1) and (2) could be directly compared with La Cartuja WWTP. Furthermore, energy efficiency of the two different sludge treatments applied could be also compared with (3 and 4). This breakdown was also performed taking into account available measured data. Figure 2 shows the main blocks of La Almozara including the exergy inputs and outputs of each stage.

Tables 7, 8, 9 shows the specific (b) and total exergy value (B) of the main internal flowstreams of the plant along the year 2010.

According to the definitions previously proposed, diverse exergy ratios could be found at global or local scale. The indexes calculation has been carried out at global scale (global process) and for each of the treatment steps, as follows:

Global process: wastewater, chemical reagents and energy consumed are the fuel (F) of La Almozara WWTP. Product (P) is depurated water, and residue (R) is digested sludge. Thus, the averaged exergy cost values obtained in 2010 were the next:

$$k^*_{prod} = \frac{B_{w,i} + B_{Cl3Fe} + B_{poly} + B_{EC}}{B_{w,o}} \quad k^*_{proc} = \frac{B_{w,i} + B_{Cl3Fe} + B_{poly} + B_{EC}}{B_{w,o} + B_{sl,d}}$$

Primary treatment (1): wastewater and electricity consumed are the fuel (F), and the Product (P) is primary decanted wastewater. Residue (R) was here purged sludge from primary decanter. As expected, the unit exergy costs of that stage are quite low, corresponding to a "light" treatment centered only on solids removal:

$$k^*_{prod,1} = \frac{B_{w,i} + B_{EC1}}{B_{w,1}}$$

$$k^*_{proc,1} = \frac{B_{w,i} + B_{EC1}}{B_{w,1} + B_{sl,1}}$$

Secondary treatment (2): F is wastewater coming from primary treatment and electricity consumed in fan blowers, P is depurated water, and R are the purged sludge from secondary decanter. As biological treatment is really produced here

Table 7 Specific and total exergy values of the main flowstreams of the La Almozara WWTP. Water flows

Month	Raw wastewater					Wastewater to secondary treatment					Treated wastewater				
	b_{IM} (kJ/kg)	$b_{s,rm}$ (kJ/kg)	b_{NP} (kJ/kg)	b_{OM} (kJ/kg)	$B_{w,e}$ (kJ)	b_{IM} (kJ/kg)	$b_{s,rm}$ (kJ/kg)	b_{NP} (kJ/kg)	b_{OM} (kJ/kg)	$B_{w,1}$ (kJ)	b_{IM} (kJ/kg)	$b_{s,rm}$ (kJ/kg)	b_{NP} (kJ/kg)	b_{OM} (kJ/kg)	$B_{w,s}$ (kJ)
Jan-10	2.55	3.67E−03	0.96	7.50	9.25E+09	2.55	2.20E−03	0.93	6.94	8.76E+09	2.55	3.01E−04	0.70	0.72	3.33E+09
Feb-10	2.51	3.15E−03	0.89	7.66	7.90E+09	2.51	1.89E−03	0.86	6.11	6.77E+09	2.51	2.74E−04	0.65	0.79	2.82E+09
Mar-10	2.51	3.50E−03	1.06	8.15	9.89E+09	2.51	2.10E−03	1.03	6.50	8.47E+09	2.51	2.60E−04	0.69	0.78	3.35E+09
Apr-10	2.52	3.19E−03	0.76	6.25	8.51E+09	2.52	1.91E−03	0.74	4.86	7.25E+09	2.52	1.92E−04	0.68	0.63	3.42E+09
May-10	2.50	3.23E−03	0.82	6.33	9.16E+09	2.50	1.94E−03	0.79	5.06	7.93E+09	2.50	2.05E−04	0.62	0.65	3.59E+09
Jun-10	2.50	2.90E−03	0.83	6.05	9.26E+09	2.50	1.74E−03	0.81	4.91	8.11E+09	2.50	2.88E−04	0.59	0.75	3.79E+09
Jul-10	2.52	2.61E−03	0.67	5.19	8.93E+09	2.52	1.56E−03	0.65	4.11	7.75E+09	2.52	2.06E−04	0.45	0.52	3.71E+09
Aug-10	2.50	2.44E−03	0.61	5.04	8.32E+09	2.50	1.46E−03	0.59	2.88	6.09E+09	2.50	1.65E−04	0.41	0.49	3.47E+09
Sep-10	2.52	3.32E−03	0.72	6.52	9.78E+09	2.52	1.99E−03	0.70	4.70	7.93E+09	2.52	2.33E−04	0.54	0.64	3.71E+09
Oct-10	2.50	3.38E−03	0.87	7.07	1.02E+10	2.50	2.03E−03	0.85	5.99	9.16E+09	2.50	3.15E−04	0.64	0.98	4.04E+09
Nov-10	2.53	4.13E−03	0.93	8.44	1.03E+10	2.53	2.48E−03	0.90	6.14	8.29E+09	2.53	3.28E−04	0.75	0.95	3.67E+09
Dec-10	2.54	4.79E−03	1.05	9.48	1.14E+10	2.54	2.87E−03	1.01	7.48	9.59E+09	2.54	3.28E−04	0.76	0.90	3.65E+09

Table 8 Specific and total exergy values of the main flowstreams of the La Almozara WWTP. Chemical reagents and residues (from water line)

Month	Chemical reagent added				Residues obtained in "water line": primary and secondary treatment							
	b_{Cl3Fe} (kJ/kg)	B_{Cl3Fe} (kJ)	b_{poly} (kJ/kg)	B_{poly} (kJ)	$b_{s,rm}$ (kJ/kg)	$B_{s,rm}$ (kJ)	b_{fat} (kJ/kg)	B_{fat} (kJ)	b_{sl2} (kJ/kg)	B_{sl2} (kJ)	b_{sl1} (kJ/kg)	B_{sl1} (kJ)
Jan-10	0.012	9.84E+06	0.030	2.56E+07	1.34E−04	1.13E+05	0.00E+00	0.00E+00	0.59	4.92E+08	1.11	9.29E+08
Feb-10	0.014	9.84E+06	0.030	2.17E+07	2.33E−04	1.66E+05	0.00E+00	0.00E+00	2.04	1.46E+09	0.95	6.78E+08
Mar-10	0.012	9.85E+06	0.030	2.57E+07	2.57E−04	2.17E+05	0.00E+00	0.00E+00	2.20	1.85E+09	1.06	8.92E+08
Apr-10	0.011	9.84E+06	0.030	2.72E+07	8.19E−05	7.31E+04	0.00E+00	0.00E+00	1.50	1.34E+09	0.96	8.59E+08
Mai-10	0.010	9.84E+06	0.030	2.89E+07	6.15E−05	5.84E+04	0.00E+00	0.00E+00	1.72	1.64E+09	0.97	9.26E+08
Jun-10	0.010	9.84E+06	0.031	3.01E+07	1.37E−04	1.35E+05	1.61E−04	1.58E+05	2.31	2.28E+09	0.88	8.64E+08
Jul-10	0.009	9.84E+06	0.031	3.25E+07	4.36E−05	4.64E+04	0.00E+00	0.00E+00	1.57	1.67E+09	0.79	8.36E+08
Aug-10	0.010	9.84E+06	0.031	3.12E+07	7.78E−05	7.94E+04	0.00E+00	0.00E+00	2.07	2.11E+09	0.74	7.51E+08
Sep-10	0.010	9.84E+06	0.031	3.06E+07	6.92E−05	6.93E+04	0.00E+00	0.00E+00	2.57	2.57E+09	1.00	1.00E+09
Oct-10	0.010	9.85E+06	0.031	2.99E+07	1.46E−04	1.43E+05	0.00E+00	0.00E+00	1.64	1.61E+09	1.02	1.00E+09
Nov-10	0.011	9.84E+06	0.030	2.64E+07	7.69E−05	6.66E+04	0.00E+00	0.00E+00	2.02	1.75E+09	1.25	1.08E+09
Dec-10	0.011	9.84E+06	0.030	2.65E+07	7.69E−05	6.69E+04	1.02E−04	8.82E+04	1.53	1.33E+09	1.44	1.26E+09

Table 9 Specific and total exergy values of the main flowstreams of the La Almozara WWTP. Sludge valorization unit

Month	B_{biogas} (kJ)	B_{hgas} (kJ)	$B_{c,gas}$ (kJ)	$B_{cool,o}$ (kJ)	$B_{cool,i}$ (kJ)	B_{el} (kJ)
Jan-10	1.69E+09	2.63E+08	7.70E+07	6.46E+08	5.46E+08	5.64E+08
Feb-10	1.56E+09	2.38E+08	6.96E+07	5.84E+08	4.93E+08	5.35E+08
Mar-10	1.79E+09	2.63E+08	7.70E+07	6.46E+08	5.46E+08	5.97E+08
Apr-10	1.82E+09	2.55E+08	7.46E+07	6.25E+08	5.28E+08	5.80E+08
May-10	1.84E+09	2.63E+08	7.70E+07	6.46E+08	5.46E+08	5.90E+08
Jun-10	1.50E+09	2.55E+08	7.46E+07	6.25E+08	5.28E+08	4.67E+08
Jul-10	1.39E+09	2.63E+08	7.70E+07	6.46E+08	5.46E+08	4.15E+08
Aug-10	1.31E+09	2.63E+08	7.70E+07	6.46E+08	5.46E+08	3.86E+08
Sep-10	1.40E+09	2.55E+08	7.46E+07	6.25E+08	5.28E+08	4.53E+08
Oct-10	1.67E+09	2.63E+08	7.70E+07	6.46E+08	5.46E+08	5.37E+08
Nov-10	1.77E+09	2.55E+08	7.46E+07	6.25E+08	5.28E+08	5.84E+08
Dec-10	1.68E+09	2.63E+08	7.70E+07	6.46E+08	5.46E+08	6.02E+08

and exergy content in organic matter is removed, lower energy efficiencies were found:

$$k^*_{prod,2} = \frac{B_{w,1} + B_{EC2}}{B_{w,2}} \quad k^*_{proc,2} = \frac{B_{w,1} + B_{EC2}}{B_{w,o} + B_{sl,2}}$$

Sludge treatment (3): F is composed by primary and secondary purged sludge, chemical dosing, electricity consumed, and heat supplied by the ICE. P is biogas produced, and R are the digested sludge. Thermochemical transformation process has similar energy efficiency than secondary treatment, as shown in the next figures:

$$k^*_{prod,3} = \frac{B_{sl,1} + B_{sl,2} + B_{poly} + B_{Cl3Fe} + B_{EC3} + B_Q}{B_{biogas}} \quad k^*_{proc,3} = \frac{B_{sl,1} + B_{sl,2} + B_{poly} + B_{Cl3Fe} + B_{EC3} + B_Q}{B_{sl,d} + B_{biogas}}$$

Sludge valorization (4): F is now biogas feeding the ICE, P is electricity produced and R is heat dissipated by the ICE in the water jacket and the chimney. Exergy efficiency varied depending on the use or not of the heat as a product (cogeneration), as it can be seen in the two exergy ratios introduced here:

$$k^*_{prod,4} = \frac{B_{biogas}}{B_{El}} \quad k^*_{proc,4} = \frac{B_{biogas}}{B_{El} + B_Q}$$

The summary of results for La Almozara WWTP, obtained after applying the explained equations for the different values, is presented in Table 10.

Table 10 Exergy efficiency parameters for La Almozara WWTP

	LA-global	LA-(1)	LA-(2)	LA-(3)	LA-(4)
k^*_{prod}	2.72	1.18	2.28	2.25	3.08
k^*_{proc}	1.98	1.06	1.36	1.24	2.01

1 Primary treatment, *2* Secondary treatment, *3* Sludge treatment, *4* Sludge valorization

4.2 Case Study 2: La Cartuja WWTP

Figure 3 shows the main exergy inputs/outputs (that is, the F-P-R definition) of the four main blocks in which La Cartuja WWTP was divided to apply the exergy analysis.

Following the methodology adopted to La Almozara, specific and total exergy flows of the main WWTP streams were represented for the year 2010 (see Tables 11, 12, 13). Note that electricity consumption related to the deodorization system was shared into the primary, secondary and sludge treatment units according to its relative power consumption.

Then exergy efficiency parameters for the main four blocks of the WWTP are presented and analyzed next:

Global process: wastewater, chemical reagents and energy consumed (electricity and diesel) are the fuel (F) of La Cartuja WWTP. Product (P) is depurated water, and residue (R) is the dry sludge plus the exhausted heat in the condenser and the chimney. Thus, lower overall efficiencies were obtained in the year 2010, as can be checked in the next figures:

$$k^*_{prod} = \frac{B_{w,i} + B_{Cl3Fe} + B_{poly} + B_{EC} + B_{go}}{B_{w,o}} \; k^*_{proc} = \frac{B_{w,i} + B_{Cl3Fe} + B_{poly} + B_{EC} + B_{go}}{B_{w,o} + B_{Q,cond} + B_{Q,c,gas}}$$

Primary treatment (1): the same as described in La Almozara WWTP. Similar exergy efficiencies were obtained in this stage, as it is shown next.

$$k^*_{prod,1} = \frac{B_{w,i} + B_{EC1}}{B_{w,1}} \; k^*_{proc,1} = \frac{B_{w,i} + B_{EC1}}{B_{w,1} + B_{sl,1}}$$

Secondary treatment (2): It is quite similar to that one presented in La Almozara, but now ion chloride was also introduced as a fuel of the biological treatment. Combined effect of deodorization plant, as well as chemical dosing increased a bit the exergy efficiency of this stage with respect to La Almozara WWTP in the year 2010. Remember that higher pollution content was entering here, then a little bit higher exergy costs are also expected.

$$k^*_{prod,2} = \frac{B_{w,1} + B_{EC2} + B_{Cl3Fe}}{B_{w,2}} \; k^*_{proc,2} = \frac{B_{w,1} + B_{EC2} + B_{Cl3Fe}}{B_{w,o} + B_{sl,2}}$$

Table 11 Specific and total exergy values of the main flowstreams of the La Almozara WWTP. Water flows

Month	Raw wastewater					Wastewater to secondary treatment					Treated wastewater				
	b_{IM} (kJ/kg)	$b_{s,rm}$ (kJ/kg)	b_{NP} (kJ/kg)	b_{OM} (kJ/kg)	$B_{w,e}$ (kJ)	b_{IM} (kJ/kg)	$b_{s,rm}$ (kJ/kg)	b_{NP} (kJ/kg)	b_{OM} (kJ/kg)	$B_{w,1}$ (kJ)	b_{IM} (kJ/kg)	$b_{s,rm}$ (kJ/kg)	b_{NP} (kJ/kg)	b_{OM} (kJ/kg)	$B_{w,s}$ (kJ)
Jan-10	2.44	5.66E−03	0.93	10.67	5.06E+10	2.44	3.72E−03	0.90	8.38	4.22E+10	2.44	3.01E−04	0.70	1.01	1.49E+10
Feb-10	2.36	6.23E−03	0.93	11.80	5.56E+10	2.36	3.70E−03	0.90	8.90	4.48E+10	2.36	1.78E−04	0.74	0.86	1.46E+10
Mar-10	2.48	5.40E−03	1.02	10.77	5.37E+10	2.48	3.99E−03	0.99	8.63	4.55E+10	2.48	2.87E−04	0.63	1.04	1.56E+10
Apr-10	2.42	4.72E−03	0.89	9.06	4.76E+10	2.42	3.54E−03	0.87	7.70	4.22E+10	2.42	2.87E−04	0.57	1.00	1.53E+10
May-10	2.39	5.30E−03	1.04	9.85	5.21E+10	2.39	3.91E−03	1.01	8.22	4.56E+10	2.39	3.42E−04	0.65	1.08	1.61E+10
Jun-10	2.37	4.70E−03	0.90	9.39	5.07E+10	2.37	3.46E−03	0.87	7.44	4.27E+10	2.37	2.74E−04	0.63	1.02	1.61E+10
Jul-10	2.40	4.11E−03	0.86	7.87	4.54E+10	2.40	1.89E−03	0.83	5.16	3.42E+10	2.40	2.19E−04	0.58	0.87	1.57E+10
Aug-10	2.42	4.04E−03	0.79	7.42	4.42E+10	2.42	2.15E−03	0.76	5.15	3.47E+10	2.42	2.33E−04	0.54	0.72	1.53E+10
Sep-10	2.42	4.86E−03	0.80	8.28	4.88E+10	2.42	2.40E−03	0.78	5.78	3.81E+10	2.42	2.60E−04	0.54	0.94	1.66E+10
Oct-10	2.41	4.59E−03	0.97	8.35	5.07E+10	2.41	2.72E−03	0.94	6.67	4.33E+10	2.41	2.19E−04	0.67	0.93	1.73E+10
Nov-10	2.40	5.25E−03	0.93	9.16	5.51E+10	2.40	3.17E−03	0.90	7.21	4.64E+10	2.40	2.87E−04	0.66	1.05	1.81E+10
Dec-10	2.45	5.52E−03	1.00	10.41	6.23E+10	2.45	3.51E−03	0.97	7.53	4.93E+10	2.45	3.83E−04	0.83	1.12	1.98E+10

Table 12 Specific and total exergy values. La Cartuja WWTP. Chemical reagents and residues (from water line)

Month	Chemical reagent added				Residues obtained in "water line": primary and secondary treatment							
	b_{Cl3Fe} (kJ/kg)	B_{Cl3Fe} (kJ)	b_{poly} (kJ/kg)	B_{poly} (kJ)	$b_{s,rm}$ (kJ/kg)	$B_{s,rm}$ (kJ)	b_{fat} (kJ/kg)	B_{fat} (kJ)	b_{sl2} (kJ/kg)	B_{sl2} (kJ)	b_{sl1} (kJ/kg)	B_{sl1} (kJ)
Jan-10	0.15	5.55E+08	0.041	1.48E+08	0.18	6.39E+08	9.87E−05	3.55E+05	4.19	1.51E+10	2.6	9.47E+09
Feb-10	0.14	5.32E+08	0.044	1.62E+08	0.08	3.10E+08	8.72E−05	3.21E+05	3.49	1.29E+10	2.2	7.96E+09
Mar-10	0.17	6.32E+08	0.054	2.04E+08	0.28	1.06E+09	9.45E−05	3.55E+05	4.74	1.78E+10	2.9	1.10E+10
Apr-10	0.16	6.27E+08	0.051	1.96E+08	0.19	7.32E+08	8.95E−05	3.44E+05	4.83	1.86E+10	2.9	1.12E+10
May-10	0.15	5.95E+08	0.047	1.85E+08	0.19	7.45E+08	9.06E−05	3.55E+05	4.76	1.87E+10	2.9	1.13E+10
Jun-10	0.15	5.94E+08	0.043	1.74E+08	0.19	7.72E+08	8.60E−05	3.44E+05	3.96	1.58E+10	2.3	9,11E+09
Jul-10	0,11	4,47E+08	0,040	1,63E+08	0,20	8,31E+08	8,72E−05	3,55E+05	3,19	1,30E+10	1,7	6,96E+09
Aug-10	0,10	4,17E+08	0,030	1,26E+08	0,14	5,69E+08	8,55E−05	3,55E+05	2,97	1,24E+10	1,5	6,44E+09
Sep-10	0,11	4,50E+08	0,034	1,42E+08	0,25	1,05E+09	8,11E−05	3,44E+05	3,51	1,49E+10	1,9	8,26E+09
Oct-10	0,10	4,54E+08	0,035	1,53E+08	0,15	6,43E+08	8,22E−05	3,55E+05	3,48	1,51E+10	2,0	8,59E+09
Nov-10	0,11	4,94E+08	0,036	1,57E+08	0,17	7,53E+08	7,80E−05	3,44E+05	3,81	1,68E+10	2,2	9,71E+09
Dec-10	0,14	6,17E+08	0,041	1,85E+08	0,18	7,96E+08	7,91E−05	3.55E+05	3.97	1.78E+10	2.3	1.02E+10

Table 13 Specific and total exergy values. La Cartuja WWTP, sludge valorization block

Month	B_{fuel}	B_{cond}	Calculated flows			Flow data sheets		
			$B_{h,air}$	$B_{h,gas}$	$B_{c,gas}$	$B_{h,air}$	$B_{h,\ gas}$	$B_{c,gas}$
	(kJ)	(kJ)	(kJ)	(kJ)	(kJ)	(kJ)	(kJ)	(kJ)
Jan-10	1.65E+08	5.75E+08	2.50E+08	1.12E+10	2.82E+09	2.79E+08	1.33E+10	3.36E+09
Feb-10	1.78E+09	5.19E+08	2.29E+08	1.02E+10	2.59E+09	2.82E+08	1.20E+10	3.03E+09
Mar-10	9.38E+07	5.75E+08	2.89E+08	1.28E+10	3.24E+09	3.56E+08	1.33E+10	3.36E+09
Apr-10	1.14E+08	5.56E+08	2.96E+08	1.31E+10	3.31E+09	3.27E+08	1.28E+10	3.25E+09
May-10	9.84E+07	5.75E+08	3.05E+08	1.33E+10	3.37E+09	3.32E+08	1.33E+10	3.36E+09
Jun-10	2.76E+07	5.56E+08	2.54E+08	1.12E+10	2.85E+09	4.32E+08	1.28E+10	3.25E+09
Jul-10	1.40E+08	5.75E+08	2.10E+08	9.29E+09	2.35E+09	2.82E+08	1.33E+10	3.36E+09
Aug-10	1.09E+07	5.75E+08	1.96E+08	8.57E+09	2.17E+09	2.72E+08	1.33E+10	3.36E+09
Sep-10	1.11E+07	5.56E+08	2.35E+08	1.03E+10	2.61E+09	3.29E+08	1.28E+10	3.25E+09
Oct-10	1.08E+07	5.75E+08	2.43E+08	1.07E+10	2.71E+09	3.60E+08	1.33E+10	3.36E+09
Nov-10	1.62E+08	5.56E+08	2.80E+08	1.24E+10	3.14E+09	3.23E+08	1.28E+10	3.25E+09
Dec-10	2.71E+08	5.75E+08	2.93E+08	1.31E+10	3.30E+09	3.61E+08	1.33E+10	3.36E+09

Table 14 Exergy efficiency parameters for La Almozara WWTP

	LC-global	LC-(1)	LC-(2)	LC-(3)	LC-(4a)	LC-(4b)
k_{prod}	3.7	1.27	2.84	2.44	9.77	9.77
k_{proc}	3.06	1.04	1.45	2.39	2.64	2.34

1 Primary treatment. *2* Secondary treatment, *3* Sludge treatment, *4* Sludge valorization

Sludge treatment (3): F is composed by primary and secondary purged sludge, chemical dosing, electricity, and hot air to fluidized bed and diesel consumed. Product P were the hot gases, and if the kiln efficiency was only separately computed, a bit higher efficiency was found:

$$k^*_{prod,3} = \frac{B_{sl,1} + B_{sl,2} + B_{poly} + B_{air} + B_{EC3} + B_{go}}{B_{hgas}}$$

$$k^*_{kiln} = \frac{B_{sl,1} + B_{sl,2} + B_{air} + B_{go}}{B_{h,gas}}$$

Sludge valorization (4): P was the electricity produced in the steam turbine Rankine cycle, which was thermodynamically modeled in order to calculate its exergy value. F consumed was always the hot gases exergy coming to the cycle, and diverse values to the exergy costs of the process could be given depending of the consideration of the heat released by the condenser and hot air produced as a power plant residue. Anyway, both exergy costs indicate the very low efficiency of the steam turbine cycle of the La Cartuja WWTP.

$$k^*_{prod,4} = \frac{B_{hgas}}{B_{El}} k^*_{proc,4a} = \frac{B_{hgas}}{B_{El} + B_{c,gas} + B_{h,air}} k^*_{proc,4b} = \frac{B_{h,gas}}{B_{El} + B_{c,gas} + B_{h,air} + B_{cond}}$$

As the exergy of dry sludge was higher in this process, usually better exergy efficiencies were found in thermal drying of the sludge, this means a lower overall efficiency of the plant, since a significant amount of available energy remains of those sludge which is not reused in the WWTP is thrown to the environment. Anyway, again it is important to remember that higher pollution means higher exergy content in residues (Table 14).

5 Results Discussion

Tables presenting the exergy cost indexes obtained for both WWTP in a monthly basis along the year 2010 are provided in the Annex Tables 15, 16, 17 and Tables 18, 19, 20 summarize the study of La Almozara and La Cartuja WWTPs respectively.

As it can be checked in those monthly tables, no high differences were found along the year for both exergy indexes, exergy cost of the product and exergy cost of the process. Only those figures related to sludge treatment varied, since destination of digested or dry residues sometimes is not monthly administrated and, in consequence, the reported amount of residue may vary. Previous results included the average value for the whole year.

Within each of the WWTPs, exergy cost of the primary treatment is lower than the exergy cost of the secondary treatment, as it can be quickly understood because the gravity separation process is much easier. Similar figures were obtained for those primary and secondary treatment in both case studies. However, La Cartuja had a bit higher values, mainly provoked by the deodorization plant and higher pollution content in raw wastewater, which does not compensates its higher capacity. Highest differences were found in the the sludge treatment and further valorization process. It is quite clear that thermochemical transformation into biogas and further cogeneration increases the exergy efficiency of the WWTP, as it is corroborated by its lower energy consumption. Main reason of that better efficiency is the lower energy quality of digested sludge with respect to dry ones. It is also noteworthy to indicate that power cycle of La Cartuja WWTP operates at design conditions, and in consequence, low efficiency arises. On the contrary, gasometer capacity of La Almozara is not enough to store biogas produced in winter season, and some gas surpluses had to be burned to the atmosphere.

6 Conclusions

Thanks to available technical information and to the continuous interaction with the plant managers, it was possible to perform, for the first time, the complete local exergy analysis of the two WWTPs of a medium-size city like Zaragoza. Despite of that, local mass, energy and exergy balances were required in most cases in order to know some exergy flows without enough information to carry out in those calculations. For instance, detailed composition of sludge at diverse plant stages was not always available. Anyway, and with all the reserves, the results were really consistent for the twelve analyzed months for both WWTPs, which could lead to further conclusions.

The detailed exergy analysis of the two WWTPs of Zaragoza included the comparison of the diverse stages of the wastewater plant: primary (solids elimination), secondary (organic matter depuration), sludge treatment and further valorization. This local analysis of the exergy efficiency of the wastewater plant located the weak points in this kind of plants: research on new methods to produce energy from organic matter content in wastewaters, and the appropriate sludge treatment should be encouraged, since overall exergy efficiency of the WWTPs is not really high taking into account that some useful energy (electricity) is required to remove an unused but usefulness energy stream.

This is an example of the application of energy-based methodologies in the field of water issues. Apart from the research in energy-efficient water technologies, the appropriate integration of water and energy is also a cornerstone in the search of a sustainable water cycle.

Acknowledgments The authors would like to thank the support given to this work, which was under the framework of the R + D+i project ENE2009-14515-CO-01, financed by the Spanish Ministry of Economy and Competitiveness. Additionally, the authors greatly acknowledge the help of the Zaragoza WWTPs plant managers (Utedeza and Depuración de Aguas del Mediterráneo) to carry out this work.

A.1 7 Annex

Detailed calculation, in a monthly base, of the exergy efficiency parameters for the two WWTP of Zaragoza city at their different stages, are shown in the following Tables A1, A2, A3, A4, A5 and A6

Table A1 Unit exergy costs (k*) and exergy efficiency of La Almozara WWTP. Black-box model

Month	Global process			
	k^*_{prod}	η_{prod}	k^*_{proc}	η_{proc}
Jan-10	2.562	0.390	2.005	0.499
Feb-10	2.563	0.390	1.829	0.547
Mar-10	2.702	0.370	1.877	0.533
Apr-10	2.315	0.432	1.700	0.588
May-10	2.381	0.420	1.705	0.587
Jun-10	2.363	0.423	1.628	0.614
Jul-10	2.351	0.425	1.698	0.589
Aug-10	2.355	0.425	1.623	0.616
Sep-10	2.530	0.395	1.656	0.604
Oct-10	2.407	0.416	1.752	0.571
Nov-10	2.602	0.384	1.813	0.552
Dec-10	2.864	0.349	2.013	0.497

Table A2 Unit exergy costs (k*) and exergy efficiency of La Almozara WWTP. Water line

Month	Primary treatment (solid removal)				Secondary (aerobic) treatment			
	k^*_{prod}	η_{prod}	k^*_{proc}	η_{proc}	k^*_{prod}	η_{prod}	k^*_{proc}	η_{proc}
Jan-10	1.061	0.942	0.960	1.042	2.645	0.378	1.939	0.516
Feb-10	1.173	0.853	1.066	0.938	2.422	0.413	1.421	0.704
Mar-10	1.174	0.852	1.062	0.942	2.546	0.393	1.447	0.691
Apr-10	1.180	0.848	1.055	0.948	2.136	0.468	1.347	0.743
May-10	1.162	0.861	1.040	0.961	2.230	0.448	1.346	0.743
Jun-10	1.153	0.867	1.042	0.959	2.174	0.460	1.223	0.818
Jul-10	1.166	0.858	1.052	0.950	2.125	0.471	1.308	0.764
Aug-10	1.384	0.723	1.232	0.812	1.798	0.556	1.011	0.989
Sep-10	1.244	0.804	1.104	0.906	2.169	0.461	1.139	0.878

(continued)

Table A2 (continued)

Month	Primary treatment (solid removal)				Secondary (aerobic) treatment			
	k^*_{prod}	η_{prod}	k^*_{proc}	η_{proc}	k^*_{prod}	η_{prod}	k^*_{proc}	η_{proc}
Oct-10	1.126	0.889	1.015	0.986	2.288	0.437	1.438	0.695
Nov-10	1.250	0.800	1.106	0.905	2.275	0.440	1.334	0.749
Dec-10	1.191	0.840	1.053	0.950	2.650	0.377	1.624	0.616

Table A3 Unit exergy costs (k*) and exergy efficiency of La Almozara WWTP. Sludge line

Month	Sludge treatment				Energy valorization: cogeneration			
	k^*_{prod}	η_{prod}	k^*_{proc}	η_{proc}	k^*_{prod}	η_{prod}	k^*_{proc}	η_{proc}
Jan-10	1.451	0.689	0.963	1.038	2.999	0.333	1.989	0.503
Feb-10	1.886	0.530	1.121	0.892	2.922	0.342	1.968	0.508
Mar-10	2.100	0.476	1.187	0.842	2.996	0.334	2.024	0.494
Apr-10	1.745	0.573	1.070	0.934	3.130	0.319	2.118	0.472
May-10	1.954	0.512	1.141	0.876	3.120	0.321	2.100	0.476
Jun-10	2.773	0.361	1.375	0.727	3.211	0.311	2.015	0.496
Jul-10	2.528	0.396	1.326	0.754	3.349	0.299	1.981	0.505
Aug-10	2.911	0.344	1.418	0.705	3.386	0.295	1.944	0.514
Sep-10	3.345	0.299	1.493	0.670	3.098	0.323	1.921	0.520
Oct-10	2.228	0.449	1.230	0.813	3.107	0.322	2.026	0.494
Nov-10	2.240	0.446	1.227	0.815	3.036	0.329	2.058	0.486
Dec-10	2.304	0.434	1.249	0.801	2.798	0.357	1.895	0.528

Table A4 Unit exergy costs (k*) and exergy efficiency. La Cartuja WWTP (general)

Month	Global process			
	k^*_{prod}	η_{prod}	k^*_{proc}	η_{proc}
Jan-10	3.974	0.252	3.237	0.309
Feb-10	4.499	0.222	3.708	0.270
Mar-10	4.053	0.247	3.256	0.307
Apr-10	3.714	0.269	2.966	0.337
May-10	3.801	0.263	3.053	0.328
Jun-10	3.702	0.270	3.056	0.327
Jul-10	3.488	0.287	2.940	0.340
Aug-10	3.423	0.292	2.904	0.344
Sep-10	3.442	0.291	2.890	0.346
Oct-10	3.428	0.292	2.881	0.347
Nov-10	3.510	0.285	2.915	0.343
Dec-10	3.586	0.279	2.998	0.334

Table A5 Unit exergy costs (k*) and exergy efficiency. La Cartuja WWTP. Water line

Month	Primary treatment (solid removal)				Secondary (aerobic) treatment			
	k^*_{prod}	η_{prod}	k^*_{proc}	η_{proc}	k^*_{prod}	η_{prod}	k^*_{proc}	η_{proc}
Jan-10	1.256	0.796	1.026	0.975	3.087	0.324	1.536	0.651
Feb-10	1.293	0.773	1.098	0.911	3.327	0.301	1.767	0.566
Mar-10	1.238	0.808	0.997	1.003	3.191	0.313	1.489	0.672
Apr-10	1.189	0.841	0.939	1.065	3.027	0.330	1.369	0.730
May-10	1.199	0.834	0.961	1.041	3.081	0.325	1.428	0.700
Jun-10	1.245	0.803	1.026	0.975	2.904	0.344	1.465	0.683
Jul-10	1.404	0.712	1.166	0.857	2.441	0.410	1.335	0.749
Aug-10	1.344	0.744	1.134	0.882	2.500	0.400	1.384	0.722
Sep-10	1.343	0.744	1.104	0.906	2.520	0.397	1.328	0.753
Oct-10	1.228	0.815	1.024	0.976	2.725	0.367	1.456	0.687
Nov-10	1.240	0.806	1.025	0.975	2.764	0.362	1.435	0.697
Dec-10	1.312	0.762	1.087	0.920	2.684	0.373	1.410	0.709

Table A6 Unit exergy costs (k*) and exergy efficiency. Cartuja WWTP. Sludge line

Month	Sludge treatment*				Energy valorization: steam turbine cycle**			
	k^*_{prod}	η_{prod}	k^*_{proc}	η_{proc}	k^*_{prod}	η_{prod}	k^*_{proc}	η_{proc}
Jan-10	2.439	0.410	1.482	0.675	13.086	0.076	2.479	0.403
Feb-10	2.445	0.409	1.468	0.681	10.047	0.100	2.349	0.426
Mar-10	2.475	0.404	1.435	0.697	15.060	0.066	2.585	0.387
Apr-10	2.495	0.401	1.592	0.628	9.927	0.101	2.387	0.419
May-10	2.463	0.406	1.653	0.605	9.651	0.104	2.367	0.423
Jun-10	2.452	0.408	1.573	0.636	8.719	0.115	2.274	0.440
Jul-10	2.452	0.408	1.728	0.579	9.081	0.110	2.234	0.448
Aug-10	2.475	0.404	1.712	0.584	9.697	0.103	2.242	0.446
Sep-10	2.479	0.403	1.642	0.609	8.547	0.117	2.238	0.447
Oct-10	2.447	0.409	1.696	0.589	7.867	0.127	2.190	0.457
Nov-10	2.348	0.426	1.775	0.563	9.597	0.104	2.355	0.425
Dec-10	2.359	0.424	1.667	0.600	9.009	0.111	2.323	0.431

*Calculated exergy flows
**Including heat delivered by the condenser as a residue (R)

References

1. Hardy, L., Garrido, A.: Análisis y evaluación de las relaciones entre el agua y la energía en España (in Spanish). Papeles del agua virtual n° 6, Fundación Marcelino Botín eds., Santander (2010)
2. Uche, J., Martínez, A., Castellano, C., Subiela V.: Life cycle analysis of water cycle in cities: an environmental analysis tool to take into account. Desalination and Water Treatment (in press) (2012)

3. Hellström D.: An exergy analysis for a wastewater treatment plant- an estimation of the consumption of physical resources. Water Environment Research **69** (1), 44–51 (1997)
4. Hellström, D.: Exergy analysis of nutrient recovery processes. Water Sci. Technol. **48**(1), 27–36 (2003)
5. Mora, C.H., Oliveira, S.: Environmental exergy analysis of wastewater treatment plants. Therm. Eng. **5**, 24–29 (2006)
6. Mora, C.H., Oliveira, S.: Exergoenvironmental evaluation of wastewater treatment process. In: Proceedings of 22th International ECOS Conference. Aug 31–Sep 3 2009, Fox do Iguacu, Brasil (2009)
7. Gallegos, A., Zaleta, A., Rangel, V.H.: On an exergy efficiency definition of a wastewater treatment plant. Int. J. Thermodynamics **6**, 169–176 (2003)
8. Martínez, A., Uche, J., Rubio, C., Carrasquer, B.: Exergy cost of water supply and water treatment technologies. Desalination Water Treat **24**, 123–131 (2010)
9. Carrasquer, B., Martinez, A., Uche, J.: An analysis of technical alternatives in a wastewater treatment plant by using sludge as energy resource. In: Proceedings of Advanced Energy Systems (AES) Conference, 19–21 Oct, Barcelona (2010)
10. Ebro River Water Authority (2011). Ebro river new management plan (Preliminary). Available at http://www.chebro.es
11. Valero, A., Lozano, M., Muñoz, M.A.: A general theory of exergy saving I. On the exergy costs. AES Congress, vol. 2–3, ASME Book H0341C, New York (1986)
12. Gaggioli, R., Petit, P.: Second Law analysis for pinpointing the true inefficiencies in final conversion systems. A.C.S, Division of Fuel Chemistry (1976). 21
13. Sussman, M.: Choosing a reference environment-state for available-energy computations. In: Proceedings of 72nd Annual Meeting. American Institute of Chemical Engineers, San Francisco (USA) (1979)
14. Ahrendts, J.: The exergy of chemically reacting systems. Technical report, VDI. Forschungsheft 579, Dusseldorf, (1977)
15. Szargut, J.: Chemical exergies of the elements. Appl. Energy **32**, 269–286 (1989)
16. Szargut J., Valero A., Stanek W., Valero, Al.: Towards an international legal reference environment. In: Proceedings of ECOS 2005, Trondheim, Norway (June 20–22), pp. 409–420 (2005)
17. Valero, A., Uche, J., Al, Valero, Martínez, A.: Physical Hydronomics: Application of the exergy analysis to the assessment of environmental costs of water bodies. The case of the inland basins of Catalonia Energy **34**, 2101–2107 (2008)
18. Martínez, A.: Exergy cost assessment of water resources: Physical Hydronomics. Ph.D. Thesis. Department of Mechanical Engineering, University of Zaragoza
19. Szargut, J.: Exergy Method, Technical and Ecological Application. WIT Press (2005)
20. Rivero, R., Garfias, M.: Standard chemical exergy of elements updates. Energy **31**, 3310–3326 (2006)
21. Tai, S., Matsushige, K., Goda, T.: Chemical exergy of organic matter in wastewater. Int. J. Environ. Stud. **27**, 301–315 (1986)
22. Martínez, A., Uche, J.: Chemical exergy assessment of organic matter in a waterflow. Energy **35**, 77–84 (2009)
23. Exergy calculator. Available at http://www.exergoecology.com/excalc. Last visit: March, 2012
24. Kotas T.J.: The exergy method of thermal plant analysis. Essex (ed.) (1985)
25. Lozano M.A, Valero A.: Determinación de la exergía para sustancias de interés industrial (in Spanish). University of Zaragoza. Department of Thermodynamics and Phisico-Chemistry (1986)
26. Lozano M.A, Valero A.: Determinación de la exergía para sustancias de interés industrial (in Spanish). University of Zaragoza. Department of Mechanical Engineering (1986)

Fuel Cells: Cogeneration of C_2 Hydrocarbons or Simultaneous Production/Separation of H_2 and C_2 Hydrocarbons

Victor José Ferreira, José Luís Figueiredo and Joaquim Luís Faria

Abstract Chemicals and energy cogeneration processes have been shown to be a promising alternative to conventional reactors and fuel cells. Solid electrolyte membrane reactors (SEMRs) have been widely studied in fuel cells applied as a chemical reactor. This chapter describes the SEMRs and important catalytic aspects on the oxidative coupling of methane (OCM) to understand how these two technological alternatives can be combined to increase the C_2 hydrocarbon yield, cogenerate electric power or produce valuable chemical compounds by using of SEMR in its two main operating modes (fuel cell and O^{2-} ion "pumping" mode).

1 Introduction

In recent years, fuel cell technologies have received much attention to provide clean and viable energy solutions [1, 2]. Fuel cells are open thermodynamic systems operating on the basis of electrochemical reactions in which reactants are consumed from an external source to produce electric power [1, 3].

Fuel cell can be applied as a chemical reactor, particularly in Solid Electrolyte Membranes Reactors (SEMRs) [4, 5]. These latter are devices constituted mainly of a membrane which allows the selective permeation of at least one of the species of a mixture to which they are exposed [5]. The reactants or the product of the reaction are supplied or removed through the membrane, providing several advantages, such as the simultaneous reaction, separation and increasing of yields towards useful chemicals and/or electric energy [5].

V. J. Ferreira (✉) · J. L. Figueiredo · J. L. Faria
LCM—Laboratory of Catalysis and Materials—Associate Laboratory LCM/LSRE,
Faculdade de Engenharia, Universidade do Porto, Rua Dr. Roberto Frias s/n 4200-465 Porto,
Portugal
e-mail: victor.ferreira@fe.up.pt

G. Ferreira (ed.), *Alternative Energies*, Advanced Structured Materials 34,
DOI: 10.1007/978-3-642-40680-5_10, © Springer-Verlag Berlin Heidelberg 2013

On the other hand, conversion of methane into higher hydrocarbons (ethane and ethylene) shows great potential as a novel use of natural gas, OCM being one of the most important routes [6]. The OCM has been intensively studied since the pioneer work of Keller and Bhasin [7]. However, there are difficulties in the direct conversion of CH_4 because it can be completely oxidised to CO, CO_2 and H_2O with limited selectivity to ethane and ethylene [8] leading to low concentrations of C_2 hydrocarbons in the stream, which makes their separation uneconomical.

It is in this direction that considerable efforts have been made to find suitable catalysts and a specific configuration for the OCM [7, 9–17]. SEMRs seem to be a suitable configuration. In such configurations electrical energy, ethane and ethylene (C_2 hydrocarbons) can be co-generated in a fuel cell applied as a chemical reactor [18–21], which is much more attractive from an energy conversion and environmental standpoint. This chapter describes operating principles and several types of fuel cells reported in literature as well as important aspects of the OCM (types of catalysts, reactors and operating modes) to understand how these two technologies can be integrated to offer an alternative to conventional reactors for OCM and fuel cells with pure water as a by-product.

Then, the chapter presents the current state of some configurations for possible industrial application about development of fuel cells applied as a chemical reactor for co-generation of C_2 hydrocarbons or the integration of the simultaneous co-production of C_2 hydrocarbons and H_2 using O^{2-} ion conductors.

2 Fuel Cells

2.1 Operating Principles of Fuel Cells

Fuel cell is a device that generates electrical energy and heat via electrochemical reaction between the fuel and an oxidant agent like oxygen. All full cells operate using the same basic principles, where the main difference is the employed electrolyte or operating temperature, efficiency, applications and costs. A fuel cell has four main parts: anode, cathode, electrolyte and external circuit. Gaseous fuel, for example hydrogen, is supplied continuously to the anode (negative electrode) and oxygen is fed continuously to the cathode (positive electrode). Depending on the electrolyte, either protons or oxide ions are transported through an ion-conductor and electron-insulator (electrolyte) while electrons travel through an external circuit to deliver electric power and heat [22]. The reactions are shown below.

Anode:

$$2H_2(g) \rightarrow 4H^+ + 4e^- \tag{1}$$

Cathode:

$$O_2 + 4H^+ + 4e \rightarrow 2H_2O \quad (2)$$

Over cell reaction:

$$2H_2(g) + O_2(g) \rightarrow 2H_2O + \text{electrical power} + \text{heat} \quad (3)$$

2.2 Types of Fuel Cells

Fuel cells are different according to the electrolyte employed or working temperature. Hence, they can be classified based on the choice of fuel and electrolyte in 6 major groups [2]. Alkaline Fuel Cell (AFC), Phosphoric Acid Fuel Cell (PAFC), Solid Oxide Fuel Cell (SOFC), Molten Carbonate Fuel Cell (MCFC), Proton Exchange Membrane Fuel Cell (PEMFC) and Direct Methanol Fuel Cell DMFC. Table 1 summarises the different types of fuel cells according to the electrolyte and operating temperature.

As mentioned above, fuel cells are usually used as an electric power generator. However, the fuel cell types shown in Table 1 can be used as a chemical reactors to produce electric energy and convert fuel to desired chemicals. The fuel cell applied as a chemical reactor consists of an external load and chemical product recovery [24]. The fuel and oxidant are supplied separately to the electrochemical cell (anode and cathode) at the fuel cell reactor and then useful chemicals and electricity are produced. Electricity is supplied to an external load, while useful chemicals are collected.

The chemical and energy cogeneration processes are similar to those involved in conventional heterogeneous catalytic processes, which use reactors such as batch reactors, plug flow reactors, continuous stirred tank reactors. In fact, the catalysts used in heterogeneous chemical processes can help to choose suitable electrocatalysts for cogeneration processes. The main advantages of the chemical energy co-generation methods over the conventional catalytic reactors are as follows:

- The production in the electrochemical reactor is essentially controlled by cell potential;
- Cogeneration processes can operate at low, moderate or high temperatures;
- No competition for the same catalytic sites, since the reactants are fed separately;
- The suitable selection of anode material and the fuel allow controlling a successful chemical generation;
- The selectivity of the process can be controlled by potential variation or type of catalyst used in the electrode.

Table 1 Operational specifications of fuel cell types [23–28]

Cell type	Temperature (°C)	Electrode material	Electrolyte	Primary Fuel	Oxidant	Issues	Partial efficiency (%)
AFC	60–90	Metal carbon	NaOH/KOH	H_2	O_2/air	CO_2 troubles	60
PAFC	160–200	Pt-on carbon	H_3PO_4	H_2 reformate	O_2/air	CO_2 sensitivity	55
SOFC	800–1,000	Ni/Y_2O_3-ZrO_2	ZrO_2 with Y_2O_3	H_2/CO/CH_4 reformate	O_2/air	Ceramics cells	60–65[a]
MCFC	600–800	Ni–Cr	$LiCO_3$-K_2CO_3	H_2/CO reformate	CO_2/O_2/air	CO_2 recycling necessary	55–65[a]
PEMFC	80–110	Pt-on carbon	Polymer membrane	H_2 reformate	O_2/air	Moisture of fuel	60
DMFC	90–120	Pt-on carbon	Polymer membrane	Methanol	O_2/air	Vapour and liquid feeds	60–70

[a] Electrical energy additional production from thermal energy co-generation is not concerned

Therefore, electrical energy cogeneration and useful chemicals have become a more attractive application not only in terms of energy conversion, but also from an environmental standpoint.

The choice of the fuel cell type for a determined cogeneration processes depends mainly on the temperature. It can be observed in Table 1 that the fuel cell types operate at low (60–200 °C) or high (600–100 °C) temperatures. Therefore, the exploratory evaluation of fuel cells as reactors for producing chemicals has been studied in these temperature ranges. For example, hydroxylamine [29], acids [30], and hydrogen peroxide [31] have been produced from inorganic reagents using different media in aqueous electrolyte fuel cells at low temperature and fuels has been converted to useful chemicals, such as 1-propanol, and cyclohexlamine in PEMFC reactors [24], which also operate at low temperature. On the other hand, electrochemical oxidation of methane to synthesis gas or C_2 hydrocarbons with electricity in SOFC reactors are very attractive alternatives to conventional catalytic reactors.

2.3 Solid Electrolyte Membrane Reactor

A solid electrolyte membrane reactor consists in a fuel cell applied as chemical reactor that has a membrane selectively permeated by at least one of the components of a mixture to which it is exposed. In this reactor at least one of the reactants or products are supplied or removed partly or wholly through a membrane. Consequently, this means that the membrane serves as the wall or as part of the reactor wall.

According to permselectivity and permeability, membranes can be classified into two types: porous and dense membranes. Porous membranes offer a high permeability to molecules but with low permselectivity. The typical gas transport mechanisms in porous membranes are: molecular diffusion and viscous flow, capillary condensation, Knudsen diffusion as well as surface diffusion. Porous membranes have been widely successfully employed in many oxidation reactions, such as oxidative coupling of methane [32], oxidative dehydrogenation of ethane [15] propane [33], etc.

On the other hand, dense membranes have a rather high permselectivity to some special species but normally a lower permeability compared to porous membranes. The transport process in a dense membrane involves oxygen diffusion or ionic jumping in the lattice. The solid oxide dense membranes can be classified into two types: solid electrolytes and mixed ion electron conductors (MIEC). Unlike MIEC, the dominant characteristic of the solid electrolytes is the ionic conductivity. In fact, it is usually two or more orders of magnitude higher than the electronic conductivity [21]. Therefore, their classification is based on the conducting ion. Many conductors have been discovered (H^+, K^+, Na^+, Cu^+, Ag^+, Li^+, O^{2-}, and F^-). However, H^+ and O^{2-} ion conductors have been mainly used in solid electrolyte membrane reactors (SEMRs) [21, 34, 35].

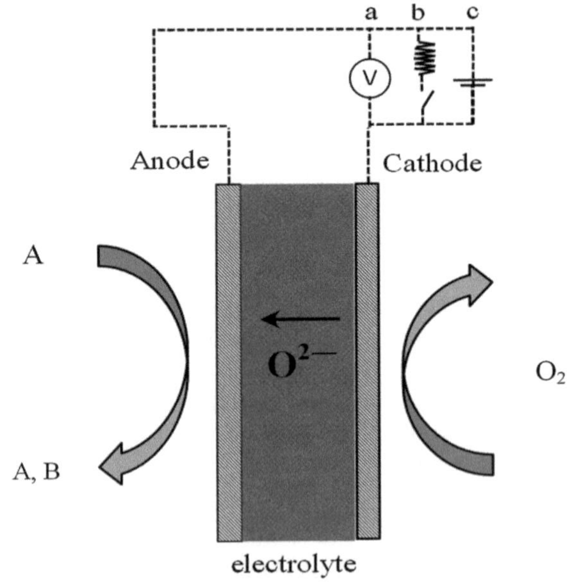

Fig. 1 Schematic diagram of a SEMR: **a** open-circuit operation; **b** closed-circuit operation in fuel cell mode; **c** closed-circuit operation in "pumping" mode

The typical configuration of a SEMR is illustrated in Fig. 1; the membrane is an oxygen-ion conductive solid electrolyte. The cathode is exposed to oxygen-containing gas, e.g., air, and the anode is exposed to the reactants, e.g., hydrocarbons. The two electrodes are connected to a circuit in three operation modes. Open-circuit operation in which the two electrodes are connected to a voltmeter (case 1a), closed-circuit connected to an external resistive load (case 1b), or to an external power source (case 1c). This last mode of operation is called electrochemical oxygen "pumping".

In the open-circuit operation, there is not net current through the electrolyte. The difference in chemical potential is converted into the open-circuit electromotive force of the cell. If the primary goal is the production of electricity the fuel cell operates in closed-circuit to convert directly chemical energy into electrical energy. This latter is also called fuel cell mode of operation. On the other hand, if the primary goal is the production of B chemicals, an external power source can be used to impose a current through the cell in the desired direction (case 1c).

These operation modes generate three basic applications of the SEMRs as shown in Fig. 1.

1. Solid Electrolyte Potentiometry (SEP),
2. SOFCs, and
3. Electrochemical oxygen "pumping".

Solid electrolyte potentiometry can be applied to study the heterogeneous catalysis by the measurement of the activity of oxygen on metal and metal oxide catalysts. The basic principle of SEP is the in situ measurement of the chemical potential difference between the two electrodes on either side of a solid electrolyte at open-circuit conditions.

SOFCs combine the concepts of a fuel cell and a chemical reactor to produce valuable chemicals from the anodic reaction with cogeneration of electrical energy instead of pure thermal energy. In SOFCs, several fuels can be used such as H_2, CO, CH_4, and CH_3OH. This is because SOFCs can be operated at high temperatures (<1,073 K).

In electrochemical oxygen "pumping", the SEMR also operate under closed-circuit conditions to carry out a reaction and produce useful chemicals rather than electrical energy. When the current is generated spontaneously and is either very low or in the undesired direction, an external power source is used to direct and control the current.

Solid electrolyte membrane reactors present several advantages such as: increased catalytic activity and selectivity, simultaneous reaction and separation in the same device, better process integration, reduced feedstock, and easy reaction rate control [36].

3 Oxidative Coupling of Methane

Production of C_2 hydrocarbons from the OCM offers a potential route for utilisation of the huge reserves of natural gas as liquid fuels, chemical and petrochemical feedstocks [37]. Methane, the main constituent of natural gas, is a stable alkane hydrocarbon and its direct conversion into C_2H_6 is a reaction with a positive variation of the Gibbs free energy:

$$2CH_4 \rightarrow C_2H_6 + H_2 \quad \Delta G^\circ = 69.06\,\text{kJ/mol} \tag{4}$$

However, this thermodynamic disadvantage can be avoided using an oxidant:

$$4CH_4 + O_2 \rightarrow 2C_2H_6 + 2H_2O \quad \Delta G^\circ = -305.6\,\text{kJ/mol} \tag{5}$$

The last reaction, the oxidative condensation of methane is known as oxidative coupling of methane. Ethane is the main product, although it can be dehydrogenated to C_2H_4. Nevertheless, in the presence of O_2 or generally of any other oxidant CH_4 can be completely oxidised to CO and CO_2:

$$2CH_4 + 3O_2 \rightarrow 2CO + 4H_2O \quad \Delta G^\circ = -543.8\,\text{kJ/mol} \tag{6}$$

$$CH_4 + 2O_2 \rightarrow CO_2 + 2H_2O \quad \Delta G^\circ = -801.1\,\text{kJ/mol} \tag{7}$$

This drawback can be overcome by using a suitable catalyst. It is known that OCM over metal oxide catalysts is a heterogeneous–homogeneous reaction. According to the accepted mechanism, the oxidative methane activation in the catalytic OCM process involves the abstraction of a H-atom from methane due to surface defects on the catalyst, leading to the formation of methyl radicals (CH_3) [37]; the two

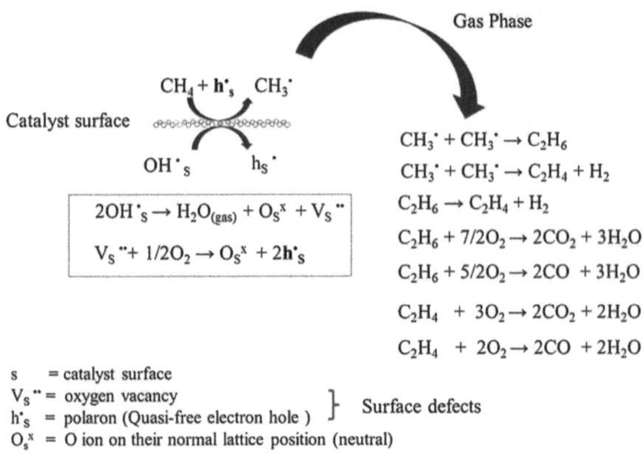

Fig. 2 OCM reaction pathways

desorbed methyl radicals are coupled (gas phase reactions) to form an ethane molecule (Fig. 2). These gas phase reactions are partly responsible for dehydrogenation C_2H_6 to C_2H_4 and formation of CO_x. In addition, the $CH_3^{\cdot}$ secondary reactions also contribute to the formation of CO_x [9].

3.1 Catalysts used for the OCM

A large number of catalysts have been evaluated for their performance in the OCM with the objective developing a catalyst highly active, selective and stable [38–43]. Oxides, such as alkaline, alkaline earth and rare earth metal oxides, single or mixed, for example perosvkites, have shown to be good catalysts for the OCM reaction. The productivity of these systems is attributed either to the catalyst basicity or the availability of active sites such as oxygen vacancies and/or other defects, which are important for CH_4 activation [44].

It has been generally established that on metal oxide catalysts the surface oxygen species may consist of lattice oxygen O^{2-}, peroxide $O_2{}^{2-}$, superoxide $O_2{}^{-}$ and besides carbonate $CO_3{}^{2-}$ and hydroxide OH^{-}. Therefore, different viewpoints concerning the reaction mechanism are centred mainly on the participation of different oxygen species in the OCM reaction. The sources of these species are the adsorbed oxygen over the catalyst (oxygen molecule and neutral atom) and the oxygen species present on the catalyst surface, such as, lattice oxygen O^{2-}, superoxide $O_2{}^{-}$ and species formed by the filling of oxygen vacancies ($Vs^{\cdot\cdot}$) by molecular oxygen to form a hole species ($h^{\cdot}s$) [45]. This reaction can be written as:

$$V_S^{\cdot\cdot} + 1/2O_2 \rightarrow O_S^x + 2h_S^{\cdot} \tag{8}$$

where 2 $\mathbf{h}^{\cdot}$ is a small polaron species, i.e., either $2(O^{-} = O_S^{\cdot})$ or $O_2{}^{2-} = (O_2)_S^{\cdot\cdot}$ [45]. The O^{-} species were shown to be the active centres on Li/MgO or Na/MgO systems [37]. Cation radii in the Li/MgO catalyst are similar ($r_{Li+} = 0.68$ Å,

$r_{Mg+2} = 0.66$ Å). They can be replaced by each other and therefore the electrical neutrality is maintained when the numbers of O^- and Li^+ ions into the oxide lattice are the same. A relationship between $[Li^+O^-]$ (active centres) and the number of $CH_3^{\cdot}$ was found [46]. These active centres are formed by oxygen vacancies ([...]) which are created between the O^{2-} and Li^+ as follows:

$$2Li^+O^{2-} + [\ldots] + \tfrac{1}{2}\, O_2 \longleftrightarrow 2Li^+O^- + O^{2-} \quad (9)$$

$$Li^+O^- + CH_4 \rightarrow Li^+OH^- + CH_3 \quad (10)$$

The regeneration of $[Li^+O^-]$ centre occurs with the reaction:

$$2Li^+OH^- + O_2 \rightarrow 2Li^+O^- + H_2O \quad (11)$$

Sodium also promotes the MgO, though it is less efficient when compared with Li. The O_2^{2-} peroxide ion was also proposed to be an active species, especially at temperatures higher than 750 °C on catalytic systems such as Na_2O_2, Na/La_2O_3 and La_2O_3, Ba/MgO [43]. On other hand, it was shown that incorporation of ions such as Mg^{2+}, Ca^{2+} and Sr^{2+} on the CeO_2 surface creates oxygen species in "interstitial" sites on the oxide surface, leading to an abundance of electrophilic oxygen species on the catalyst surface specially O_2^{2-} [47]. Instead, superoxide species O_2^- were observed on other catalysts such as LaOF [48], Ba/Nd_2O_3 and Y_2O_3-CaO catalysts [41].

In the search for a suitable catalyst, especially a stable catalyst at high temperatures, Mn-Na_2WO_4/SiO_2 catalyst has shown to have this important property. This fact has been confirmed by different research groups [39, 42, 49, 50] indicating a catalytic performance (CH_4 conversions of 20–30 % at C_2 selectivities of 70–80 %) suitable for a practical application. Amorphous SiO_2 is the support used in this type of catalyst and incipient wetness is the method used by almost all researches to incorporate the others components on the catalyst. However, different research groups have reported that in the synthesis the phase transformation form amorphous SiO_2 to the inert α-cristobalite is necessary to reduce activity [51].

Different studies have tried to understand the surface structure and the active centre responsible for the good catalytic performance of this catalyst for the OCM reaction. It has been reported in literature [52] that these studies have revealed that the best catalyst performance is only achieved when Mn, Na or K and W are present. A synergistic effect seems to be present. However, in other studies where Mn, Na and W are substituted for other metals and in case of W also for non-metals, an active catalyst is also obtained indicating a possible structure responsible for the activity. In addition, detailed variations of the support material with a using of different materials or the SiO_2 precursor with additional doping of the Mn-Na_2WO_4/SiO_2 catalyst were also performed [52] in order to contribute to the fundamental understanding of the OCM and the function of a catalyst in this reaction.

3.2 Operating Mode of the OCM

In conventional reactors two different operating modes of the OCM have been studied:

- Sequential feeding methane and oxygen (the reactants are not together) and;
- Co-feeding where both reactants coincide in the same time–space over the catalyst.

Keller and Bhasin [7] tested both operating modes. In the sequential feeding, reducible metallic oxide was utilised as the O_2 source. Then, the reduced metallic oxide is deoxidised using a gaseous oxygen stream. The typical reactions are:

$$2CH_4 + MO_{2+x} \rightarrow C_2H_4 + 2H_2O + MO_x \quad (12)$$

$$MO_x + O_2 \rightarrow MO_{2+x} \quad (13)$$

That sequential combination is the methane coupling reaction:

$$2CH_4 + O_2 \rightarrow C_2H_4 + 2H_2O \quad (14)$$

In this route, methane and the coupling products are not in contact with gaseous oxygen and therefore the total oxidation reactions do not occur. However, methane conversion and C_2 hydrocarbons selectivities are not constant (unsteady state). Therefore, CH_4 conversion and selectivity vary in the time. On the other hand, the amount of the catalyst and the reaction rate in (12) must be optimised.

Regarding the Co-feeding of CH_4 and O_2 in reactor inlet, different operation modes to contact both reactants on the catalyst surface were investigated. For that, some reactors have been designed to mix a hydrocarbon with a powerful oxidant reagent under safety conditions and to control the generated energy from the process.

3.2.1 Fixed Bed Reactors

Several fixed-bed reactor designs have been proposed to control the high exothermicity exhibited in the most of the catalyst performed in the OCM reaction. Thin-bed reactor [16], sintered metal packing [17] and monolith reactor [40] have been used at laboratory-scale and their catalytic performances have allowed assuming projections at industrial-scale units, such as multitubular and multistage adiabatic fixed beds. In fixed bed reactors, the temperature control and heat management is crucial. Since, large exothermicity steep axial and radial temperature gradients can occur in the reactor bed. On the other hand, inlet concentration of oxygen is limited to 20 % in order to keep the hot-spot temperature below 1,000 °C.

3.2.2 Fluidised Bed Reactors

The fluidised bed reactor has been designated by different authors as the best reactor concept for OCM reaction. Unlike any fixed bed reactor, the fluidised bed reactor allows managing of the huge reaction heat and presents the ability to operate isothermally and avoids a temperature runaway at the same time to continuously re-circulate or even to change deactivated catalyst. There have been a number of investigations in the 1990s around the fluidised bed reactor concept [11, 38, 53–56]. However, all of these studies showed that only a yield of less than 19.4 % was accessible in a fluidised bed reactor, and thus, limiting its yield to the similar restriction as in the case of a fixed bed reactor.

3.2.3 Membrane Reactors

It is known that the membrane reactors present a potential to advance the process industry by enhancing selectivity and yield, reducing energy consumption, improving operation safety, and miniaturising the reactor system. Membrane reactors made of inert porous or dense membrane tube with a regular OCM catalyst have been studied as oxygen distributor to improve the C_2 hydrocarbon yield [12, 57]. On the other hand, studies on OCM in inorganic membrane reactors have not only leaded to a breakthrough in obtaining a higher C_2, but also improve the understanding of the reaction mechanism. For instance, catalytically active membrane was used to change the OCM reaction mechanism and minimise the presence of the gas-phase oxygen in the methane stream [58–60].

Nevertheless, this membrane reactors presents inability to obtain high C_2 hydrocarbon yields. This because of problems inherent to the membrane reactors, such as: poor membrane surface catalytic properties and unfavourable reactor configuration. To solve this problem, it was necessary designing news of membrane reactors with a large permeation surface area to volume ratio.

All these studies involving the different oxygen species present on the catalyst surface and reactors used in the OCM have showed that the selectivity to C_2 hydrocarbons decreases as the overall conversion of methane increases. Thus the C_2 yield, which is the product of selectivity and the conversion, is usually limited to about 25 %. Several innovative reactor design approaches have been proposed and very high yields (>60 %) were reported [61, 62]. Nevertheless, the problem of low yields *per pass* remains unsolved.

The study of the types and state of oxygen to supply to the OCM reaction has also been of crucial importance to try achieving industrially acceptable C_2 hydrocarbon yields. An interesting alternative is the electrochemically supplied oxygen (O^{2-}) by the use of SEMRs. It has been described in detail in excellent reviews [21, 63, 64].

In general it was reported that the use of O^{2-} SEMRs improved the catalytic properties of the electrodes; though the C_2 hydrocarbon yields were mostly between 1 and 10 % [21]. However, these studies provided new possible alternatives to increase R&D activities to convert methane into useful chemicals or the cogeneration of electric energy and useful chemicals.

4 Co-generation of Solid Oxide Fuel Cell Reactor: Oxidative Coupling of Methane

Selective oxidation of methane in SOFCs was studied to cogenerate electric power and C_2 hydrocarbons [21]. Solid oxide fuel cell type reactor is more attractive than other conventional reactors due to an achievement of high selectivity and possible use in electricity generation. In addition, it can be used at high temperatures.

As mentioned above, a SOFC is a SEMR operating in closed circuit (fuel cell mode). The power electrical and valuable chemicals such ethylene can be produced when the ionic charge through the electrolyte is balanced by the flow electronic charge through an outside circuit (case b in Fig. 1).

Pujare and Sammellsn were the first to report the use of SOFCs for co-generation of C_2 hydrocarbons and electrical power [65]. They obtained high C_2 hydrocarbon selectivity (>90 %) and relatively low methane conversion. In a SOFC the electrolyte material is interposed between two thin electrodes (porous anode and cathode). Therefore, direct chemical combustion is prevented by the electrolyte that separates the fuel (CH_4) and from oxidant (O_2). The oxygen is reduced to oxygen ion O^{2-} on the cathode as lattice oxygen and migrates across electrolyte, which also serves as a barrier to gas diffusion. The permeated oxygen is then activated on the anode catalyst to react with methane and form valuable C_2 hydrocarbons.

Many studies have been reported focusing the catalyst preparation, characterisation and reactor performance test. Conventional solid electrolytes used in most studies are O_2 ion conductors like yttria stabilised zirconia (YSZ) [66, 67] or CeO_2 [68]. Catalysts such as KF, $BaCO_3$, NaCl/MnO_2, Sm_2O_3 deposited on Au-electrode were tested by Otsuka et al. [69, 70] and they found that the most active and selective catalyst was $BaCO_3$ on Au, which showed high selectivity to C_2 hydrocarbons, but further studies were need to stabilise the catalytic activity.

In addition, anode catalysts for an oxidative coupling of methane in SOFC become an important goal to improve the conversion, selectivity, yield and stability of the reactor. Guo et al. [71] combined 1 wt%Sr/La_2O_3 and Bi_2O_3-Ag as the catalyst-electrode and they showed that the fuel cell was operated for an extended period of time with selectivity to C_2 hydrocarbons and CH_4 conversion relativity constant. Wiyaratn et al. [72], reported a possibility in simultaneous generation of electrical energy and C_2 hydrocarbons over Au/$La_{1-x}Sr_xMnO_3$ nanocomposites in an SOFC reactor. They showed that Au nanoparticles could significantly improve the catalytic performance of the composite to be used in a partial oxidation fixed bed reactor and an SOFC reactor.

Membranes were also used in SOFCs for the OCM reaction. Guo et al. [71] studied the effect of the Sr/La_2O_3-Bi_2O_3-Ag-YSZ membrane composition and revealed that an increase in the generated current was accompanied by a decrease in C_2 selectivity and an increase in CH_4 conversion. On the other hand, fuel cell type temperature-programmed desorption (FC-TPD) technique was used to investigate oxygen species. This study allowed predict the kinetic parameters of the reactions at the anode of cogeneration of chemical production in [18, 73, 74].

Kiatkittipong W et al. [19] suggested that SOFC for oxidative coupling of methane is a good reactor for C_2 hydrocarbon production and cogeneration of electric power. However, the obtained electricity was far from a typical SOFC which solely for electricity generation. Several conventional fixed bed reactor (FBR), porous membrane reactor (PMR) and mixed ionic and electronic conducting membrane reactor (MIEMR) were compared with solid oxide fuel cell reactor [14]. This latter improved C_2 selectivity compared to FBR. However, PMR was superior to the other reactors at low temperature (<877 °C) while MIEMR was attractive at high temperature (>877 °C).

5 Simultaneous Production/Separation of H_2 and C_2 Hydrocarbons in SEMRs Using O^{2-} Ion Conductors

In most of the studies in which the OCM reaction was conducted in SEMRs using O^{2-} ion conductors, the reacting O^{2-} ions were electrochemically supplied to the active catalyst film from the O_2 coming from the gas phase or the air, and hence, the C_2 hydrocarbons were the unique valuable products. Nevertheless, there have been very few studies to apply the OCM reaction as a process by integration of the simultaneous co-production of another product of great economic interest, H_2, using a SEMR.

A firsts study was reported recently [75]. In this process the O_2 for the OCM is not directly supplied from gas phase, and hence, the active O^{2-} for the OCM is produced in situ from the electrolysis process. The O^{2-} ions migrate to electrode through a solid electrolyte to react with CH_4 as shown in (Fig. 3).

Theoretically the electrode must be selective to methane coupling reaction with the subsequent formation of C_2 hydrocarbons. Consequently, the hydrogen production rate from water electrolysis is enhanced because of using an oxygen ion conducting membrane to remove in situ of the simultaneous produced oxygen. However, it has not was reported an electrode highly active e selective to form C_2 hydrocarbons.

In previously studies, a usual solution was to prepare an anodic electrode by mixing a metal (preferably inactive like Au) with metal oxide [69]. Nevertheless, when oxygen ions reach the metal–electrolyte–gas boundary of the anode, methane can react with oxygen on the metal as well as on the oxide surface, and metals are poorly selective catalysts. Also, if the O^{2-} flux is high and the metal electrode is relatively inactive, a significant fraction of the transported O^{2-} may combine to

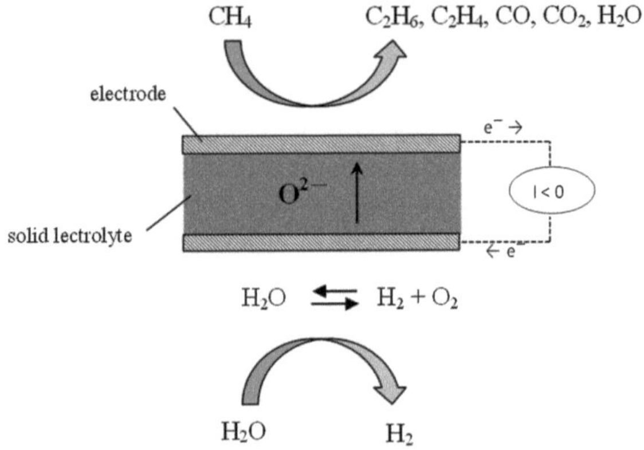

Fig. 3 Schematic diagram of the simultaneous production of H_2 and C_2 hydrocarbons in an oxygen-transporting membrane reactor

form O_2 [21]. A possible solution test is to use conducting oxides, i.e., perovskite materials as electrodes with the hope that they would serve both purposes (methane coupling and electrical conduction) [21].

However, the efficiency of this type of system can be improved by coupling electro-catalysis with conventional heterogeneous catalysis. In this case a catalyst active and selective towards C_2 hydrocarbons is located next to electrode in which the OCM reaction occurs to enhance the reaction between the O_2 molecules and methane increasing the overall efficiency of the process for the production of H_2 and C_2 hydrocarbons (Fig. 4).

This technique was applied by Carvaca et al. [76], which proposed a single chamber solid electrolyte membrane plus fixed bed reactor configuration to study the simultaneous production of H_2 and C_2 hydrocarbons in a humidified CH_4 atmosphere. They used Yttria-stabilised Zirconia (SZY) as O^{2-} conducting solid electrolyte. Platinum and silver were used as electrodes and the silver electrode was located in contact to the synthesised Ce-Na_2WO_4/SiO_2 catalyst. The results revealed, under closed circuit and in the O^{2-} ions "pumping" mode, a high production of H_2 and C_2 hydrocarbons when compared to results without catalyst.

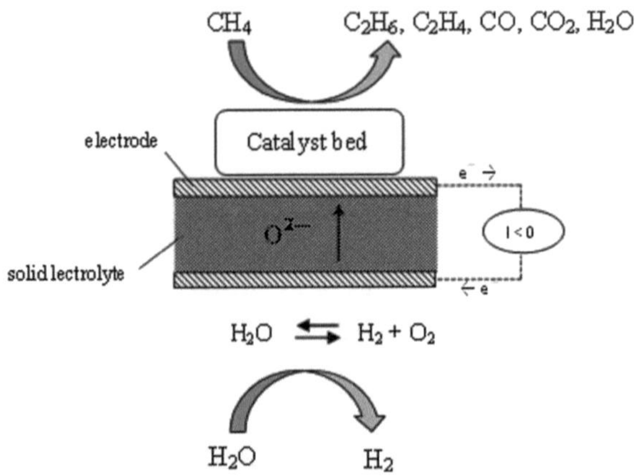

Fig. 4 Schematic diagram of the simultaneous production of H_2 and C_2 hydrocarbons in an oxygen-transporting membrane reactor + catalyst bed

This type of process offers two important advantages. Firstly, the possibilities of combining an exothermic reaction (oxidative coupling) with an endothermic reaction (steam electrolysis) by heat transfer. The second, applying a same current to produce simultaneously H_2 and C_2 hydrocarbons, which may be satisfactory for an economically feasible production of ethylene, and hence for the further implementation of this process.

6 Future Frontiers

One alternative route for the production of C_2 hydrocarbons is from natural gas via oxidative coupling of methane. So far, the process is economically unfeasible. However, recent studies suggest that this may be overcome by producing not only C_2 hydrocarbons and electricity but also simultaneous production/separation of H_2 and C_2 hydrocarbons, using the heat from the very exothermic coupling reaction. Solid fuel cell reactors really offer this interesting application. Thus, electro cogeneration and simultaneous production of H_2 and C_2 hydrocarbons have a major driven development towards commercialisation.

However, no effort is generally made to optimise the solid fuel cell reactor, applied as a cogenerator of electric power and C_2 hydrocarbons, and the working conditions, to scale-up the fuel cell. Certainly fuel cells are recently being tentatively introduced in the market because they appear as important alternative for meeting future energy requirements. The commercial progress of fuel cells has been very slow, mainly due to the fact that energy is being obtained at smaller cost with conventional technologies. Therefore, it is not strange that scientific and economic aspects relating to electrocogeneration, from C_2 hydrocarbons, in a solid fuel cell, has been the object of only limited number of studies.

The catalysts and electrodes to use in solid fuel cell reactor is the major obstacle for its application. Although, plenty of efforts have been focused on price-reasonable of synthesis, active and stability, no true breakthrough has been reported yet. Thus, development of new catalysts, increasing their activity by researching of their chemical, textural and morphological properties (particle size, electric and ionic conductivity, dispersion on the support, etc.) and reduced cost of catalyst are important mission and challenge in the cogeneration and production of C_2 hydrocarbons using solid fuel cell reactor for opening the opportunity in commercialisation. In addition, the understanding in the reactions of fuel cell reactor including thermodynamic (e.g., entropy, cooling effect, hot spot) is required to develop the optimal condition for enhancing fuel cell activity.

Acknowledgments The authors acknowledge to ACENET/0001/2007 and project PEst-C/EQB/LA0020/2011, financed by FEDER through COMPETE—Programa Operacional Factores de Competitividade and by FCT—Fundação para a Ciência e a Tecnologia. VJF gratefully acknowledges a PhD grant (SFRH/BD/33647/2009) by FCT.

References

1. Boudghene Stambouli, A., Traversa, E.: Fuel cells, an alternative to standard sources of energy. Renew. Sustain. Energy Rev. **6**(3), 295–304 (2002)
2. Kirubakaran, A., Jain, S., Nema, R.K.: A review on fuel cell technologies and power electronic interface. Renew. Sustain. Energy Rev. **13**(9), 2430–2440 (2009)
3. Mekhilef, S., Saidur, R., Safari, A.: Comparative study of different fuel cell technologies. Renew. Sustain. Energy Rev. **16**(1), 981–989 (2012)
4. Vayenas, C.G.: Catalytic and electrocatalytic reactions in solid oxide fuel cells. Solid State Ionics 28–30, Part 2(0), 1521–1539 (1988)
5. Athanassiou, C., Pekridis, G., Kaklidis, N., Kalimeri, K., Vartzoka, S., Marnellos, G.: Hydrogen production in solid electrolyte membrane reactors (SEMRs). Int. J. Hydrogen Energy **32**(1), 38–54 (2007)
6. Guo, X.-M., Hidajat, K., Ching, C.-B., Chen, H.-F.: Oxidative coupling of Methane in a Solid Oxide Membrane Reactor. Ind. Eng. Chem. Res. **36**(9), 3576–3582 (1997)
7. Keller, G.E., Bhasin, M.M.: Synthesis of ethylene via oxidative coupling of methane: I. Determination of active catalysts. J. Catal. **73**(1), 9–19 (1982)
8. Lapeña-Rey, N., Middleton, P.H.: The selective oxidation of methane to ethane and ethylene in a solid oxide electrolyte reactor. Appl. Catal. A **240**(1–2), 207–222 (2003)
9. Guczi, L., Van Santen, R.A., Sarma, K.V.: Low-temperature coupling of Methane. Catal. Rev. **38**(2), 249–296 (1996)
10. Chua, Y.T., Mohamed, A.R., Bhatia, S.: Oxidative coupling of methane for the production of ethylene over sodium-tungsten-manganese-supported-silica catalyst (Na-W-Mn/SiO2). Appl. Catal. A **343**(1–2), 142–148 (2008)
11. Andorf, R., Mleczko, L., Schweer, D., Baerns, M.: Oxidative coupling of methane in a bubbling fluidized bed reactor. Can. J. Chem. Eng. **69**(4), 891–897 (1991)
12. Coronas, J., Menendez, M., Santamaria, J.: Development of ceramic membrane reactors with a non-uniform permeation pattern. Application to methane oxidative coupling. Chem. Eng. Sci. 49(24, Part A), 4749–4757 (1994)
13. Ji, S., Xiao, T., Li, S., Chou, L., Zhang, B., Xu, C., Hou, R., York, A.P.E., Green, M.L.H.: Surface WO_4 tetrahedron: the essence of the oxidative coupling of methane over M–W–Mn/SiO_2 catalysts. J. Catal. **220**(1), 47–56 (2003)
14. Kiatkittipong, W., Tagawa, T., Goto, S., Assabumrungrat, S., Silpasup, K., Praserthdam, P.: Comparative study of oxidative coupling of methane modeling in various types of reactor. Chem. Eng. J. **115**(1–2), 63–71 (2005)
15. Klose, F., Wolff, T., Thomas, S., Seidel-Morgenstern, A.: Operation modes of packed-bed membrane reactors in the catalytic oxidation of hydrocarbons. Appl. Catal. A **257**(2), 193–199 (2004)
16. Leyshon, D.W.: Thin bed reactor for conversion of Methane to higher Hydrocarbons. In: Holmen, A., Jens, K.J., Kolboe, S. (eds.) Studies in Surface Science and Catalysis, vol. 61, pp. 497–507. Elsevier, Amsterdam (1991). doi:10.1016/s0167-2991(08)60116-0
17. van Looij, F., Mulder, A., Boon, A.Q.M., Scheepens, J.F., Geus, J.W.: Fixed bed catalytic reactors based on sintered metals. In: Guczi, L., Solymosi, F., Tétényi, P. (eds.) Studies in Surface Science and Catalysis, vol. 75, pp. 1377–1389. Elsevier, Amsterdam (1993). doi:10.1016/s0167-2991(08)64458-4
18. Kiatkittipong, W., Tagawa, T., Goto, S., Assabumrungrat, S., Praserthdam, P.: Oxidative coupling of Methane in the LSM/YSZ/LaAlO SOFC Reactor. J. Chem. Eng. Jpn. **37**(12), 1461–1470 (2004)
19. Kiatkittipong, W., Goto, S., Tagawa, T., Assabumrungrat, S., Praserthdam, P.: Simulation of oxidative coupling of Methane in Solid Oxide Fuel Cell type reactor for C2 Hydrocarbon and electricity co-generation. J. Chem. Eng. Jpn. **38**(10), 841–848 (2005)

20. Garagounis, I., Kyriakou, V., Anagnostou, C., Bourganis, V., Papachristou, I., Stoukides, M.: Solid electrolytes: applications in heterogeneous catalysis and chemical cogeneration. Ind. Eng. Chem. Res. **50**(2), 431–472 (2010)
21. Stoukides, M.: Solid-electrolyte membrane reactors: current experience and future outlook. Catal. Rev.—Sci. Eng. **42**(1–2), 1–70 (2000)
22. Ormerod, R.M.: Solid oxide fuel cells. Chem. Soc. Rev. **32**(1), 17–28 (2003)
23. Zainoodin, A.M., Kamarudin, S.K., Daud, W.R.W.: Electrode in direct methanol fuel cells. Int. J. Hydrogen Energy **35**(10), 4606–4621 (2010)
24. Alcaide, F., Cabot, P.-L., Brillas, E.: Fuel cells for chemicals and energy cogeneration. J. Power Sources **153**(1), 47–60 (2006)
25. McLean, G.F., Niet, T., Prince-Richard, S., Djilali, N.: An assessment of alkaline fuel cell technology. Int. J. Hydrogen Energy **27**(5), 507–526 (2002)
26. Neergat, M., Shukla, A.K.: A high-performance phosphoric acid fuel cell. J. Power Sources **102**(1–2), 317–321 (2001)
27. Zhu, W.Z., Deevi, S.C.: A review on the status of anode materials for solid oxide fuel cells. Mater. Sci. Eng., A **362**(1–2), 228–239 (2003)
28. Stambouli, A.B., Traversa, E.: Solid oxide fuel cells (SOFCs): a review of an environmentally clean and efficient source of energy. Renew. Sustain. Energy Rev. **6**(5), 433–455 (2002)
29. Langer, S.H., Pate, K.T.: Electrogenerative reduction of nitric oxide. Nature **284**(5755), 434–435 (1980)
30. Otsuka, K., Sawada, H., Yamanaka, I.: A hydrogen-nitric oxide cell for the synthesis of hydroxylamine. J. Electrochem. Soc. **143**(11), 3491–3497 (1996)
31. Alcaide, F., Brillas, E., Cabot, P.L., Casado, J.: Electrogeneration of hydroperoxide ion using an alkaline fuel cell. J. Electrochem. Soc. **145**(10), 3444–3449 (1998)
32. Ramachandra, A.M., Lu, Y., Ma, Y.H., Moser, W.R., Dixon, A.G.: Oxidative coupling of methane in porous Vycor membrane reactors. J. Membr. Sci. **116**(2), 253–264 (1996)
33. Kölsch, P., Noack, M., Schäfer, R., Georgi, G., Omorjan, R., Caro, J.: Development of a membrane reactor for the partial oxidation of hydrocarbons: direct oxidation of propane to acrolein. J. Membr. Sci. **198**(1), 119–128 (2002)
34. Marnellos, G., Stoukides, M.: Catalytic studies in electrochemical membrane reactors. Solid State Ionics **175**(1–4), 597–603 (2004)
35. Munder, B., Ye, Y., Rihko-Struckmann, L., Sundmacher, K.: Solid electrolyte membrane reactor for controlled partial oxidation of hydrocarbons: Model and experimental validation. Catal. Today **104**(2–4), 138–148 (2005)
36. Coronas, J., Santamaría, J.: Separations using zeolite membranes. Sep. Purif. Methods **28**(2), 127–177 (1999)
37. Lunsford, J.H.: The catalytic conversion of methane to higher hydrocarbons. Catal. Today **6**(3), 235–259 (1990)
38. Mleczko, L., Pannek, U., Rothaemel, M., Baerns, M.: Oxidative coupling of methane over a La_2O_3/CaO catalyst. Optimization of reaction conditions in a bubbling fluidized-bed reactor. Can. J. Chem. Eng. **74**(2), 279–287 (1996)
39. Lin, J.Z., Gu, J.F., Yang, D.X., Zhang, C.W., Yang, Y.L., Chu, Y.L., Li, S.B.: Stability test of W-Mn/SiO_2 catalyst for oxidative coupling of methane. Shiyou Huagong **24**, 293–298 (1995)
40. Aigler, J.M., Lunsford, J.H.: Oxidative dimerization of methane over MgO and Li^+/MgO monoliths. Appl. Catal. 70(1), 29–42 (1991)
41. Osada, Y., Koike, S., Fukushima, T., Ogasawara, S., Shikada, T., Ikariya, T.: Oxidative coupling of methane over Y_2O_3CaO catalysts. Appl. Catal. **59**(1), 59–74 (1990)
42. Sergei, P., Ping, Q., Jack, H.L.: Elementary reactions in the oxidative coupling of Methane over Mn/Na_2WO_4/SiO_2 and Mn/Na_2WO_4/MgO catalysts. Catal **179**(1), 222–230 (1998)
43. Sinev, M.Y., Korchak, V.N., Krylov, O.V.: Highly selective ethane formation by reduction of BaO/sub 2/with methane. Kinet Catal (Engl Transl); (United States) 27(5) (1987). Translated from Kinet Katal 27(5), 1274 (Sep–Oct 1986) (Medium: X; Size: Pages: 1110)

44. Voskresenskaya, E.N., Roguleva, V.G., Anshits, A.G.: Oxidant activation over structural defects of oxide catalysts in oxidative methane coupling. Catal. Rev. **37**(1), 101–143 (1995)
45. Gellings, P.J., Bouwmeester, H.J.M.: Solid state aspects of oxidation catalysis. Catal. Today **58**(1), 1–53 (2000)
46. Driscoll, D.J., Martir, W., Wang, J.-X., Lunsford, J.: The production of gas phase methyl radicals over lithium-promoted MgO. In: Che, M., Bond, G.C. (eds.) Studies in Surface Science and Catalysis, vol. 21, pp. 403–408. Elsevier, Amsterdam (1985). doi:10.1016/s0167-2991(08)64940-x
47. Ferreira, V.J., Tavares, P., Figueiredo, J.L., Faria, J.L.: Effect of Mg, Ca, and Sr on CeO_2 based catalysts for the oxidative coupling of methane: investigation on the oxygen species responsible for catalytic performance. Ind. Eng. Chem. Res. **51**(32), 10535–10541 (2012)
48. Weng, W., Chen, M., Wan, H., Liao, Y.: High-temperature in situ FTIR spectroscopy study of LaOF and BaF_2/LaOF catalysts for methane oxidative coupling. Catal. Lett. **53**(1–2), 43–50 (1998)
49. Wang, X.L., Zhang, J.N., Yang, D.X., Zhang, C.W., Lin, J.Z., Li, S.B.: Oxidative coupling of methane over W-Mn/Sio_2 catalyst in a bench-scale stainless steel fluidized-bed reactor. Shiyou Huagong **26**, 361–367 (1997)
50. Simon, U., Görke, O., Berthold, A., Arndt, S., Schomäcker, R., Schubert, H.: Fluidized bed processing of sodium tungsten manganese catalysts for the oxidative coupling of methane. Chem. Eng. J. **168**(3), 1352–1359 (2011)
51. Palermo, A., Holgado Vazquez, J.P., Lee, A.F., Tikhov, M.S., Lambert, R.M.: Critical influence of the amorphous silica-to-cristobalite phase transition on the performance of Mn/Na_2WO_4/SiO_2 catalysts for the oxidative coupling of methane. J. Catal. **177**(2), 259–266 (1998)
52. Arndt, S., Otremba, T., Simon, U., Yildiz, M., Schubert, H., Schomäcker, R.: Mn–Na_2WO_4/SiO_2 as catalyst for the oxidative coupling of methane. What is really known? Appl. Catal. A **425–426**, 53–61 (2012)
53. Mleczko, L., Pannek, U., Niemi, V.M., Hiltunen, J.: Oxidative coupling of methane in a fluidized-bed reactor over a highly active and selective catalyst. Ind. Eng. Chem. Res. **35**(1), 54–61 (1996)
54. Pannek, U., Mleczko, L.: Comprehensive model of oxidative coupling of methane in a fluidized-bed reactor. Chem. Eng. Sci. **51**(14), 3575–3590 (1996)
55. Pannek, U., Mleczko, L.: Reaction engineering simulations of oxidative coupling of methane in a circulating fluidized-bed reactor. Chem. Eng. Technol. **21**(10), 811–821 (1998)
56. Edwards, J.H., Tyler, R.J., White, S.D.: Oxidative coupling of methane over lithium-promoted magnesium oxide catalysts in fixed-bed and fluidized-bed reactors. Energy Fuels **4**(1), 85–93 (1990)
57. Lu, Y., Dixon, A.G., Moser, W.R., Ma, Y.H., Balachandran, U.: Oxygen-permeable dense membrane reactor for the oxidative coupling of methane. J. Membr. Sci. **170**(1), 27–34 (2000)
58. Nozaki, T., Fujimoto, K.: Oxide ion transport for selective oxidative coupling of methane with new membrane reactor. AIChE J. **40**(5), 870–877 (1994)
59. Zeng, Y., Lin, Y.S., Swartz, S.L.: Perovskite-type ceramic membrane: synthesis, oxygen permeation and membrane reactor performance for oxidative coupling of methane. J. Membr. Sci. **150**(1), 87–98 (1998)
60. Shao, Z., Dong, H., Xiong, G., Cong, Y., Yang, W.: Performance of a mixed-conducting ceramic membrane reactor with high oxygen permeability for methane conversion. J. Membr. Sci. **183**(2), 181–192 (2001)
61. Jiang, Y., Yentekakis, I.V., Vayenas, C.G.: Methane to ethylene with 85 percent yield in a gas recycle electrocatalytic reactor-separator. Science **264**(5165), 1563–1566 (1994)
62. Tonkovich, A.L., Carr, R.W., Aris, R.: Enhanced C2 yields from methane oxidative coupling by means of a separative chemical reactor. Science **262**(5131), 221–223 (1993)

63. Garagounis, I., Kyriakou, V., Anagnostou, C., Bourganis, V., Papachristou, I., Stoukides, M.: Solid electrolytes: applications in heterogeneous catalysis and chemical cogeneration. Ind. Eng. Chem. Res. **50**(2), 431–472 (2011)
64. Stoukides, M.: Methane conversion to C2 hydrocarbons in solid electrolyte membrane reactors. Res. Chem. Intermed. **32**(3–4), 187–204 (2006)
65. Pujare, N.U., Sammells, A.F.: Methane activation to C2 hydrocarbon species in solid oxide fuel cell. J. Electrochem. Soc. **135**(10), 2544–2545 (1988)
66. Tsiplakides, D., Neophytides, S., Vayenas, C.G.: Investigation of electrochemical promotion using temperature-programmed desorption and work function measurements. Solid State Ionics **136–137**, 839–847 (2000)
67. Vernoux, P., Gaillard, F., Bultel, L., Siebert, E., Primet, M.: Electrochemical Promotion of Propane and Propene Oxidation on Pt/YSZ. J. Catal. **208**(2), 412–421 (2002)
68. Petrolekas, P.D., Balomenou, S., Vayenas, C.G.: Electrochemical promotion of ethylene oxidation on Pt catalyst films deposited on CeO_2. J. Electrochem. Soc. **145**(4), 1202–1206 (1998)
69. Otsuka, K., Yokoyama, S., Morikawa, A.: Catalytic activity- and selectivity-control for oxidative coupling of methane by oxygen-pumping through yttria-stabilized zirconia. Chem. Lett. **14**(3), 319–322 (1985)
70. Otsuka, K., Suga, K., Yamanaka, I.: Oxidative coupling of methane applying a solid oxide fuel cell system. Catal. Today **6**(4), 587–592 (1990)
71. Guo, X.-M., Hidajat, K., Ching, C.-B.: An experimental study of oxidative coupling of methane in a solid oxide fuel cell with 1 wt%Sr/La2O3-Bi2O3-Ag-YSZ membrane. Korean J. Chem. Eng. **15**(5), 469–473 (1998)
72. Wiyaratn, W., Appamana, W., Charojrochkul, S., Kaewkuekool, S., Assabumrungrat, S.: Au/La1-xSrxMnO$_3$ nanocomposite for chemical-energy cogeneration in solid oxide fuel cell reactor. J. Ind. Eng. Chem. **18**(5), 1819–1823 (2012)
73. Kiatkittipong, W., Tagawa, T., Goto, S., Assabumrungrat, S., Praserthdam, P.: TPD study in LSM/YSZ/LaAlO system for the use of fuel cell type reactor. Solid State Ionics **166**(1–2), 127–136 (2004)
74. Tagawa, T., Kuroyanagi, K., Goto, S., Assabumrungrat, S., Praserthdam, P.: Selective oxidation of methane in an SOFC-type reactor: effect of applied potential. Chem. Eng. J. **93**(1), 3–9 (2003)
75. Caravaca, A., de Lucas-Consuegra, A., González-Cobos, J., Valverde, J.L., Dorado, F.: Simultaneous production of H_2 and C_2 hydrocarbons by gas phase electrocatalysis. Appl. Catal. B **113–114**, 192–200 (2012)
76. Caravaca, A., Ferreira, V.J., de Lucas-Consuegra, A., Figueiredo, J.L., Faria, J.L., Valverde, J.L., Dorado, F.: Simultaneous production of H_2 and C_2 hydrocarbons by using a novel configuration solid-electrolyte+ fixed bed reactor. Int. J. Hydrogen Energy. **38**, 3111–3122 (2013)

Solar Thermal Energy Use in EU-27 Countries: Evolution and Promotion

María P. del Pablo-Romero, Antonio Sánchez-Braza and Enrique Lerma

Abstract Growth in the use of renewable energies in the 27 European Union (EU-27) countries over the past decade has been remarkable. Among these energies is solar thermal energy (STE). The average annual growth rate for the EU-27 countries in installed STE capacity in the period 2005–2011 was 20.82 %. While the use of STE has reached almost 43 % of National Renewable Energy Action Plan targets for the EU-27 countries for 2020, it is still nevertheless far from being used at its full potential, mainly because of barriers that hinder the installation of STE-based technologies in homes. Over recent years, various measures have been adopted in the EU to enhance and promote the use of STE. One measure has been to implement legislative regulations that require the housing construction industry to installation this type of energy in new homes. Another alternative has been to promote STE use via three forms of incentives: tax incentives, non-refundable grants and favourable lines of finance. Twenty-five of the 27 EU countries have used subsidies to promote solar thermal energy, 13 have used tax measures, 8 have used financial incentives and only 4 have adopted feed-in tariffs, with those countries with the most implemented measures being those with greatest amount of installed STE technology.

M. P. del Pablo-Romero (✉) · A. Sánchez-Braza
Chair of Energy Economics and the Environment Economic Analysis Department,
University of Seville, Ramon y Cajal 1 41018 Seville, Spain
e-mail: mpablorom@us.es

A. Sánchez-Braza
e-mail: asb@us.es

E. Lerma
University of Seville, Ramon y Cajal 1 41018 Seville, Spain
e-mail: enrlerrod@alum.us.es

G. Ferreira (ed.), *Alternative Energies*, Advanced Structured Materials 34,
DOI: 10.1007/978-3-642-40680-5_11, © Springer-Verlag Berlin Heidelberg 2013

1 Introduction

In 2008, around the 47 % of the worldwide final energy consumption was used for heating purposes, which was higher than the total energy used for electricity (17 %) and transport (27 %) combined [43]. More specifically, the residential sector used 41 % of the global final energy for heat, with most of this used for cooking, and space and water heating. On the other hand, industry used 44 % of the world's fuel for heat, mostly for boilers, smelters and chemical processes for the production of goods. Globally, fossil fuels (including coal) make up two-thirds of all end-use fuel for the provision of heat and therefore are still the most predominant fuels used in most countries. In Europe, empirical data from final energy consumption shows that heat makes up about half (48 %) of the total consumption of fossil fuels [26].

The large share of heat in final energy demand, and the considerable use of fossil fuels explain the substantial contribution that renewable energy-derived heat—or more specifically, STE—could make in meeting climate change and energy security objectives. In 2010, according to the International Energy Agency (IEA) [22] the share of renewable energy in heat total production was 10 %, which is expected to increase to 19 % by 2030. In line with Stryi-Hipp et al. [39], renewable heating and cooling (H&C) is the sleeping giant of the energy sector, which for a long time has been poorly perceived and underestimated. As pointed out by the European Renewable Energy Council (EREC) [13], the renewable H&C market, comprising residential and industrial biomass as well as solar thermal and geothermal applications, is predicted to undergo considerable growth. Together, these energies are expected to account for approximately 21 and 45 % of the total final energy consumption in Europe in 2030 and 2050, respectively.

STE can contribute significantly to the global energy need for heat. STE-based H&C, which are available in a wide range of technologies, from established forms of domestic hot water heaters to those just entering the demonstration phase, can provide low-carbon emission energy from the capture of solar energy, which is available in abundance throughout much of the world.

By the end of 2010, the solar thermal collector capacity in operation worldwide amounted to 195.8 GWth, corresponding to 279.7 million square metres of installed collectors [44]. The vast majority of this capacity is installed in China (60 %), Europe (18 %), and the United States (7.5 %). The early development of this technology on a large scale, took place in the 1960s in Australia, Japan and Israel [21], although strong growth in this sector, mainly in China and Europe, has occurred over the past decade. In Europe, the market size more than tripled between 2002 and 2008 [3]. Nevertheless, although Europe is the second largest world zone in terms of solar thermal capacity and has shown strong growth in recent years, the fact is that it still only represents a relatively small percentage of the overall provision of energy for H&C. According to Eurostat [15], the share of renewable energy sources (RES) in H&C reached about 11.9 % in 2008 in Europe, with biomass representing 11.4 % of heat consumption, followed by geothermal (0.3 %) and solar thermal (0.2 %).

The existence of technical, economic and even institutional barriers has slowed the development of STE use in Europe [30]. As outlined in Pablo-Romero et al. [34], two main technical barriers have hindered its implementation, these being the calorific losses of storage systems and technical difficulties related to compatibility with some building construction materials [24, 41]. Despite these technical limitations, the most significant barriers appear to be economic, basically related to the initial cost of implementing STE-based systems, which usually makes them uncompetitive compared with heat produced from fossil fuels. According to Stryi-Hipp et al. [39], the costs for STE systems range between 3 Eurocents per kWh for small solar domestic hot water (DHW) thermosiphon systems in southern Europe and 19 Eurocents per kWh for small and collective solar DHW in central and northern Europe. In contrast, according to Eurostat [15], natural gas and electricity costs for domestic consumers in the EU range from 2.76 to 11.65 Eurocents per kWh for natural gas and between 8.74 and 29.75 Eurocents per kWh for electricity.

Despite these costs, the use of STE systems entails clear environmental benefits. According to Solangi et al. [38], they do not exhaust natural resources and do not generate CO_2 emissions or waste products. Furthermore, they generate other advantages, such as increasing national energy independence and allowing a greater diversification of energy sources, which in turn increases the security of the energy supply [42]. The advantages that these systems present, in spite of barriers that impede their use for heating or hot water purposes, have encouraged countries to develop a range of policies to stimulate the implementation of STE-based technology in the public and private sector.

The present work undertakes a complete review of the measures taken to date to promote low-temperature STE use in the EU-27, and analyses the specific policies developed for that purpose. This analysis allows the effects of these actions to be assessed in relation to the objectives set for 2010 and foreseeable outcomes related to objectives established for 2020.

The main information sources that we have taken into account in the writing of this review, which are far from exhaustive, are a range of reports reviewing these measures at a global level [7, 36, 38, 41], as well as the Intelligent Energy Europe report [25], the EuroACE report [12], the main EU Climate Policy Tracker 2011 report and Financing Renewable Energy in the European Energy Market of Ecofys [6, 18]. We also used EU Member States' progress reports from 2011 of overall renewable energy policy developments [13], as well as the Taxes in Europe database published by the European Commission [14] and the Energy Efficiency Policies and Measures database of the International Energy Agency [20]. Further to this, to study the manner in which these policy measures have been implemented in each EU-27 country, the EU Directives relating to this issue and their implementation into national legislation have also been analysed.

This work is structured as follows. Section 2 analyses the present situation of STE use in the EU-27 region and current objectives that have been established.

Section 3 covers legislative regulations related to the implementation of this energy. Section 4 examines the subsidies adopted to promote STE use, Sect. 5 analyses tax incentives for the same purpose, and Sect. 6 considers financial incentives and other measures. Section 7 discusses the effects of these policies, following which conclusions are presented.

2 Objectives and Current Situation

The EU is one of the most active developers of energy security assessment methods and assurance measures. Its initiatives are expressed in the form of "Directives" to increase the percentage that RES contribute to total energy demand. Most evidence of the use of RES can be found in the electricity sector. EU Directive 2001/77/EC required Member States to increase the share of RES in the electricity sector by using national support instruments. In contrast, no legislative framework at the EU level was available in the heating sector before Directive 2009/28/EC was implemented.

The main EU strategy to reduce energy consumption in the household and services sectors with respect to heating has been oriented towards promoting more energy-efficient building designs. As pointed out in Cansino et al. [7], the most important EU policies for the household sector concerning improved energy efficiency are the Energy Performance of Buildings Directive 2010 (EPBD), The Energy Services Directive, and the Eco-design Directive. Other Directives from the building sector warranting specific mention are Directive CPD 89/106/EEC (Construction Products Directive) and the EU Directive of 21 December 1988 on the approximation of laws, regulations and administrative provisions of Member States relating to construction products. These Directives have been implemented into the national legislation of each country.[1] Nevertheless, as the EPBD only asks Member States to set requirement levels, each country has set its own national levels which, according to Panek [35], have led to significant disparity in the implementation of EPBD standards.

Besides policy measures aimed at improving the energy efficiency of buildings, EU authorities have also promoted the use of RES for H&C uses to reduce primary energy dependency and the stress of demand on primary energy resources.

On February 14, 2006, the European Parliament adopted a report suggesting that the RES for H&C share of total energy consumption should be increased up to 20 % by 2020. Later, the 2009/28/EC Directive on the promotion of the use of energy from renewable sources set an objective that RES should account for at least 20 % of the EU's final energy consumption by 2020. This Directive sets mandatory national targets for each Member State for the overall share of RES in

[1] Country Reports on EPBD implementation can be found in BUILD UP, the European web portal for energy efficiency in buildings: http://www.buildup.eu/home.

gross final energy consumption as well as a mandatory share of 10 % RES in the transport sector.

Following the implementation of Directive 2009/28/EC, every Member State was required to develop its own National Renewable Energy Action Plan that fixes specific objectives concerning the use of RES for each energy sector, including the heating sector. These plans provide detailed roadmaps of how each Member State expects to reach its legally binding 2020 target for the share of RES in their final energy consumption.

According to the NREAPs provided by each country, 21.3 % of EU-27 heating consumption in 2020 is expected to come from RES, with biomass representing 17.2 %, STE 1.2 %, heat pumps from aerothermal and hydrothermal energy 1.6 %, and geothermal energy the remainder. These percentages are lower than that estimated by the national RES industry roadmaps [9], where renewable energy-based H&C is expected to account for 23.5 % of total H&C consumption in 2020.

Table 1 shows NREAP targets for STE use set down in the NREAPs for the 27 EU countries, with the last row in that table showing that STE consumption in 2020 is expected to amount to 5,526.03 Ktoe. Italy, France, Germany and Spain have set the highest targets, while Romania, Finland and Estonia do not include STE targets at all in their plans, and Bulgaria, Denmark, The Netherlands, Sweden, and United Kingdom provide extremely low targets. Also shown in the table (column 3) is the STE consumption estimated by Member State national RES industries. Twenty-two of the 27 EU countries expect to reach or exceed their domestic STE targets for 2020. STE consumption in 2020 is expected account for 8,868.5 Ktoe in the EU-27, which is 60 % higher than the NREAP estimate. In this case, Italy, Spain Germany and France expect to consume the most energy of this form, compared with Denmark, Ireland, Romania and Estonia, which have no expected consumption. Moreover, the report by Weiss and Biermayr [45] concerning the potential of STE use in Europe pointed out that it is feasible for an installed capacity three times greater than the NREAP target to be reached.

Table 1 also presents data on the installed solar thermal energy capacity in EU-27 countries for the years 2001, 2005, and 2011. The total installed STE capacity in EU-27 in 2011 was 27,545,011 MWh, even though the market is still in the very early stages of development given that only about 0.3 % of the heat demand in Europe is provided by STE [39]. Germany has always been the country with the highest installed capacity, accounting for 10,663,800 MWh in 2011, which represents almost 38 % of the EU-27 total capacity of STE. Thereafter follows (based on 2011 figures) Austria, Greece, Italy, Spain and France, with all of these countries (including Germany) representing about 81 % of EU total capacity. The evolution of uptake over these years shows that Greece and France have become less important players in this field. Figure 1 shows the share of installed capacity of STE in the EU-27 countries.

In terms of capacity of low-temperature STE installed per inhabitant, at the end of 2011 the EU-27 had a capacity of 78.34 kWth/1,000 inhabitants. As shown in Fig. 2, EU-27 countries with an above average installed capacity are led by

Table 1 Solar thermal energy targets and installed capacity in EU-27 countries

	Solar thermal consum. NREAP (Ktoe)	Solar thermal consum. National roadmap (Ktoe)	Installed capacity (MWh) 2001	Installed capacity (MWh) 2005	Installed capacity (MWh) 2011	Annual growth rate 2005–2011 (%)	Percentage of 2020 objective already installed NREAP (%)	Percentage of 2020 objective already installed National roadmap (%)
Germany	686.0	1,225.0	2,883,335	4,976,300	10,663,800	13.54	133.69	74.86
Austria	269.0	912.0	1,659,672	1,817,060	3,336,430	10.66	106.67	31.46
Greece	355.0	582.0	1,965,040	2,133,040	2,862,595	5.03	69.35	42.30
Italy	1,586.0	1,455.0	254,135	476,385	2,052,716	27.56	11.13	12.13
Spain	644.0	1,298.0	176,568	382,900	1,914,913	30.77	25.57	12.69
France	927.0	927.0	462,000	626,920	1,659,000	17.61	15.39	15.39
Poland	506.0	319.0	–	85,568	632,561	39.57	10.75	17.05
Portugal	160.0	236.0	147,630	87,640	613,773	38.32	32.99	22.37
The Netherlands	23.0	361.0	231,560	434,301	585,445	5.10	218.91	13.95
Czech Rep.	22.0	55.0	–	61,551	554,378	44.24	216.71	86.68
Cyprus	90.4	96.0	–	350,140	489,591	5.75	46.54	43.86
Denmark	16.0	–	201,446	243,264	435,681	10.20	234.18	–
UK	34.0	743.0	123,144	140,812	414,767	19.73	104.91	4.80
Sweden	6.0	250.0	130,291	180,530	332,290	10.70	476.28	11.43
Belgium	5.7	31.7	25,518	55,685	291,513	31.77	439.83	79.09
Slovenia	21.0	44.0	–	73,822	132,331	10.22	54.19	25.86
Ireland	20.0	–	2,310	3,500	121,590	80.64	52.28	–
Slovak Rep.	30.0	42.0	–	44,919	102,445	14.73	29.37	20.98
Hungary	82.0	55.0	–	34,300	89,384	17.31	9.37	13.98
Romania	–	–	–	–	86,100	12.22	–	–
Bulgaria	21.0	40.0	–	–	79,450	17.35	32.54	17.08

(continued)

Table 1 (continued)

	Solar thermal consum. NREAP (Ktoe)	Solar thermal consum. National roadmap (Ktoe)	Installed capacity (MWh) 2001	Installed capacity (MWh) 2005	Installed capacity (MWh) 2011	Annual growth rate 2005–2011 (%)	Percentage of 2020 objective already installed NREAP (%)	Percentage of 2020 objective already installed National roadmap (%)
Malta	2.7	9.8	–	13,552	33,287	16.16	103.72	29.21
Finland	–	142.0	33,285	10,332	27,666	17.84	–	1.68
Luxembourg	8.1	14.0	700	9,380	18,976	12.46	20.15	11.66
Latvia	2	11.0	–	1,855	7,945	27.44	34.16	6.21
Lithuania	9.0	20.0	–	1,505	4,375	19.46	4.18	1.88
Estonia	–	–	–	574	2,009	23.22	–	–
EU-27	5,526.0	8,868.5	8,296,634	12,245,835	27,545,011	20.82	42.73	26.81

Source EurObserv'ER [17], EREC [9], Eurostat [16]

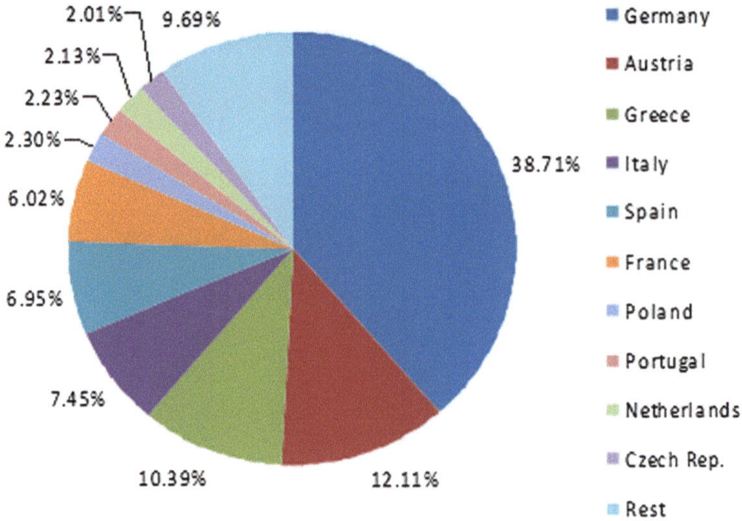

Fig. 1 Share of solar thermal energy installed capacity in EU-27 (2011) (*Source* Own elaboration from EurObserv'ER [17] data)

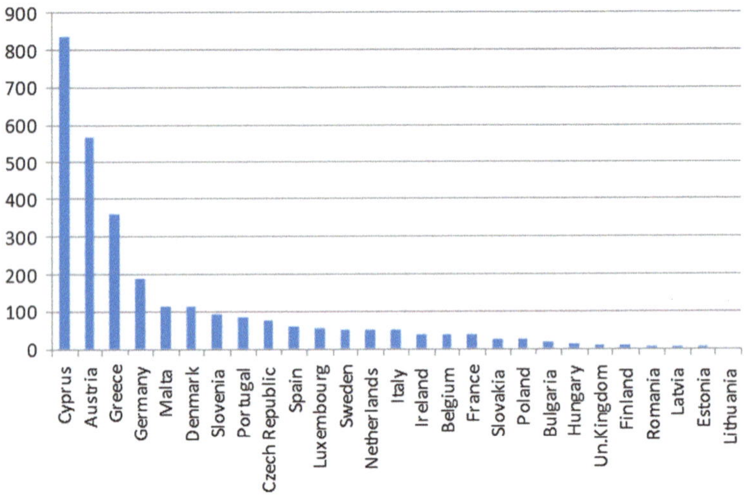

Fig. 2 Solar thermal installed capacity in EU-27 countries (2011) (kWth/1,000 inhabitants) (*Source* Own elaboration from EurObserv'ER [17] data)

Cypruswith 834 kWth/1,000 inhabitants, followed by Austria (567 kWth/ 1,000 inhab.), Greece (361 kWth/1,000 inhab.), Germany (186 kWth/ 1,000 inhab.) and Malta (114 kWth/1,000 inhab.). Countries with a below-average installed solar thermal capacity include Lithuania (2 kWth/1,000 inhab.), Estonia

(2 kWth/1,000 inhab.), Latvia (5 kWth/1,000 inhab.), Romania (5 kWth/1,000 inhab.), Finland (7 kWth/1,000 inhab.), and the UK (9 kWth/1,000 inhab.).

The average annual growth rate in installed STE capacity in the period 2005–2011 was 20.82 % for the EU-27 (Table 1). Ireland had the highest percentage growth rate, although its initial installed capacity was very low. The Czech Republic, Poland and Portugal also had annual growth rates around 40 %, while that for Spain, Italy and Belgium was around 30 %. The lowest rates correspond to Greece and Cyprus (5 %), both of which had a high initial level of installed capacity per capita. The market size in the EU-27 countries more than tripled between 2002 and 2008. However, according to EurObserv'ER [17] data, the 2008 economic slowdown resulted in decreases of 10 % in 2009, 13.4 % in 2010 and 9.3 % in 2011. As pointed out in ESTIF [10], some markets, particularly in southern European countries such as Italy, Spain, Portugal and Greece, have been traversing difficult economic times in recent years, while others, such as Germany and Poland have continued to experience impressive growth.

Table 1 also outlines the percentages of 2020 objectives already installed in 2011 with respect to NREAP and national RES industry roadmap targets, respectively. The last row shows that the UE-27 has already reached almost 43 % of the NREAP target and 27 % of the national RES industry roadmap target. Overall, some countries have already reached their NREAP target, while others are far from doing so. Among the countries which still have considerable progress to make before meeting their targets are Italy, Spain, France and Poland, which set very high targets for themselves. It is worth noting, that no country has yet reached the national RES industry roadmap previsions.

3 Solar Energy Regulations

Eight countries in the EU-27 have introduced quantitative targets in their national regulations for the integration of RES in buildings. Spain was the first European country to introduce the obligatory use of RES in buildings. Under their Technical Building Code (Royal Decree 314/2006), new and rehabilitated buildings must, among other energy requirements, incorporate STE technology. A minimum energy contribution is established depending on the climatic zone and the sanitary hot-water demand [34].

According to the information provided in the Ecofys report [18], Bulgaria has also established quantitative requirements for the use of RES. Since the Energy from Renewable Sources' Act came into effect (May 2011), new buildings and buildings under renovation require that at least 15 % RES be used. Nevertheless, Bulgaria's renewable energy act does not clarify how this requirement must be implemented, monitored or controlled. Greece has also recently introduced building regulations concerning the use of RES. From January 2011, 60 % of hot water consumption in new buildings should be provided by solar panels. In Portugal, Decree 80/2006 requires at least 1 m^2 of solar thermal panels to be installed in buildings for each conventional inhabitant of the building (limited to 50 % of total available rooftop surface area) for the production of sanitary hot water. In Slovenia, regulations concerning energy efficiency in new buildings and those undergoing major renovation require that a minimum of 25 % of the total energy demand of a building must be covered by the installation of RES-based systems such as solar hot water. Also in Sweden there is building obligations.

In Germany, the Promotion of Renewable Energies in the Heat Sector Act requires building owners to use RES for H&C purposes in new buildings with an effective surface area greater than 50 m^2. Depending on the technology used, RES must provide between 15 and 50 % of the H&C demand of the said building. If STE is used, 15 % for all new buildings or 0.04 m^2 of solar collectors per m^2 of total effective area for single/double family houses must be installed.

Law 244/2007 in Italy set a mandatory quota of energy consumption in new buildings to be covered by RES starting from January 1, 2009. However, the starting date has been postponed several times; in 2011, a new legislative decree postponed the obligation to comply through until May 31, 2012. It is anticipated that 20 % of the energy consumption for H&C plus water heating in new buildings should be covered by RES, with this increasing to 35 % from January 1, 2014.

In addition to these national regulations, some countries have adopted targets in regional or even local regulations for the integration of RES in buildings. The city of Barcelona has pioneered the implementation of city-based solar energy regulations in Europe, with its Solar Thermal Ordinance which came into effect in August, 2000. After this, many Spanish town councils approved bylaws in their jurisdictions. As pointed out in Pablo-Romero et al. [34], currently more than 50 municipal bylaws have been approved that cover around 20 % of the Spanish population. Other cities in Europe have also introduced solar energy-based regulations in their ordinances. According to ESTIF [11], following the example of the town of Carugate in Italy, local solar energy-based obligations are enforced in a number of small municipalities around Milan. In Ireland, a number of progressive local authorities introduced building energy standards which require a substantial increase in the energy performance of new buildings (between 40 and 60 % reduction in energy usage). In Portugal, the new Portuguese buildings code requires building owners to install STE systems or some other form of renewable energy that provides a similar energy saving. At a regional level, Denmark has established targets for the use of renewable energies [2]. Municipalities are obliged to establish heating plans based on feasibility studies with alternative/RES systems with the aim of reducing dependency on oil-based heating systems.

In Slovakia, there is no regulation concerning the minimum share of RES utilisation in buildings, but there is an obligation to evaluate the possibility of RES utilisation in new large buildings. And finally, in Lithuania, one directive has been passed concerning the energy efficiency of buildings, which fixes the minimum required share of RES in the heating of buildings; this is due to commence in 2014.

4 Subsidies

STE is one of the main technologies commonly supported through public subsidies in the EU-27. As an instrument of energy policy, subsidies have the interesting property that they are easy for public authorities to manage. A simple scheme based on an application, a check-comparison with standard legal requirements, and a possible review by authorities is all that is required to implement this type of policy measure. Usually, the level of subsidy is fixed as a percentage of the total cost of the investment, with legal maximums settled in nominal terms. Subsidies consisting of a flat-rate for aid are less common (as in Belgium or Ireland, among others).

The granting of subsidies is not done automatically, but is subject to public sector verification. Often, installers are required to be licensed by the responsible administration in order for the investment is subsidized. When subsidies are applied, they may vary depending on whether the beneficiaries are from the public or private sector, or, if they are from the latter, if they are households or companies. Sometimes the benefit is conditional on the new installation replacing a previous one that did not use RES (this is the case in Denmark, for example). When different levels of government are involved (e.g. central and regional), the subsidy is usually co-financed by the governments implicated. This occurs in Austria, Belgium, Italy and Spain, among others.

Almost all EU countries, to a greater or lesser extent, provide specific subsidies for STE technologies for H&C. The technology that is mainly promoted is that of solar collectors. However, the nature of the direct subsidy scheme can be very different between countries. In most cases, these subsidies are involved in promotion programs that are renewed annually. In other cases, the subsidies are included in specific programs with a specific date of expiration (that can or cannot be renewed upon completion). In any case, it should be noted that both the number of promotion programs as well as the amount of funding allocated have decreased in most EU countries during the past year, as these measures are very sensitive to the economic situation of the day.

Table 2 offers an overview of subsidies and direct grant schemes for EU-27 countries. A relationship can be seen between the type of scheme eligible for subsidy in each country and its endogenous nature.

5 Tax Incentives

In addition to subsidies, STE-based systems are often promoted through a range of tax incentives. The main tax incentives used by the EU-27 countries are deductions, exemptions and reduced tax rates. Table 3 provides an overview of the use of these tax incentives.

Table 2 Overview of subsidies to promote RES Solar Thermal H&C in the EU-27

Member state	Subsidies/Direct grants
Austria	Depending on collector size: Collectors $\leq$ 100 m^2: €100/m^2 for standard collectors, €150/m^2 for vacuum collectors. The support is granted "de-minimis" and is limited to a maximum of 30 % of the environmentally-relevant costs
	Collectors > 100 m^2: "De-minimis": 20 % of all environmentally-relevant investment costs, and up to 40 % of additional costs
	In both cases, costs may not exceed €200,000 over three fiscal years
Belgium	Flemish Region: €75/m^2 (minimum €525–maximum €1,500). In 2012, this premium was scheduled to increase to €200/m^2
	Walloon Region: €1,500 per installation with a surface of 2–4 m^2 + €100 for every additional m^2
	Brussels Capital Region: €3,000 per installation with a surface of 2–4 m^2 + €200 for every additional m^2 (limited in any event to 30 % of the invoice for the supply and installation)
Bulgaria	Included in general measures for all RES: Incentive grants of 15–20 % of the loan amount were provided to borrowers for solar installations for heat generation
Cyprus	Central solar active domestic hot water systems: 45 % of the eligible budget under the restriction of maximum eligible expenditures (maximum €20,000)
	Central solar domestic H&C systems: 55 % of the eligible budget under the restriction of maximum eligible expenditures (maximum €20,000 for heating systems and €50,000 for H&C systems)
Czech republic	Solar thermal collectors, hot water production only: €2,270 for family homes, €1,030 (per dwelling unit) for multiple-dwelling houses
	Solar thermal collectors, hot water production and pre-heating: €3,300 for family homes, €1,445 (per dwelling unit) for multiple-dwelling houses
	RES projects: enterprises can apply for investment grants for RES projects. The minimum subsidy is €20,640. The maximum is a percentage of the eligible costs: in general, small enterprises are entitled to 50 %, medium-sized enterprises to 40 % and large enterprises to 30 %
Denmark	Subsidies for purchasing and installing approved heating systems to replace scrapped oil-fired boilers by solar heating in combination with a connection to district heating: €1,345 for the establishment of district heating units and 25 % of investment costs (given for single unit houses)
Estonia	Currently there is no direct subsidy scheme for solar thermal RES for H&C in Estonia. Nevertheless, the Environmental Investment Centre organizes financing of various environmental projects for environmental investments and for the development of projects supporting sustainability and restoration of the environment
Finland	Energy grants for households: up to 25 % of eligible costs (materials and equipment only)

(continued)

Table 2 (continued)

Member state	Subsidies/Direct grants
France	The amounts and the type of subsidies for investment costs vary between regions, and can reach up to 50 % (for households)
Germany	Solar collectors < 40 m^2: €120/m^2 for existing buildings. The base support level for new buildings (building application after January 1, 2009) is 25 % lower
Greece	STE-based installations can benefit from cash grants, which cover part of the expense for the investment project, and a leasing subsidy which covers part of the payable installments relating to the use of new mechanical and other equipment. Depending of the geographical areas, the subsidies can cover between 20 and 40 % of the eligible costs
	For new installations of large STE-based systems, subsidies are set depending on the fulfillment of some requirements. These subsidies can cover between 40 and 50 % of the eligible costs
Hungary	Subsidies of up to 50 % of the investment cost are provided for STE-based installations for the public sector. Industrial installations can expect support of up to 50–60 % of the subsidy rate, while municipal buildings can receive up to 85 %
Ireland	Home owners can apply for a grant of €250/m^2 for flat plate collectors and €300/m^2 for evacuated tube collectors, up to a maximum of 6 m^2. In December 2011, this scheme was re-launched. For solar heating systems, the fixed cash grant based on this new scheme is €800
Italy	A 30 % subsidy is offered to public building owners and municipal gas distributors for investments in solar heating systems
Latvia	Investment projects involving the construction of new combined heat and power (CHP), utilizing RES can be supported: the beneficiary must contribute at least 25 % of the amount of the total eligible cost
Lithuania	Currently there is no direct subsidy scheme for STE-based RES for H&C in Lithuania. By and large, for all RES, the Special Program for Climate Change Fund can provide support for RES projects in the form of subsidies. These subsidies are granted to persons or legal entities not engaged in economic and commercial activities, rural communities, managers of the public entities (with the exception of income-generating projects), legal persons engaged in economic and commercial activities (with the exception of income-generating projects). Persons may be granted a subsidy solely for the implementation of small-scale projects
Luxembourg	For STE-based systems for domestic hot water, subsidies of 50 % up to a maximum of €3,000 in single-family homes and up to a maximum of €15,000 in the case of a multi-family building. For systems also used for space heating, the subsidy increases to a maximum of €5,000 for single-family homes and to a maximum of €15,000 per multi-family building. An additional €300 is available for combined wood- and STE-based heating systems
	For companies, a STE-based system with or without heating support is eligible for a subsidy of 33 % of the investment cost
	Municipalities may benefit from incentives up to 40 % of the cost of energy planning or renovation projects
Malta	Energy installations related to the use of STE-based water heaters and collectors receive capital grants of 40 % of the investment costs to a maximum of €560

(continued)

Table 2 (continued)

Member state	Subsidies/Direct grants
The Netherlands	In 2008, one subsidy scheme was established to stimulate the installation of STE-based boiler heating systems in households; this was applied to existing dwellings only (built before 1 January 2008) and targets private owners and social housing associations. The grant was set at €200/GJ for boilers with a solar collector surface smaller than 6 m^2, and €180/GJ for larger boilers
	The scheme was discontinued in December 2011 and there are currently no new (similar) schemes expected
Poland	Support for the purchase and installation of solar panels for hot water heating in buildings assigned or used for residential purposes: 45 % of the costs that qualify under the scheme. Business activities do not qualify for subsidies
	Support has also been provided for projects in which solar collectors are aggregated across hundreds of installations within the territory covered by one municipality: 70–85 % of the qualified costs
Portugal	In 2009, a special program to promote the installation of solar thermal panels in homes and buildings was launched. The program applied to the purchase of a solar thermal "kit", comprising panels and ancillary equipment, installation, yearly maintenance for six years, and a six-year guarantee. Up to 65 % of the investment amount was subsidized. The program expired on December 31, 2009
	In June 2010, a support plan to subsidies investments on solar thermal installations was established for small- and medium-sized businesses, as well as social entities and sporting associations. Investments could be financed by up to 40 % for medium-sized businesses, and up to 45 % for small businesses
	In July 2010, another scheme was launched, targeting social entities and sporting associations. The amount subsidized depended on the region in question: northern regions could receive benefits of up to 70 % of the invested amount, whereas Lisbon and the Algarve could receive up to 50 %
	Both programs expired in late 2010
Romania	Subsidies are aimed both at residential installations, as well as installations in the public and administrative sector
	The residential part of the programme is targeted at the owners of detached houses and the co-owners of semi-detached houses. The amount subsidised is up to €1,430 for solar collectors
	For installations in the public and administrative sector, the subsidised amount is up to 90 % of system costs
Slovakia	Households that install solar panels are eligible for a subsidy under specific criteria
	Amount of subsidy for solar collectors: €200/m^2 for up to 8 m^2; €50/m^2 for installations above 8 m^2; €300/m^2 for apartment houses, if area of solar collectors for one apartment is less than 3 m^2
	Requirements for supported solar collectors are the following: efficiency should be at least 525 kWh/m^2 per year for installations completed as of 2010; and certificate of Solar Keymark (issued in EU) is necessary

(continued)

Table 2 (continued)

Member state	Subsidies/Direct grants
Slovenia	Promotion of solar collectors in households: flat plate systems may receive up to 25 % of the recognized investment costs, but no more than €150/m^2; vacuum systems up to 25 % of the recognized investment costs, but no more than €200/m^2. The size of the STE-based system that receives the subsidies is not capped. This support measure requires the use of certified installers
Spain	Subsidies are distributed by regional governments that have been assigned this task by the Federal Government. The requirements have remained the same, but each region decides what kind of installation is the most suitable to receive subsidies. The users of STE-based installations (households, firms and public sector) can benefit from 35 to 70 % of all environmentally-relevant investment costs, depending on the region
Sweden	Permanent residences and non-commercial buildings receive a subsidy of €0.275/kWh per year. The maximum annual subsidy is €825 for one-family homes, and €550 for individual apartments. The grant is limited to a maximum of €27,500 per property
	Commercial buildings are also eligible for the €0.275/kWh subsidy, which is granted based on the total annual energy production. The maximum grant is 30 % of the total costs for material and labour, and is paid in the form of a tax refund
	These subsidies for solar heating were in force until 2011
UK	The Low-Carbon Buildings Program provides grants for the installation of micro-generation technologies in a range of buildings that include households, community organizations, and the public, private and non-profit sectors. STE-based hot water systems may receive an overall maximum of €467 or 30 % of the relevant eligible costs. This program was extended to April 2011
	The Renewable Heat Premium (RHI) payment provided vouchers to householders to assist them in the purchase of renewable heating technologies prior to the extension of the RHI scheme to cover the domestic sector. The scheme specifically targets those households that are not connected to the gas grid. Applicants need to ensure that basic energy efficiency measures are in place to qualify and that Micro-generation Certified Scheme (MCS) certified equipment is used and installed by MCS certified installers (or equivalent standard). Subsidy levels for STE-based systems are €300/unit

Source Own elaboration

Table 3 Member States that use tax incentives to promote solar thermal energy use in the EU-27

Member State	Deductions	Exemptions	Reduced tax rates
Austria	✔		
Belgium	✔		
Bulgaria			
Cyprus			
Czech Republic		✔	
Denmark		✔	
Estonia			
Finland		✔	
France	✔		✔
Germany			
Greece	✔		
Hungary			
Ireland			
Italy	✔		✔
Latvia			
Lithuania			
Luxembourg			
Malta			
The Netherlands	✔		
Poland			
Portugal	✔		✔
Romania			
Slovakia			
Slovenia			
Spain	✔		
Sweden		✔	
UK	✔		✔

Source Own elaboration

5.1 Deductions

At present, nine European countries (Austria, Belgium, France, Greece, Italy, The Netherlands, Portugal, Spain and the UK) offer different direct tax deductions to encourage the use of STE.

The Austrian Income Tax Act defines energy saving measures as special expenses for which tax allowances can be reclaimed. These measures also include expenses for heat pumps, solar thermal and bioenergy systems. These expenses can be deducted from the taxable income.

In Belgium, a tax reduction of 40 % of the investment cost was introduced on personal income tax, with a maximum of €2,830 per installation. The measure can be accumulated with an investment premium. The regulation makes support conditional on the use of certified equipment and certified installers. France also has a tax deduction scheme dedicated to promote RES in the H&C of private households. Tax deductions cover 13–45 % of costs of eligible equipment and can be combined with other public fiscal incentives. The tax deduction has a ceiling of €8,000 for individuals and €16,000 for married or registered couples. There is also a supplement of €400 per dependant persons. Deductions are available both for new and renovated houses.

In Greece, a 20 % deduction of up to €700 is available on personal income tax for money spent on the installation of RES, such as solar panel systems. In Italy, personal income tax deductions up to a total of 55 % of the investment outlaid on solar thermal systems, spread over 10 years, can be obtained. To stimulate investments in RES, The Netherlands has implemented a scheme that allows Dutch companies investing in RES (including those related to STE) a deduction of 44 % on such investments from their fiscal profit up to a national maximum of €110 million per year.

Portugal established a support program for private persons to finance solar cooling plants; this was provided in the form of a 30 % tax reduction on the investment or up to a limit of €803 (valid in general for all renewable energy investments). This measure was not continued after 2011.

In Spain, a reduction in personal income tax is permitted for up to 20 % of the investment made for works carried out in the home to replace electricity, water, gas or other supply installations with those that use RES. The installation of solar panels is specifically considered as a way to contribute to the production of the sanitary hot water demand in homes. The maximum allowance per tax period is €1,350. When the amount is greater than that allowance, the deduction can be applied across successive tax periods, up to a total of €4,000. Furthermore, for investments in environmentally friendly installations, the deduction for companies, whose financial year coincides with the natural year, is 2 % for investments made in 2010 and 2011, and 8 % for investments made in 2012.

In the UK, the full cost of an investment in designated energy-saving technology plants and machinery can be written off against the taxable profits of the period in which the investment is made. This scheme is open to all businesses that pay UK corporation or income tax, regardless of their size, sector or location.

5.2 Exemptions

Four countries (Czech Republic, Denmark, Finland and Sweden) have implemented tax exemptions to promote RES for H&C, including the installation of STE-based systems.

In the Czech Republic, the Property Tax Act stipulates that property tax consists of a tax on land and a tax on buildings. Land that serves solely for the purpose of improving the environment (RES projects meet this criteria) is exempted from the land tax. Furthermore, the Act stipulates that buildings in which a fossil fuel heating system was replaced by a RES system using solar energy are exempt from the tax on buildings for a period of five years from the year the change occurred.

The Income Tax Act of the Czech Republic stipulates that income generated from solar installations is exempt from taxes.

In Denmark, the use of RES for heating is supported through tax exemptions. Solar heating plants are exempt from both energy and CO_2 emissions taxes. In Finland, taxes on heating are based on the net carbon emissions from the input fuels. These are defined as zero for RES, meaning that solar heating plants are exempt from CO_2 emissions taxes. Sweden has a similar exemption from both energy and CO_2 emissions taxes.

5.3 Reduced Tax Rates

While the use of reduced tax rates to promote RES is an instrument employed largely to promote the use of RES such as biofuel, only four countries (France, Italy, Portugal and the UK) have introduced reduced value-added tax (VAT) rates on components and materials required for eligible H&C systems based on solar energy technologies.

In France, tax deductions are available for energy-saving equipment for private or collective dwellings. Generally, equipment is eligible for tax deduction if purchased and installed at the principal dwelling. Installation costs are not included in the eligible costs, but this measure is complemented by a reduced VAT of 5.5 % on material and installation costs. This extra incentive does not cover the installation of solar panels with a production capacity larger than 3 kW.

Consumers in Italy can also benefit from a reduced VAT (10 % instead of the usual 20 %) in the case of the refurbishment of a house when this includes the installation of solar thermal systems. In Portugal, a reduced VAT rate of 13 % for purchased equipment was applicable to all renewable energy products. However, this program expired at the end of 2011.

Finally, in the UK, a reduced VAT of 5 % is charged on certain energy-saving materials if these are used in non-business buildings or village halls. This reduced VAT covers the installation of solar panels, as well as all installations (including conventional systems) if they are carried out in the sole or main residence of a person over 60 years of age.

6 Other Support Measures: Low-Interest Loans and Feed-in Tariffs

Some countries have also implemented other measures to promote STE use. These include financial support measures and systems of support for the power generated by STE-based technology.

6.1 Low-Interest Loans

Financial support measures are also used to promote STE use, with several EU-27 countries having offered the possibility of reduced-interest loans to fund the implementation of systems based on STE. This measure has only recently been introduced but has had significant uptake in a very short time. In this way, a total of eight countries (France, Germany, Malta, Portugal, Slovakia, Slovenia, Spain and the UK) have made financial aid programs available under a variety of forms.

France has implemented a zero-interest loan program with no restriction on the budget dedicated to it. The loan is available to all land owners for the financing of construction works for energy saving actions in their principal dwelling or in a dwelling leased to tenants. The loan is capped at €20,000 or €30,000, depending on the type of work undertaken. The pay-back time of the loan is 10 years.

Germany has offered low-interest loans since 2007 for the financing of solar thermal plants for H&C. It offers long-term, low-interest loans with fixed interest rates that are 1.04–1.98 % lower than standard lending rates and a grace period before payments commence. The Maltese government has also made soft loans available for investments in renewable energy, including STE. These loans are only available for owners of tourist accommodations. In Portugal, low interest rate loans have been available from 2007 to 2013; these include a 5 year term with a two-year grace period and up to a maximum amount of €750,000 for the installation of systems using any type of RES for H&C.

In Slovakia, loans are granted at below-market interest rates. In this case, the beneficiaries are private companies that undertake investments in solar thermal energy systems for H&C.

Slovenia also has a low-interest rate loan program to finance projects based on the use of RES for H&C. These loans can cover between 50 and 90 % of the predicted investment cost.

In Spain, individuals, small- and medium-sized businesses, groups of property owners, town councils and other legal entities, with the exception of large companies, can apply for an 11 year loan at an interest rate of Euribor +0.30 % with a 1 year grace period to finance solar thermal energy projects.

The UK has a zero-interest loan program offered under the Energy Efficiency Loans Scheme, which is mainly directed at small- and medium-sized businesses. Further support is available under the Home Loans Scheme (launched in Scotland), which offers interest-free loans to the domestic sector for the installation of renewable heat and electricity technologies.

6.2 Feed-in Tariffs

Feed-in tariffs are used in a small number of EU-27 countries, comprising France, Luxembourg, The Netherlands and the UK.

In order to support renewable energy use in H&C, France created a “heating fund” in 2008. Varying between regions, project size, and sector (agricultural/ industrial, public sector), this fund provides operators of H&C installations with a feed-in premium. Eligible projects include those using STE. Installations must comply with certain levels of production and environmental performance to benefit from the incentive. The budget for this fund was €1.2 billion for the period 2008–2013.

Within the regulations for Luxembourg’s feed-in tariff, an additional premium for the use of solar energy heat was introduced. This premium is paid for each MWh of commercialized heat produced, with amounts ranging from €37/MWh to €42/MWh. Commencing in 2012, RES for heating in The Netherlands became eligible for support under a scheme called SDE+, in which heat produced by renewable energy projects was subject to receipt of a feed-in premium. Eligible technologies include heat derived from STE.

Finally, to help meet the UK’s overall RES targets, the government there launched the Renewable Heat Incentive, which consisted of a feed-in tariff for heat produced from RES. These tariffs vary depending on the installation’s size and the technology employed. The renewable heat incentives began operating in April 2011.

7 Discussion

Table 4 summarizes government-implemented measures that have been used to promote solar thermal energy use by the EU-27 countries, and the main outcomes derived for the period 2005–2011. Column 2 shows the countries which have implemented regulations promoting the use of solar energy, Column 3 those that have used subsidies, Column 4 shows those that have implemented tax incentives, Column 5 financial measures and Column 6 Feed-in tariffs. Column 7 shows the installed capacity, Column 8 the annual growth rate and finally, Column 9 the percentage of installed capacity related to NREAP target.

In general it can be seen that measures designed to promote STE use have not been implemented in the same way by all 27 EU countries. Notable differences can be seen in terms both of the regulations and in the fiscal and financial measures employed, which in most cases are related to the specific features of a given country given that STE-generated heat must be used locally [23]. In this sense, large differences in solar radiation energy can be found not only within the 27 EU Member States, but also within different regions of the same country; these differences are determined by latitude, continentally, terrain and local climatic variations. According to Šúri et al. [40], the highest potential for solar-based electricity generation, and therefore for STE, is in Portugal, Malta, Cyprus, most parts of Spain, Italy, Croatia, southern France and Corsica, and Greece. In contrast, Scotland, northern Sweden and Finland have the least favourable conditions for the implementation of STE-based technologies.

Table 4 Summary of policy measures to promote solar thermal energy use in EU-27 countries. Improvements in national targets

	Regulation	Subsidies	Tax incentives	Financial incentives	Feed-in tariffs	Installed capacity (MWh) 2011	Annual growth rate 2005–2011 (%)	Objective reached NREAP (%)
Germany	✔	✔		✔		10,663,800	13.54	133.69
Austria		✔	✔			3,336,430	10.66	106.67
Greece	✔	✔	✔			2,862,595	5.03	69.35
Italy	✔	✔	✔			2,052,716	27.56	11.13
Spain	✔	✔	✔	✔		1,914,913	30.77	25.57
France		✔	✔	✔	✔	1,659,000	17.61	15.39
Poland		✔				632,561	39.57	10.75
Portugal	✔	✔	✔	✔		613,773	38.32	32.99
The Netherlands		✔	✔		✔	585,445	5.10	218.91
Czech Rep.		✔	✔			554,378	44.24	216.71
Cyprus		✔				489,591	5.75	46.54
Denmark	✔	✔	✔			435,681	10.20	234.18
UK		✔	✔	✔	✔	414,767	19.73	104.91
Sweden		✔	✔			332,290	10.70	476.28
Belgium		✔	✔			291,513	31.77	439.83
Slovenia	✔	✔		✔		132,331	10.22	54.19
Ireland	✔	✔				121,590	80.64	52.28
Slovak Rep.		✔		✔		102,445	14.73	29.37
Hungary		✔				89,384	17.31	9.37
Romania		✔				86,100	12.22	–
Bulgaria	✔	✔				79,450	17.35	32.54
Malta		✔		✔		33,287	16.16	103.72
Finland		✔	✔			27,666	17.84	–
Luxembourg		✔			✔	18,976	12.46	20.15
Latvia		✔				7,945	27.44	34.16
Lithuania						4,375	19.46	4.18
Estonia						2,009	23.22	–

Source Own elaboration

It is worth noting that most of the EU-27 countries that have implemented solar regulations have high solar radiation levels in general. Furthermore, four of the five countries with the largest STE capacity have implemented strategies to promote its use. These regulations seem to have had considerable impact. In Spain, for example, most of the solar energy-based installations that have been installed are directly linked to the Technical Building Code regulations, which accounted for 83 % of new installations in 2010 [4]. Likewise, in Greece, recent regulatory changes seem to have kick-started new growth in this sector. In spite of efforts at a governmental level, the strategies implemented have not been sufficient to reach the established objectives. For example, regulations to encourage the installation of solar energy-based systems normally only apply to new or rehabilitated buildings, which constitute a small proportion of the total number of buildings. So the effects of this approach are limited. Impediments to the full-scale implementation of RES are accentuated even more so when the deep crisis affecting the construction sector in countries as Spain, Greece or Portugal is taken into account. In this sense, the impact of legislation changes will be limited whilst the construction sector is in decline [19].

Although regulations implemented across the EU-27 countries do not appear to be sufficient for all of them to reach their NREAP targets, the regulations may be a useful tool for local governments, as through them, use of the most suitable RES for a particular territory can be promoted, thus keeping in line with the European Directive on Renewable Energy. In this sense, as pointed in Klessmann et al. [28], the implementation of solar energy use at a local level (for example, the case of Barcelona) is showing good progress.

The second of the instruments used to promote STE in the EU-27 countries is the concession of subsidies. As Table 4 shows, subsidies are the most widely used instrument employed by these countries, with only Estonia and Lithuania not ever having offered such financial incentives. According to Cansino et al. [7], the main reason for the use of subsidies is that, by reducing the high costs of investment in a straightforward manner, they encourage the adoption of specific technologies that are usually capital intensive. This is in tune with the results of Sarzynski et al. [36] for the USA, where the importance of the concession of subsidies for the establishment of new technological systems is encouraged due to the reduced costs involved. It is perhaps for this reason that, currently, subsidies are the main global instrument used for promoting the implementation of solar-based energy [41].

Subsidies also have the advantage that governments can adapt them to their own territorial characteristics, permitting each government to promote solar energy implementation in the way they best see fit; as such, there is no common pattern for the way subsidies are used. Nevertheless, although governments can regulate subsidies based on the specific characteristics of each EU country, they also introduce an important segmentation of the market and an excessively broad and complex policy development that may hinder the development of these systems.

Further to this, subsidies have the disadvantage of being closely linked to budgetary resources and therefore to budgetary constraints. In most EU countries, subsidies for renewable energy for heat have been lowered, as in the cases of Germany and Spain, or even terminated, as seen in the lower Austrian region and the Czech Republic. Also, Portugal's solar energy market has been affected by the nation's financial crisis [10]. The effects of such "stop-and-go" policy measures include significant market contraction, with attendant business closures and job losses. Therefore, in an ideal situation, promotion measures should be stable over time. However, continuity in supporting such measures seems difficult in the current context of budget cuts. In this way, the EU-27 needs to be pro-active by providing grants and support facilities for the coordination of programs and counteracting imbalances between different countries in order to apply these incentives when they are needed. Moreover, according to Krozer [29], through the use of RES, Europe has saved more money than was provided in terms of publically-funded support.

In addition to the EU support, Member States need to look for new, budget-independent STE support [5]. Among these are systems of support for the power generated by STE, over a limited time, which would simultaneously stimulate the development of more efficient installations. However, to the present time, feed-in tariffs have only had a minor impact on STE promotion given that only four countries have used this approach (Table 4).

Besides subsidies, EU countries have also adopted tax deductions, although to a lesser extent. Thirteen countries have adopted tax measures; however, as shown in Table 4, such measures have only really been adopted by countries with adequate economic capacity to do so. In this sense, and in contrast to that noted in Sarzynski et al. [36], the use of tax incentives does not appear to have less capacity to stimulate the implementation of solar installations than do subsidies. These authors pointed out that those states in the USA that have established income tax incentives, have not demonstrated an enhanced capacity to develop solar technologies. According to Pablo-Romero et al. [34], several reasons may be behind this occurrence. First, tax deductions involve the need to make a previous investment, for which there may not be sufficient liquidity. Second, tax deductions are not sufficiently attractive because they are quantitatively meagre. And finally, for tax incentives to have any effect, they must remain in operation for a sufficient period time.

Very few studies, however, have examined specifically the effectiveness of these measures on the adoption of RES. Most econometric studies assessing the effectiveness of renewable energy policies to date have focused on state-level policies in the United States, particularly Renewable Portfolio Standards (RPS) [1, 8, 33, 37, 46]. With respect to European countries, the studies of Marques et al. [31, 32] are noteworthy. Most of these cited studies, however, address policy outcomes containing a binary variable. To our knowledge, the only study to apply rigorous econometric methods to this problem in Europe is the study by Jenner et al. [27]; however, that study refers only to electric renewable energy and the effect of feed-in tariffs. Thus, it is not possible to adequately assess the effectiveness of tax measures on renewable energy, and specifically on that concerning STE.

Finally, low-interest loans have only been used sparingly by EU countries. Of the countries which have undergone a major increase in the percentage of STE,

Germany, Spain and France are the only ones that have used such loans to encourage uptake of solar energy technologies. However, as stated in Cansino et al. [7], this type of approach has the advantage that it can reduce the average cost per unit and can be easily implemented by banking institutions. Furthermore, these incentives do not lead to substantial budget increases if adequate arrangements are established with private banks. Nevertheless, public financial support is most likely necessary due to the present financial situation of many banks.

8 Conclusion

The overall use of energy for H&C systems by the household and services sectors in EU-27 countries represents a large proportion of total energy use, thereby justifying the substantial contribution that renewable energy-derived heat could make in meeting climate change and energy security objectives. Solar energy-based heat production could contribute significantly to the global energy need for heat; however even though Europe is the second largest world zone in terms of STE capacity, and has had an annual growth rate of 20.82 % in recent years, the fact is that it still only represents a small percentage of the total energy required for H&C and still far from its maximum potential.

After analyzing the energy policies of EU-27 countries, we identified 25 countries that have adopted measures to promote the use of STE, corroborating the opinion of those experts who maintain that the increased use of STE can only be achieved if it is accompanied by increased support from government authorities, due to the existence of technical, economic and even institutional barriers.

The most widespread measure to promote STE use is the provision of subsidies. Only Estonia and Lithuania have not ever offered subsidies. In our opinion, subsidies encourage the adoption of specific technologies by reducing the high costs of investment and can be adapted to the specific characteristics of each country and region therein. Nevertheless, subsidies are conditioned by budgetary constraints, meaning that the continuity of support for such measures is made difficult within the current global financial context. As such, EU-27 countries need to be pro-active in looking for new, budget-independent support for the on-going implementation of STE.

Thirteen EU-27 countries have adopted tax measures, with tax deductions used in nine of these, and exemptions and reduced tax rates used in four. These measures seem to have been relatively successful, but an analysis that examines specifically their effectiveness with respect to the adoption of STE is recommended. Their continuity also depends on the severity of budgetary constraints.

Low-interest loans have only been used by eight countries. However, the establishment of this type of action may be appropriate in the context of budgetary constraints, even though state financial support is also likely necessary due to the financial situation of many banks at present. Feed-in tariffs also play only a minor role in promoting STE uptake, although such tariffs could be a worthwhile instrument to promote greater efficiency of the installations.

Finally, nine countries have implemented regulations specifically promoting the uptake and use of solar energy. These regulations seem to have had considerable impact, particularly at a local level; however, their effects have not been sufficient for established objectives to be reached, as the regulations normally only apply to installations in new or rehabilitated buildings, which constitute a small proportion of the total number of buildings that could be brought under the umbrella of such programs.

Acknowledgments The authors wish to acknowledge the financial support received by the Cátedra de la Economía de la Energía y del Medio Ambiente (Chair of Energy and Environmental Economics) from the Fundació Roger Torné.

References

1. Adelaja, S., Hailu, Y.G.: Effects of Renewable Portfolio Standards and Other State Policies on Wind Industry Development in the U.S. Michigan State University, East Lansing, MI (2008)
2. Annunziata, E., Frey, M., Rizzi, F.: Towards nearly zero-energy buildings: the state-of-art of national regulations in Europe. Energy J. (2013). doi:0.1016/j. 2012.11.049
3. Arvizu D. et al.: Direct solar energy. In: O. Edenhofer et al. (eds.) IPCC Special Report on Renewable Energy Sources and Climate Change Mitigation. Cambridge University Press, Cambridge and New York (2011)
4. ASIT: La actualidad de la industria solar térmica. Asociación Solar de la Industria Térmica, Madrid (2012)
5. Buerger, V., Klinsli, S., Lehr, U., et al.: Policies to support renewable energies in the heat market. Energ. Policy **36**, 3150–3159 (2008)
6. De Jager, J., Klessmann, C., Stricker, E., et al.: Financing Renewable Energy in the European Energy Market. Ecofys, Belgium (2011)
7. Cansino, J.M., Pablo-Romero, M.P., Román, R., et al.: Promoting renewable energy sources for heating and cooling in EU-27 countries. Energ. Policy **39**(6), 3803–3812 (2011)
8. Carley, S.: State renewable energy electricity policies: an empirical evaluation of effectiveness. Energ. Policy **38**(8), 3071–3081 (2009)
9. EREC: Mapping Renewable Energy Pathways towards 2020. EREC, Belgium (2011)
10. ESTIF: Solar Thermal Markets in Europe. Trends and Market Statistics 2011. European Solar Thermal Industry Federation, Brussels (2012)
11. ESTIF: Best Practice Regulations for Solar Thermal. ESTIF, Bruxelles (2007)
12. EuroACE: Current Financial and Fiscal Incentive Programs for Sustainable Energy in Buildings from Across Europe. EuroACE, Belgium (2009)
13. European Commission: Members States' progress reports on 2011. European Commission, Belgium (2013a)
14. European Commission: Taxes in Europe. Database. European Comissión, Belgium (2013b)
15. Eurostat: EU Energy in Figures: Statistical Pocketbook. European Commission, Belgium (2012)
16. Eurostat: Data on 'Energy statistics'. http://epp.eurostat.ec.europa.eu/portal/page/portal/energy/data/database (2013)
17. EurObserv'ER: EurObserv'ER database. EurObserv'ER (2013)
18. Höhne, N., Geurts, F., Teckenburg, E., et al.: EU Climate Policy Tracker. Ecofys, Belgium (2011)

19. Fantidis, J.G., Bandekas, D.V., Potolias, C., et al.: Financial analysis of solar water heating systems during the depression: case study of Greece. Inzinerine Ekonomika (Eng. Econ.) **23**(1), 33–40 (2012)
20. IEA: Energy Efficiency Policies and Measures. Database. IEA, Paris (2013)
21. IEA: Solar Energy Technology Perspectives. IEA, Paris (2013)
22. IEA: World Energy Outlook 2012. Renewable Energy Outlook. OECD/IEA, Paris (2012)
23. IEA: Renewables for Heating And Cooling. OECD/IEA, Paris (2007)
24. IEA: Barriers to Technology Diffusion: The Case of Solar Thermal Technologies. OECD/IEA, Paris (2006)
25. Intelligent Energy Europe: Re-shaping Renewable Energy Policy Country Profiles. Intelligent Energy Europe, Utrecht (2010)
26. Jäger-Waldau, A., Szabó, M., Monforti-Ferrario, F., et al.: Renewable Energy Snapshots. Joint Research Centre, European Commission, Luxembourg (2011)
27. Jenner, S., Groba, F., Indvik, J.: Assessing the strength and effectiveness of renewable electricity feed-in tariffs in European Union countries. Energ. Policy **52**, 385–401 (2013)
28. Klessmann, C., Held, A., Rathmann, M., et al.: Status and perspectives of renewable energy policy and deployment in the European Union—what is needed to reach the 2020 targets? Energ. Policy **39**, 7637–7657 (2011)
29. Krozer, Y.: Cost and benefit of renewable energy in the European Union. Renew. Energ. **50**, 68–73 (2013)
30. Labriet, M., Cabal, H., Lechón, Y., et al.: The implementation of the EU renewable directive in Spain: strategies and challenges. Energ. Policy **38**(5), 2272–2281 (2010)
31. Marques, A.C., Fuinhas, J.A., Pires, J.: Motivations driving renewable energy in European countries: a panel data approach. Energ. Policy **38**(11), 6877–6885 (2010)
32. Marques, A., Fuinhas, J., Manso, J.: A quantile approach to identify factors promoting renewable energy in European countries. Environ. Resour. Econ. **49**(3), 351–366 (2011)
33. Menz, F.C., Vachon, S.: The effectiveness of different policy regimes for promoting wind power: experiences from the states. Energ Policy **34**(14), 1786–1796 (2006)
34. Pablo-Romero, M.P., Sanchez-Braza, A., Perez, M.: Incentives to promote solar thermal energy in Spain. Renew. Sust. Energ. Rev. **22**, 198–208 (2013)
35. Panek, A.: Impact, Compliance and Control of Legislation: Summary Report. ASIEPI, Poland (2010)
36. Sarzynski, A., Larrieu, J., Shrimali, G.: The impact of state financial incentives on market deployment of solar technology. Energ. Policy **46**(1), 550–557 (2012)
37. Shrimali, G., Kneifel, J.: Are government policies effective in promoting deployment of renewable electricity resources? Energ. Policy **39**(9), 4726–4741 (2011)
38. Solangi, K.H., Islam, M.R., Saidur, R., et al.: A review on global solar energy policy. Renew. Sust. Energ. Rev. **15**(4), 2149–2163 (2011)
39. Stryi-Hipp, G., Weiss, W., Mugnier, D., et al.: Strategic Research Priorities for Solar Thermal Technology European Technology Platform on Renewable Heating and Cooling. RHC, ESTIF and Renewable Energy House, Belgium (2012)
40. Šúri, M., Huld, T.A., Dunlop, E.D., et al.: Potential of solar electricity generation in the European Union Member States and candidate countries. Sol. Energ. **81**, 1295–1305 (2007)
41. Timilsina, G.R., Kurdgelashvili, L., Narbel, P.A.: Solar energy: markets, economics and policies. Renew. Sust. Energ. Rev. **16**(1), 449–465 (2012)
42. Tsoutsos, T., Frantzeskaki, N., Gekas, V.: Environmental impacts from the solar energy technologies. Energ. Policy **33**(3), 289–96 (2005)
43. Veerapen, J., Beerepoot, M.: Co-Generation and Renewables. OECD/IEA, Paris (2011)
44. Weiss, W., Mauthner, F.: Solar Heat Worldwide. IEA Solar Heating and Cooling Programme, Graz (2012)
45. Weiss, W., Biermayr, P.: Potential of Solar Thermal in Europe. ESTIF, Belgium (2012)
46. Yin, H., Powers, N.: Do state renewable portfolio standards promote in-state renewable generation? Energ. Policy **38**(2), 1140–1149 (2009)

Environmental Performance of Applying Alternative Energies to the Collection, Transport and MBT Plant Within an Integrated MSW Management System

David Zambrana Vásquez, Alfonso Aranda Usón, Ignacio Zabalza Bribián and Germán Ferreira

Abstract This study extends the environmental assessment of Municipal Solid Waste (MSW) management strategies using Life Cycle Assessment (LCA) methodology with the objective of evaluating the environmental implications of applying alternative energies to the collection, transport and operation of a Mechanical–Biological Treatment (MBT) plant within an integrated MSW management system. To this end, the environmental implications of the use of alternative energies in two stages of the MSW management system were taken into account: (1) collection and transportation systems for the residual household waste of MSW; and (2) the MBT plant operation including the recovery of its residual fraction. As a case study, the MSW management system of the Ecocity Valdespartera and the MBT plant in Zaragoza have been evaluated. These sites are located in the Autonomous Community of Aragon (Spain). In this study, different scenarios of alternative energy supply from renewable energy sources were evaluated at each stage of the management system. Impact assessment for each of the scenarios considered the following six impact categories: (1) acidification (kg SO_2 eq.); (2) global warming (100 years) (kg CO_2 eq.); (3) eutrophication (kg PO_4 eq.); (4) photochemical oxidation (kg C_2H_4 eq.); (5) abiotic depletion (kg Sb eq.); and (6) ozone layer depletion (kg CFC-11 eq.). These categories are contained in the CML 2 baseline 2,000 impact assessment method V2.05. The software Simapro V. 7.3.2 was also used. Results show that when alternative energy supply scenarios from renewable energy sources (RES) are included in both the collection system and the operation of the MBT, environmental benefits can be achieved in comparison to current state of affairs.

D. Z. Vásquez · A. A. Usón (✉) · I. Z. Bribián · G. Ferreira (✉)
CIRCE—Research Centre for Energy Resources and Consumption, University of Zaragoza, Mariano Esquillor 15 50018 Zaragoza, Spain
e-mail: alaranda@unizar.es

G. Ferreira
e-mail: germanf@unizar.es

D. Z. Vásquez
e-mail: zambrana@unizar.es

G. Ferreira (ed.), *Alternative Energies*, Advanced Structured Materials 34,
DOI: 10.1007/978-3-642-40680-5_12,

In this scenario, the avoided emissions are greater than those generated in most of the impact categories under study. The results identify scientific and technical processes that can be used to promote fundamental changes in the management of upstream flows of MSW in MBT plants and in its operation.

List of symbols

Variables

E_a	CO_2-eq. emissions avoided, (t)
E_g	CO_2-eq. emissions generated, (t)

Greek Symbols

β	Difference between generated and avoided CO_2-eq. emissions, (t)

Subscripts

i	MSW management system
j	Scenario for MSW treatment and recovery methods
x	Subsystem or activity within the MSW management system i
y	MSW treatment and recovery method considered in scenario j

Acronyms

LCA	Life Cycle Assessment
GWP	Global Warming Potential
IMS	Integrated Management System
IPCC	Intergovernmental Panel on Climate Change
LCI	Life Cycle Inventory
MBT	Mechanical–Biological Treatment
MSW	Municipal Solid Waste
RDF	Refuse Derived Fuel
tCO_2-eq	Tonnes of CO_2 equivalent
HHV	Higher Heating Value
LHV	Lower Heating Value
tkm	Tonnes per kilometre

1 Introduction

A significant problem faced by the administrations of various countries in the European Union (EU) and in the rest of the world is an increase in the generation of industrial and urban waste. Municipal solid waste (MSW) consists of all waste generated in urban and municipal environments, i.e., non-hazardous waste generated in homes, shops, offices and services, including waste from the cleaning of public places, parks, and recreational areas, dead pets, furniture, appliances, waste and debris from minor construction work and home repairs [1, 2]. In 2010, more than 250 million tonnes of MSW were generated in the EU-27 countries [3].

From the point of view of the waste management systems the collection stage is essential for the sustainable development of urban centres. Considering that the collection and transportation can amount to more than 70 % of the total budget of MSW management [4–6]; the energy consumption plays an important role in this budget. This energy consumption is also related to the emission of gaseous pollutants (e.g., CO_2 emissions) and contributes to the environmental impact of collection.

The collection, sorting, transportation, treatment, and ecological recycling of MSW can provide solutions that are sustainable in the medium- and long-term, taking into account the demand needs of urban development [7]. Collection systems for the separated and non-separated fractions of domestic waste at its source vary by country, region, and city [8]. Studies in the literature have focused on various aspects of the environmental impact of different waste collection systems using LCA methodology [9–17].

In Spain, the majority of MSW is collected in surface containers made of high-density materials designed to collect glass, light packaging, and paper/cardboard with separate containers for door-to-door collection of the residual household waste of MSW; in certain cases, organic waste and packaging are collected separately [18]. Other methods of waste collection, such as stationary and mobile pneumatic waste collection systems, have been implemented in certain cases; these have been studied by several authors [9, 14]. Ecoembalajes España S. A. (Ecoembes) reported that during the year 2010 the distribution of collection methods in Spain, which is linked to the chosen separation model, was as follows: 93.1, 0.6, 0.7 and 5.6 %, respectively, for surface containers or high-density material containers placed in public places, pneumatic waste collection systems, separate door-to-door collection, and underground containers [19].

On the other hand, during the last decade, MBT plants in European countries have begun to pursue important technological alternatives in managing MSW to meet the goals set forth by Directive 2008/98/EC [20–22]. Mechanical-Biological Treatment (MBT) plants use mechanical and other physical processes in combination with biological processes to recover the recyclable fractions of MSW and to obtain usable products, e.g., biogas and compost, from the organic fraction [23]. As a consequence of the mechanical processing associated with the treatment of MSW in MBT plants, the energy consumption related to these processes can amount more than the 90 % of the total energy consumption of the plant. In this context, accurate assessment of the environmental implications of both systems is essential in planning and promoting waste management methods. Such assessment would help reduce the environmental impact of lower the consumption of energy resources and reduce associated economic costs.

In an MBT plant, two types of flows, upstream and downstream, can be identified; these represent inflows (e.g., mixed household waste) and outflows (e.g., the residual fraction), respectively. In many countries, the residual fraction refused by

MBT plants is deposited in landfill without performing any material or energy recovery process [24]. A number of authors have studied the main features of MBT plants and have identified the flows and processes associated with the upstream and downstream flows of these plants [25–27]. Thus, study of the environmental implications of the residual fraction, as well as of alternative methods for material and energy recovery, can be useful in generating innovative strategies for managing the upstream waste flows of MBT plants and providing for the selection, collection, and transportation of MSW [24, 26, 28].

Studies in the existing literature focus on various aspects of the environmental impact generated at particular stages of MSW management. Among these studies, several discuss collection and final treatment systems, considering an MBT plant, and the Life Cycle Assessment methodology (LCA) [9, 10, 15, 16, 18, 29–33]. Although these publications represent specific case studies that address the methods and stages of MSW management, e.g., the design and operation of MBT plants [29, 32], the estimates reported vary considerably, indicating the need for further research in the use of alternative energies in these systems.

The objective of the current study is to analyse the environmental performance of applying alternative energies to the collection, transport and MBT plant within an integrated MSW management system. To this end, an environmental assessment method based on LCA was applied, taking into account two stages of the system: (1) management of the upstream waste flows of the MBT plant (collection and transportation systems for the residual household waste of MSW); and (2) the operation of the MBT plant itself within an integrated system aimed at recovering the MSW value in an environmentally friendly manner. The MBT plant in Zaragoza and the MSW management system of Ecocity Valdespartera, both of which are located in the Autonomous Community of Aragon in Spain, have been evaluated as a case study. In contrast to the approach adopted in this study, previous work has not focused attention on the use of alternative energies in both systems and its environmental implications.

2 Methodology

The methodology proposed is based on the method described by Aranda Usón et al. [34], which focuses on the assessment of environmental impact in terms of CO_2 equivalent emissions (CO_2 eq.) for a management system for end-of-life tyres (ELT).

Equations 1–3 and Table 1 summarise the methodology used in terms of CO_2-eq. emissions, which corresponds to the impact category of global warming or *global warming potential*. It should be noted that the methodology can be replicated in terms of other impact categories such as, for example, acidification (SO_2 eq.), eutrophication (PO_4 eq.), and ozone layer depletion (CFC-11 eq.), among others. Table 1 shows a matrix for general analysis.

Table 1 Net CO_2-eq. emissions—relationship matrix of MSW management systems and recovery scenarios

MSW recovery scenario j↓	1	2	3	←MSW i management system
1	β_{11}	β_{12}	β_{13}	
2	β_{21}	β_{22}	β_{23}	

This matrix represents the difference between the amount of CO_2-eq. emissions generated by an MSW management system i and the amount of CO_2-eq. emissions avoided in a recovery scenario j.

The elements of the matrix that are presented in Table 1, β_{ij} are calculated as follows:

$$\beta_{ij} = E_{g_i} - E_{a_j} \tag{1}$$

$$E_{g_i} = \sum_{x=1}^{x=n} E_{g_x} \tag{2}$$

$$E_{a_j} = \sum_{y=1}^{y=m} E_{a_y}, \tag{3}$$

where E_{gx} represents the CO_2-eq. emissions generated by n subsystems of the ith MSW management system and E_{ay} represents the CO_2-eq. emissions avoided by m valorisation methods considered by the recovery stage j. The matrix elements β_{ij} may have positive or negative values. If the values are positive, the emissions generated are greater than the emissions avoided in each particular scenario. If the values of the matrix elements are negative or zero, the emissions avoided are greater than or equal to the emissions generated. It is worth mentioning that a lower negative value in the matrix indicates a higher net profit in terms of CO_2-eq. emissions. The amount of CO_2-eq. emissions generated $\left(E_{gi}\right)$ includes CO_2-eq. emissions caused by the collection system; these are integrated with CO_2-eq. emissions associated with the transportation of the residual waste from collection points to the central collection station and from there to the MBT plant with CO_2-eq. emissions due to the operation of the MBT plant and with CO_2-eq. emissions associated with the processing operations in the material and energy recovery plants and the final disposal methods.

LCA methodology has been used to evaluate the environmental impacts of each of the stages under consideration. This provides a structured analysis of inputs and outputs at each stage of the life cycle of products and services [35]. The Society of Environmental Toxicology and Chemistry (SETAC)[1] defines LCA as an objective process for evaluating the environmental burden associated with a product, process, or activity. This evaluation is achieved by performing the following steps:

[1] Society of Environmental Toxicology and Chemistry: www.setac.org.

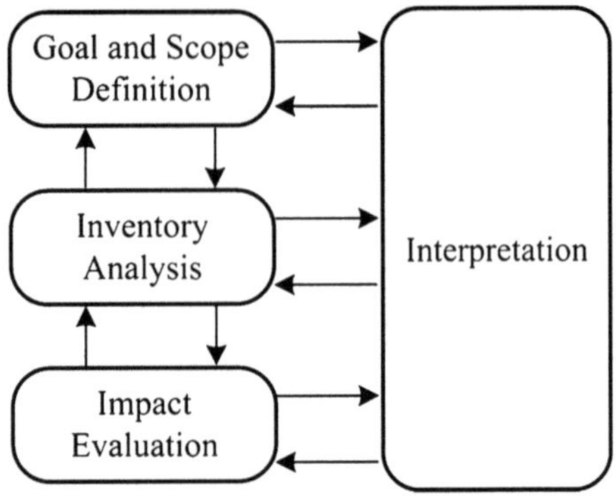

Fig. 1 Main phases of an LCA study (*Source* [37])

(1) identifying the energy, materials, and all types of waste that are released to the environment; (2) determining the environmental impact of energy consumption and of the materials released to the environment; (3) evaluating and implementing environmental improvement practises. The methodology described above is standardised by the ISO 14040:2006 [36, 37]. The assessment process is divided into four basic steps: (1) defining the goal and scope of the analysis; (2) inventory analysis; (3) impact assessment; and (4) interpretation. Figure 1 shows the main phases of the LCA study and illustrates the dynamic character and the relationship of the four phases.

ISO 14040:2006 prescribes a clear definition of the objective and scope from the beginning of the LCA studies, including system boundaries and functional units. After this stage, inventory analysis is performed by data collection within system boundaries. In this research, environmental impacts were determined from a *midpoint*-level approach. A *midpoint* impact category indicator is considered to be a point resulting from the cause-effect (environmental mechanism) relationship of a particular impact category with respect to the causative factor (a set of conditions that can lead to an impact) and an *endpoint*-level impact category indicator [35]. In other words, *midpoint* methodologies define environmental effects (e.g., acidification, global warming) without identifying the damage caused to human health and to ecosystem quality, as do *endpoint* methodologies.

Taking into account the phases of impact assessment (classification, characterisation, normalisation, and weighting) at the *midpoint* level, the characterisation factors applied to each impact category correspond to those proposed by the CML method [38]. The CML 2 baseline 2000 V2.05 method was used to quantify and compare the potential environmental impacts of the life cycle inventory (LCI) using the software tool SimaPro V. 7.3.2 [39]. A brief description of the ten impact categories considered by the CML 2 method is given in the studies of Banar et al. [40], Bravo et al. [41]. Of these ten impact categories, Cleary [42] states that the most relevant ones for MSW management are (1) acidification (kg SO_2 eq.); (2) global warming (100 years) (kg CO_2 eq.); (3) eutrophication (kg PO_4 eq.); (4) photochemical oxidation (kg C_2H_4 eq.); (5) abiotic depletion (kg Sb eq.); and

(6) ozone layer depletion (CFC-11 eq. kg.). In this research, the six above-mentioned impact categories were considered. The impact category of global warming is among the most-used categories in the studies reviewed by Cleary [42]; in various studies, it is used to quantify the environmental impact at different stages of an MSW management system [9, 15, 16, 31, 33]. Thus, the results obtained here are compared with those reported by other authors, considering primarily the global warming impact category that summarises greenhouse effect gas emissions in terms of CO_2-eq. emissions.

2.1 Goal and Scope

2.1.1 Objective

The main objective of this study is to analyse the environmental performance of applying alternative energies to the MSW management system by applying the proposed environmental assessment method, considering two stages of the system: (1) the collection and transportation system for the residual household waste of MSW; and (2) the operation of the MBT plant itself.

2.1.2 Functional Unit

In this study, the functional unit is 1 tonne of MSW collected and treated in the MBT plant under study.

2.2 Target Area and Quality Data

The MBT plant in Zaragoza receives the "all-in-one" or residual household waste of MSW from the selective collection of 62 municipalities spread across four regions of the Autonomous Community of Aragon: Zaragoza (23), Ribera Baja del Ebro (10), Campo de Cariñena (14), and Campo de Belchite (15). According to the Integrated Waste Management Plan of Aragón (G.I.R.A.) for 2009–2015, a programme of selective collection and recycling establishes the selective collection of glass, paper and cardboard packaging, light packaging, mixed fraction, drug packaging and remnants, pesticide containers, used industrial oils, used tyres, electrical and electronic equipment, and batteries. This collection is carried out by different Integrated Management Systems (IMS) that are authorised to make such collections; it also uses additional collection systems, such as door-to-door collection in the large centres of waste production and clean points [1].

Ecocity Valdespartera is located within the study area described in the city of Zaragoza. Ecocity Valdespartera has the first pneumatic waste collection system employed in the Autonomous Community of Aragon. This system is designed to collect the residual household waste of MSW, light packaging, and other organic waste generated by 9,687 households, schools, shops, bars, and restaurants that represent in total 30,000 inhabitants.

2.3 System Description

Figure 2 shows the stages included in the MSW management system under study.

A diagram of the input and output flows of the MBT plant in Zaragoza can be seen in the study of Aranda Usón et al. [34]; and a diagram of the pneumatic waste collection system in Ecocity Valdespartera was described by Aranda Usón et al. [43]. The system includes all activities and processes involved in the collection of the residual household waste of MSW at its source, the system for transportation of the waste to the MBT plant, the MBT plant operation, and the current scenario of material and energy recovery from the output flows e.g., the residual fraction.

The pneumatic waste collection system of Ecocity Valdespartera collects the residual household waste of MSW, light packaging, and other organic waste generated in residential and commercial areas. The system includes an

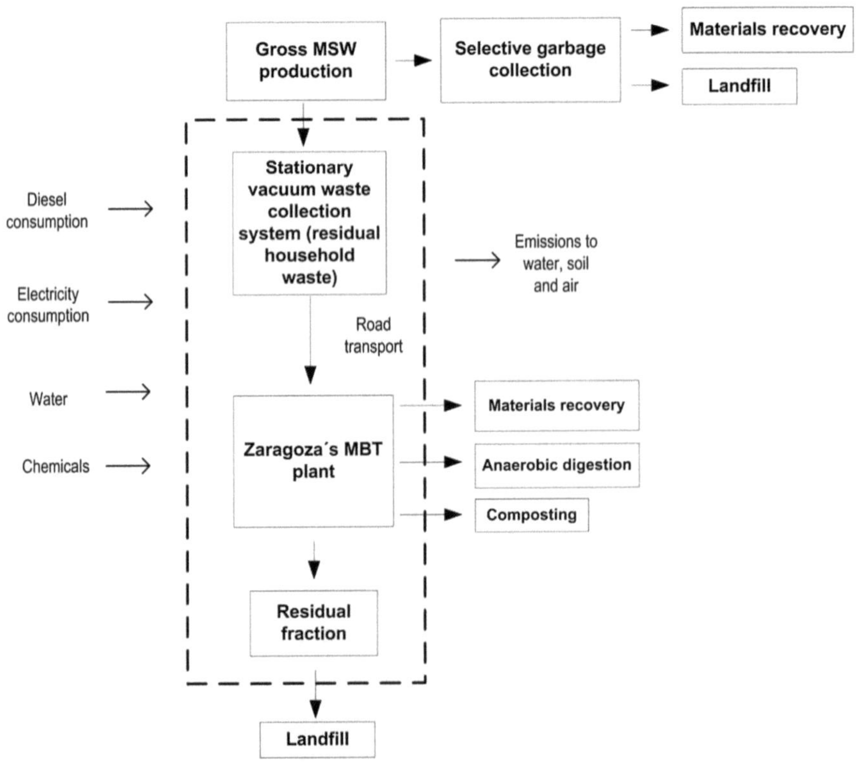

Fig. 2 Current MSW management system used in the study area

underground network of special pipes, underground storage containers, boxes, and inlets and chutes that are located inside and outside the blocks of buildings. The system was designed to have a maximum waste collection capacity corresponding to the volume of waste generated at 100 % occupancy of Ecocity Valdespartera. Several interior and exterior collection points are available to collect, as a starting point, the residual household waste of MSW and light packaging. The system is designed to collect the organic fraction of waste at its source separately from the mixed fraction. After users place bags of waste inside the boxes, the boxes are stored in underground containers, which are later transported by a pneumatic system to the central collection station according to a collection schedule. The central collection station is located at the centre of the collection network and represents the heart of the system. The central station features turbo-extractors, cyclones, waste compressors, air cleaning filters, and general equipment such as conveyors, cranes, compressed air equipment, and automated control systems. This collection method consumes electricity for collection of household waste and for compaction inside a container according to the residue fraction. When the waste container is full, it is transported by a truck as a single load to the MBT plant in Zaragoza. Energy consumption depends mainly on the programme and collection schedule; thus, it is related to the frequency of operation of the system, the amount of waste collected, and the distance from the collection points to the central station. Transportation of the waste from the central station to the MBT plant involves diesel consumption of the truck for a distance of 19.8 km.

The MBT plant in Zaragoza uses a mechanical pre-treatment system that consists of four processing lines for the residual household waste of MSW, a line for light packaging, four anaerobic digesters that have a rotating mixing system operated by agitation, closed tunnels with forced ventilation, and maturation platforms for final stabilisation. At the starting point, MSW undergoes manual selection, during which bulky items, such as electronics, furniture, and hazardous waste, are removed by operators before mechanical treatment. Using rotary screens, the waste stream is separated into two or three fractions of different sizes. Then, with the use of sieves, the waste stream is separated into fractions based on the shape and physical behaviour of its components. From the coarse fraction after mechanical separation, cardboard, newspaper, plastic and glass containers, aluminium cans, and other products can be selected. From the fine fraction and part of the intermediate fraction, the organic fractions are mechanically discarded; these fractions will be subjected to a stabilisation process through composting with or without a digestion phase. To recover the most suitable materials for recycling from the intermediate fractions obtained from the screening, inorganic waste streams pass through automatic separation operations that involve the use of electromagnets, induction by Foucault currents, infrared radiation, and other methods. The rejected waste from the described operations constitutes part of the residual fraction. Currently, at the MBT plant in Zaragoza, the rejected fraction is compacted into bales or bundles and deposited in landfill without energy recovery.

2.4 System Boundary Selection

Within the boundaries of the MSW management system under study, attention was paid to the activities of the collection and transportation system for the residual household waste of MSW, and to the operation of the MBT plant. To evaluate the environmental implications of each of these stages, the following system boundaries were selected:

- Components weighing more than 1 % of the final weight of the product are considered.
- Components representing less than 1 % of the total economic value of the product are not considered.
- No stages that contribute less than 1 % of the inventory analysis or less than 1 % of the total environmental impact are considered.

The boundaries of the analysed systems (processes, manufacturing, waste transportation and processing, and the inputs and outputs considered) are as follows:

- Second-order boundaries are defined considering the phases of production and the production of energy and raw materials for each component.
- Third-order boundaries are defined considering the infrastructure and production of materials needed for implementation. However, this study does not account for machinery or staff.

2.5 Life Cycle Inventory

Table 2 shows (1) an estimate of the major fractions of MSW generated annually in Ecocity Valdespartera; these include light packaging, residual household waste of MSW, paper/cardboard, and glass, allowing for a ratio of 100 % occupancy and an average ratio of MSW generation of 0.438 t/year [34]; and (2) the actual generation of these fractions in the year 2011.

Table 3 shows the inventory of the pneumatic waste collection system at load factors of 13 and 100 %. It should be noted that the current occupancy ratio in Ecocity Valdespartera is less than 50 %.

The load factor of a collection system represents the relationship between the current load and the estimated full design load capacity, considering the latter to be the MSW generation corresponding to 100 % occupancy of the residential area served by the system. Thus, comparing the data shown in Table 2 regarding MSW generation for 100 % occupancy and MSW generation in 2011, it is possible to estimate a load factor of approximately 13 %.

Table 4 shows the inventory of main inputs and outputs for treating 1 tonne of MSW at the MBT plant under study.

Table 2 Composition of MSW generated per year for 100 % occupancy and MSW generated in 2011

Fraction	MSW generated at 100 % occupancy (t/year)	MSW generated in 2011 (t/year)
Light packaging	2,969	397.20
Mixed fraction	16,278	2,178.00
Paper/cardboard	3,010	402.72
Glass	1,261	168.72
Total	23,518	3,146.64

Table 3 Main LCI of the MSW pneumatic waste collection system at load factors of 13 and 100 %

	Input	Unit	Load factor %	
			13	100
Collection	Electricity consumption	kWh/year	517,315	962,327
Infrastructure	Number of inlets and chutes	Unit	487	487
	Pipe network	km	13	13
	Central collection station	m^3	6,304	6,304
	Number of turbo-extractors (110 kW)	Unit	8	8
	Number of compressors	Unit	4	4
	Number of containers	Unit	8	8
	Number of cranes	Unit	1	1
	Number of spreaders	Unit	1	1
	Number of rotary separators	Unit	4	4
	Number of cyclones	Unit	4	4
Transportation	Transportation to Zaragoza's MBT plant	tkm/year	50,989	381,081

Table 4 Main LCI of the pneumatic waste collection system at a load factor of 100 % and for the treatment of 1 tonne of MSW in the MBT plant under study

Stage	Input	Output	Unit		Source
MBT plant	Water		t/t MSW	0.536	[29]
	Electricity		kWh/t MSW	80	[32]
	Diesel		kg/t MSW	0.8	[32]
	Sulphuric acid		kg/t MSW	0.789	[29]
	NAOH		kg/t MSW	0.095	[29]
	Chlorine		kg/t MSW	0.406	[29]
		Residual fraction	t/t MSW	0.519	[34]
		Compost	t/t MSW	0.120	[34]
		Biogas	t/t MSW	0.020	[34]
		Paper/cardboard	kg/t MSW	30.69	[34]
		Plastic	kg/t MSW	5.15	[34]
		Metals	kg/t MSW	29.42	[34]

The composition of the residual fraction deposited in the MBT plant landfill in Zaragoza, as well as the material fractions and by-products obtained by mechanical and biological treatment operations, were estimated using the methodology proposed by Aranda Usón et al. [34], taking into account the composition of the residual household waste of MSW that enters the plant. Recycling materials decreases emissions by reducing the use of raw materials in the manufacture of new products [32]. Indeed, producing paper and cardboard from recycled materials requires less energy than does manufacturing the same products from raw materials [10]. The amounts of material and energy recovered in the fractions were estimated taking into account the different outflows for 1 tonne of residual household waste of MSW processed at the MBT plant and its corresponding recovery rate. In the case of the MBT plant in Zaragoza, these recovery rates correspond to 78.36, 18.82, 4.79, and 79.51 % for organic material, paper/cardboard, plastics, and metals, respectively [34]. The electrical energy inputs were estimated considering the electricity generation mix by fuels in Spain shown in Table 5.

Table 5 National electricity generation mix by fuel in Spain, 2011 (*Source* [48])

Source	Unit (GWh)
Regular regime	
Conventional hydroelectric power	25,260
Hydroelectric pumped-storage	2,315
Nuclear	57,649
Anthracite	7,782
Black lignite	3,973
Tar	28,583
Steel gas	985
Natural gas	51,357
Petroleum products	1,714
Special regime	
Hydroelectric	5,332
Wind	42,007
Photovoltaic	7,019
Thermo-solar	1,777
Coal	511
Steel gas	183
Natural gas	31,806
Petroleum products	2,691
Biomass	2,936
Biogas	875
Renewable M. S. W	631
Non-renewable M. S.W	631
Other sources	171
Total	276,188

3 Results and Discussion

3.1 Alternative Energy Supply from RES Sources for the Collection and Transportation of MSW

The pneumatic waste collection system used in the case study is compared with a hypothetical conventional system of collection that uses surface containers suitable for the collected fractions in a way that is consistent with the urbanism policies of Ecocity Valdespartera. Both systems under study have been described by Aranda Usón et al. [43]. The hypothetical system was designed based on the facilities and collection services for MSW used by the city of Zaragoza [1]. The considerations and results published by Taghizadeh and Safavian [44] for the design of conventional MSW collection and transportation systems were also taken into account. The hypothetical system uses surface containers made of high-density materials and located in public places for the residual household waste of MSW, glass, light packaging, and paper/cardboard. A side-loading collection truck collects the fractions from surface and door-to-door collection containers on a specific day according to a weekly schedule and taking into account the daily generation of MSW. The complete collection service includes cleaning of the containers by a washing vehicle and a transportation distance of 19.8 km between Ecocity Valdespartera and Zaragoza's MBT.

The pneumatic waste collection system of Ecocity Valdespartera is currently operational with a load factor of approximately 13 %. From this baseline, two scenarios were studied, considering each system at load factors of 13 and 100 %. Table 6 shows a comparison of the collection and transportation systems in terms of acidification (kg SO_2 eq.), global warming (100 years) (kg CO_2 eq.), eutrophication (kg PO_4 eq.), photochemical oxidation (kg C_2H_4 eq.), abiotic depletion (kg Sb eq.), and ozone layer depletion (kg CFC-11 eq.).

The results for the conventional system and the pneumatic system, which are shown in Table 6, indicate that there is a difference in CO_2-eq. emissions for both systems, depending on whether they operate at load factors of 13 or 100 %. At low load, the pneumatic system operation has higher emission values in the first five impact categories than the conventional system. In terms of ozone layer depletion, however, the value remains constant. The results also show that neither collection system shows an advantage at low loads compared to a load factor of 100 %. This is result is observed because the pneumatic waste collection system is designed to achieve maximum performance when operating with a load factor of 100 %, this conclusion that was also reached by Punkkinen et al. [15]. With respect to the conventional system, studies reviewed in the literature show a range of emissions, for example, in the category of global warming, from 40 to 300 kg CO_2 eq./t, taking into account the distance travelled, both for collection and transportation to the MBT plant, the type of collection truck, the average truck load, the distance between the containers, and the type of driving [9, 15].

Table 6 Comparison of collection systems for the baseline scenario for treatment and recovery

Recovery scenario j↓	Conventional system	Pneumatic system	←MSW i management system
Acidification kg SO_2 eq./t			
Load factor 13 %	0.48	0.98	
Load factor 100 %	0.41	0.35	
Global warming (100 years) kg CO_2 eq./t			
Load factor 13 %	201.21	278.50	
Load factor 100 %	192.65	173.39	
Eutrophication kg PO_4 eq./t			
Load factor 13 %	0.10	0.27	
Load factor 100 %	0.05	0.04	
Photochemical oxidation kg C_2H_4 eq./t			
Load factor 13 %	0.05	0.08	
Load factor 100 %	0.04	0.04	
Abiotic depletion kg Sb eq./t			
Load factor 13 %	0.44	1.09	
Load factor 100 %	0.38	0.27	
Ozone layer depletion CFC-11 eq./t			
Load factor 13 %	0.01	0.01	
Load factor 100 %	0.01	1.43E − 03	

In the following, the effect of installing a photovoltaic solar panel on the flat roof of the transfer station of the pneumatic system as an alternative renewable energy supply is studied. The use of two biodiesel blends (B10 and B20) as alternative energy supplies in the side-loading collection trucks used to empty the containers located in public places in the conventional system is also studied.

First, given that electricity consumption associated with the operation of a pneumatic waste collection system generates a significant impact per tonne of MSW collected, the following scenario includes the installation of photovoltaic solar panels as an alternative fuel source to reduce the consumption of electricity, considering the electricity mix of Table 5, which, in turn, affects the reduction of emissions to the atmosphere. The system was designed taking into account the unique availability of the flat roof of the central station of the pneumatic waste collection system. This roof has a usable area of 600 m^2. To establish the climatic data of the chosen site, the angle values of the day, hour angle, solar declination, and height for the chosen position and the PVGIS radiation database for that location were taken into account. As a result, an approximate value of 105,838 kWh/year generated by the photovoltaic system was obtained. A nominal power of 72 kW was considered. The photovoltaic system, which uses monocrystalline silicon technology, was characterised using the Ecoinvent 2.2 database and the features presented in the study of Bravo et al. [41]. The system inventory includes solar cells, a photovoltaic solar panel, wiring and an AC/DC converter for a typical installation on deck with 30° inclination facing south. The inventory was estimated per unit of m^2 of installation from the study by Fthenakis and Kim [45].

Table 7 Comparison of the pneumatic system with a photovoltaic installation

Recovery Scenario j↓	Pneumatic system FV	Pneumatic system	←MSW i management system
Acidification kg SO_2 eq./t			
Load factor 13 %	0.89	0.98	
Load factor 100 %	0.34	0.35	
Global warming (100 years) kg CO_2 eq./t			
Load factor 13 %	266.27	278.50	
Load factor 100 %	171.23	173.39	
Eutrophication kg PO_4 eq./t			
Load factor 13 %	0.26	0.27	
Load factor 100 %	0.04	0.04	
Photochemical oxidation kg C_2H_4 eq./t			
Load factor 13 %	0.08	0.08	
Load factor 100 %	0.04	0.04	
Abiotic depletion kg Sb eq./t			
Load factor 13 %	0.99	1.09	
Load factor 100 %	0.25	0.27	
Ozone layer depletion CFC-11 eq./t			
Load factor 13 %	0.01	0.01	
Load factor 100 %	1.32E − 03	1.43E − 03	

Table 7 shows the effect of installation of the photovoltaic system on the pneumatic system. There is a reduction of the indicators in each impact category; however, because a greater surface for the installation of a solar generator of higher power is not available, lowering of the emissions in the case of CO_2 eq. ranges between 4 and 1 % for load factors of 13 and 100 %, respectively. No data are available from similar studies considering the installation of a photovoltaic system for MSW pneumatic waste collection systems.

For conventional waste collection, the use of biodiesel B10 and B20 blends prepared from sunflower oil in side-loading collection trucks was analysed. The inventory used for this scenario is presented in the study of Gómez et al. [13], in which the use of biodiesel blends in side-loading collection trucks in the city of Zaragoza was analysed. The inventory from the Ecoinvent 2.2 database for the biodiesel lifecycle to the consumer and the energy required for the production of biodiesel from the study [46] were also considered. The effect of using biodiesel blends on diesel vehicle emissions by heavy vehicles, corresponding to Euro 3 technology, was obtained from the 2009 Guide for Road Transportation Emissions of the European Environment Agency; this study indicates that almost no engine modifications are needed for B10 and B20 blends. It was further assumed that the energy consumption of the truck remains constant [13] when functioning with diesel–biodiesel blends. Thus, when the percentage of biodiesel is greater, the consumption in the truck is modified as a result of the different calorific value of this fuel. The truck corresponds to a MSW collection truck of 21 tonnes, as described in the Ecoinvent 2.2 database. Fuel consumption and fuel emissions are

Table 8 Comparison of the conventional system using B10 and B20 biodiesel blends

Recovery scenario j↓	Conventional system	System conventional B10	System conventional B20	←MSW i management system
Acidification kg SO_2 eq./t				
Load factor 13 %	0.48	0.48	0.49	
Load factor 100 %	0.41	0.42	0.42	
Global warming (100 years) kg CO $_2$eq./t				
Load factor 13 %	201.21	201.74	202.03	
Load factor 100 %	192.65	193.08	193.27	
Eutrophication kg PO_4 eq./t				
Load factor 13 %	0.10	0.11	0.12	
Load factor 100 %	0.05	0.06	0.07	
Photochemical oxidation kg C_2H_4 eq./t				
Load factor 13 %	0.05	0.05	0.05	
Load factor 100 %	0.04	0.05	0.05	
Abiotic depletion kg Sb eq./t				
Load factor 13 %	0.44	0.43	0.41	
Load factor 100 %	0.38	0.36	0.34	
Ozone layer depletion CFC-11 eq./t				
Load factor 13 %	0.01	0.01	0.01	
Load factor 100 %	0.01	5.19E − 03	4.71E − 03	

also included, taking into consideration driving stops, abrasion of tires, abrasion from the road, and truck infrastructure.

Table 8 shows the influence of the use of B10 and B20 biodiesel blends on the transportation of surface containers by side-loading waste collection trucks. The baseline scenario for the MBT plant operation was used for this calculation. The data show that there is an increase in CO_2 eq. emissions when the fuel in the trucks is changed; also, the eutrophication impact category tends to increase compared to the other categories, abiotic depletion tends to decrease, and other categories tend to remain constant with slight variations. The finding of increased emissions, primarily of CO_2 eq., agrees with the results of previous studies; from a comprehensive analysis of biodiesel production, the cultivation and transportation processes directly affect CO_2 eq. emissions, even considering that 10 and 20 % of the emissions related to the use of diesel fuel in the case study are recorded as emissions avoided.

3.2 Alternative Energy Supply from ERS for MBT Plant Operation

A recovery scenario that includes the use of Residual Derived Fuel (RDF) from the residual fraction of the MBT plant in Zaragoza was evaluated from the baseline scenario. The G.I.R.A. 2009–2015 [1] does not consider energy recovery by

Table 9 Comparison of two scenarios for the recovery and management systems under study

Recovery Scenario j↓	G1	G2	←MSW i management system
Acidification kg SO_2 eq./t			
R1	0.34	0.41	
R2	−1.55	−1.48	
Global warming (100 years) kg CO_2 eq./t			
R1	171.23	191.31	
R2	−117.18	−97.10	
Eutrophication kg PO_4 eq./t			
R1	0.04	−2.45	
R2	−2.47	0.07	
Photochemical oxidation kg C_2H_4 eq./t			
R1	0.04	0.05	
R2	−0.10	−0.09	
Abiotic depletion kg Sb eq./t			
R1	0.99	0.33	
R2	−14.66	−14.59	
Ozone layer depletion CFC-11 eq./t			
R1	1.32E − 03	4.39E − 03	
R2	1.31E − 03	4.39E − 03	

incineration as a final treatment option for the outflows of MBT plants, specifically, of the residual fraction. For the case study of this paper, the impact of the use of 100 % of the residual fraction in a cement plant in the town of Morata de Jalón, near the city of Zaragoza, will be analysed. The composition and higher calorific value of the residual fraction was determined to be 17,929.24 kJ/kg using the methodology described by Aranda Usón et al. [34]. Considering the Ecoinvent 2.2 database and a HCV of 34,000 kJ/kg for the pet coke used in the cement plant, the emissions avoided per tonne of waste treated at the MBT plant were estimated. Treatment and transportation to the cement plant, which is 76 km from the MBT plant, were also considered; however, quantification of the emissions from the incineration of RDF in the cement plant was not performed due to the difficulty of quantifying a ratio of contribution for CO_2 emissions from the production of clinker [47].

Table 9 shows the results of the comparison of the baseline scenario (R1) and the RDF recovery scenario (R2) with the pneumatic system at a load factor of 100 % with the photovoltaic installation (G1) and the conventional system at a load factor of 100 % using a B20 biodiesel blend (G2).

For the recovery scenarios under study, it can be observed that the emissions avoided are greater than those generated in almost all impact categories when energy recovery in a cement plant is used as a final disposal alternative for the residual fraction; in eutrophication, the emissions avoided are greater than those generated only in the pneumatic waste collection system scenario.

Given the scenarios under study, it should be noted that the MBT plant in Zaragoza currently uses biogas produced in their facilities to create electricity and

Table 10 Comparison of two scenarios for the recovery and management systems under study

Recovery scenario j↓	G1	G2	←MSW i management system
Acidification kg SO_2 eq./t			
R1FV	0.31	0.38	
R2FV	−1.58	−1.50	
Global warming (100 years) kg CO_2 eq./t			
R1FV	168.21	188.29	
R2FV	−120.20	−100.12	
Eutrophication kg PO_4 eq./t			
R1FV	0.04	0.06	
R2FV	−2.48	−2.45	
Photochemical oxidation kg C_2H_4 eq./t			
R1FV	0.04	0.05	
R2FV	−0.10	−0.10	
Abiotic depletion kg Sb eq./t			
R1FV	0.23	0.31	
R2FV	−14.69	−14.61	
Ozone layer depletion CFC-11 eq./t			
R1FV	1.32E − 03	4.39E − 03	
R2FV	1.31E − 03	4.39E − 03	

heat for self-sufficiency [34]. However, the impact on emissions of an additional photovoltaic system on the roof of the main building of the MBT plant will be studied. The usable surface is 10,450 m^2, and the features of the photovoltaic installation are similar to those described in paragraph 4.3.2, thus providing a nominal power of 1,254 kW. Annual production is estimated at 1,843.3 MWh/year. Table 10 shows the impact of the installation of the photovoltaic plant in the scenarios of Table 9, including photovoltaic installation in the R1 (R1FV) scenario and in the R2 (R2FV) scenario for G1 and G2 collection.

For the recovery scenarios under consideration, the emissions avoided are greater than those generated in all impact categories, except ozone layer depletion. The latter is notably close to zero when a photovoltaic installation is included in the MBT plant to supply additional electrical energy.

4 Conclusions

An analysis of the benefits of using alternative energy sources such as solar photovoltaic and biofuels for transportation in the collection system was performed. Regarding the use of biofuels for transportation, it can be concluded that such variables as production processes and transportation of biodiesel constrain the net benefits in terms of emissions (e.g., CO_2 eq.) of the use of such material in collection trucks and MSW transportation in conventional systems that use B10 and B20 biodiesel blends.

On the other hand, the benefits of using alternative energy source as solar photovoltaic in the operation of the MBT plant considering two different scenarios of recovery of the residual fraction.

Decisions regarding investments in the application of alternative energies to the existing collection methods and operation of the treatment plants should be justified in terms of environmental, technological, and economic sustainability. A line of research to be developed from this study will focus on a study of the life cycle costs of each of the alternatives discussed in this article.

Acknowledgments This article was developed from results obtained in the framework of the ECOURBAN project, "Methodology for energy and environmental impact assessment and the ecodesign of urban areas", co-financed by the Spanish Ministry for Science and Innovation (Spanish National Plan for Scientific Research, Development and Technological Innovation 2008–2011—Ref. number ENE2010-19850) and coordinated by CIRCE.

References

1. Gobierno de Aragón (2009) Gestión Integral de los Residuos de Aragón (Integrated Waste Management Plan of Aragón). G.I.R.A. (2009–2015). In: Aragón, Gd. (ed.) ORDEN de 22 de Abril de 2009 del Consejero de Medio Ambiente, Spain (Decree of April 22, 2009 the Minister of the Environment, Spain, 2009)
2. Kreith, F.: Handbook of solid waste management. McGraw-Hill, New York (1994)
3. Eurostat Data Centre on Waste. Municipal waste (env_wasmun). Available from http://epp.eurostat.ec.europa.eu/portal/page/portal/waste/data/database (2012)
4. Tavares, G., Zsigraiova, Z., Semiao, V., Carvalho, M.G.: Optimisation of MSW collection routes for minimum fuel consumption using 3D GIS modelling. Waste Manage. (Oxford) **29**, 1176–1185 (2009)
5. United Nations Environment Programme. Assessment of Current Waste Management System and Gaps therein Developing Integrated Solid Waste Management Plan. Training Manual 2. Division of Technology, Industry and Economics. International Environmental Technology Centre, Osaka/Shiga, Japan (2009)
6. de Oliveira Simonetto, E., Borenstein, D.: A decision support system for the operational planning of solid waste collection. Waste Manage. **27**, 1286–1297 (2007)
7. Council of Europe. Management of municipal solid waste in Europe. Resolution 1543 Committee on the Environment Agriculture and Local and Regional Affairs (2007)
8. Eurostat. Environmental statistics and accounts in Europe. Luxembourg: Publications Office of the European Union (2010)
9. Eisted, R., Larsen, A.W., Christensen, T.H.: Collection, transfer and transport of waste: accounting of greenhouse gases and global warming contribution. Waste Manage. Res. **27**, 738–745 (2009)
10. Feo, G.D., Malvano, C.: The use of LCA in selecting the best MSW management system. Waste Manage. (Oxford) **29**, 1901–1915 (2009)
11. Gallardo, A., Bovea, M.D., Colomer, F.J., Prades, M.: Analysis of collection systems for sorted household waste in Spain. Waste Manage. (Oxford) **32**, 1623–1633 (2012)
12. Gallardo, A., Bovea, M.D., Colomer, F.J., Prades, M., Carlos, M.: Comparison of different collection systems for sorted household waste in Spain. Waste Manage. (Oxford) **30**, 2430–2439 (2010)

13. Gómez, M., Sebastián, F., Royo, J.: Biodiesel versus natural gas fuelled engines to improve contribution of captative fleets to air quality. In: Proceedings of 16th European Biomass Conference and Exhibition BC&E. EU BC&E Conference, Valencia, Spain (2008)
14. Iriarte, A., Gabarrell, X., Rieradevall, J.: LCA of selective waste collection systems in dense urban areas. Waste Manage. (Oxford) **29**, 903–914 (2009)
15. Punkkinen, H., Merta, E., Teerioja, N., Moliis, K., Kuvaja, E.: Environmental sustainability comparison of a hypothetical pneumatic waste collection system and a door-to-door system. Waste Manage. (Oxford) **32**, 1775–1781 (2012)
16. Teerioja, N., Moliis, K., Kuvaja, E., Ollikainen, M., Punkkinen, H., Merta, E.: Pneumatic vs. door-to-door waste collection systems in existing urban areas: a comparison of economic performance. Waste Manage. (Oxford) **32**, 1782–1791 (2012)
17. Wäger, P.A., Hischier, R., Eugster, M.: Environmental impacts of the Swiss collection and recovery systems for Waste Electrical and Electronic Equipment (WEEE): a follow-up. Sci. Total Environ. **409**, 1746–1756 (2011)
18. Bovea M.D., Ibáñez-Forés, V., Gallardo A., Colomer-Mendoza. FJ.: Environmental assessment of alternative municipal solid waste management strategies: a Spanish case study. Waste Manage. 30, 2383–2395 (2010)
19. Ecoembalajes España, S.A. (ECOEMBES). Recogida selectiva de Envases Ligeros (Selective collection of Light Packaging). http://www.ecoembes.com (2012)
20. European Parliament C. Directive 2008/98/EC of the European Parliament and of the Council of 19 November 2008 on waste and repealing certain Directives (Text with EEA relevance). Official J. Eur. Union (OJ L 312), pp. 3–30 (2008)
21. Pires, A., Martinho, G., Chang, N.-B.: Solid waste management in European countries: a review of systems analysis techniques. J. Environ. Manage. **92**, 1033–1050 (2011)
22. Tintner, J., Smidt, E., Böhm, K., Binner, E.: Investigations of biological processes in Austrian MBT plants. Waste Manage. (Oxford) **30**, 1903–1907 (2010)
23. Lornage, R., Redon, E., Lagier, T., Hébé, I., Carré, J.: Performance of a low-cost MBT prior to landfilling: study of the biological treatment of size-reduced MSW without mechanical sorting. Waste Manage. (Oxford) **27**, 1755–1764 (2007)
24. Aranda Usón, A., Ferreira, G., Zambrana Vásquez, D., Zabalza Bribián, I., Llera Sastresa, E.: Estimation of the energy content of the residual fraction refused by MBT plants: a case study in Zaragoza's MBT plant. J. Cleaner Prod. **20**, 38–46 (2012)
25. Montejo, C., Ramos, P., Costa, C., Márquez, M.C.: Analysis of the presence of improper materials in the composting process performed in ten MBT plants. Bioresour. Technol. **101**, 8267–8272 (2010)
26. Nithikul, J., Karthikeyan, O.P., Visvanathan, C.: Reject management from a Mechanical Biological Treatment plant in Bangkok, Thailand. Resour. Conserv. Recycl. **55**, 417–422 (2011)
27. Velis, C.A., Longhurst, P.J., Drew, G.H., Smith, R., Pollard, S.J.T.: Biodrying for mechanical-biological treatment of wastes: a review of process science and engineering. Bioresour. Technol. **100**, 2747–2761 (2009)
28. Montejo, C., Costa, C., Ramos, P., Márquez, MdC: Analysis and comparison of municipal solid waste and reject fraction as fuels for incineration plants. Appl. Therm. Eng. **31**, 2135–2140 (2011)
29. Abeliotis, K., Kalogeropoulos, A., Lasaridi, K.: Life Cycle Assessment of the MBT plant in Ano Liossia, Athens, Greece. Waste Manag. **32**, 213–219 (2012)
30. Gentil, E.C., Damgaard, A., Hauschild, M., Finnveden, G., Eriksson, O., Thorneloe, S., et al.: Models for waste life cycle assessment: review of technical assumptions. Waste Manage. **30**, 2636–2648 (2010)
31. Mohareb, E.A., MacLean, H.L., Kennedy, C.A.: Greenhouse gas emissions from waste management–assessment of quantification methods. J. Air Waste Manag. Assoc. **61**, 480–493 (2011)

32. Papageorgiou, A., Barton, J.R., Karagiannidis, A.: Assessment of the greenhouse effect impact of technologies used for energy recovery from municipal waste: a case for England. J. Environ. Manage. **90**, 2999–3012 (2009)
33. Zhao, W., der Voet, Ev, Zhang, Y., Huppes, G.: Life cycle assessment of municipal solid waste management with regard to greenhouse gas emissions: Case study of Tianjin. China. Sci. Total Environ. **407**, 1517–1526 (2009)
34. Aranda Usón, A., Ferreira, G., Zabalza Bribián, I., Zambrana Vásquez, D.: Study of the environmental performance of end-of-life tyre recycling through a simplified mathematical approach. Therm. Sci. (2012)
35. Tukker, A.: Life cycle assessment as a tool in environmental impact assessment. Environ. Impact Assess. Rev. **20**, 435–456 (2000)
36. Guinee, J., Gorrée, M., Heijungs, R., Huppes, G., Kleijn, R., de Koning, A.: Life cycle assessment—an operational guide to the ISO standards. Centre of Environmental Sciences (CML), Leiden University (2001)
37. International Organization for Standardization: Environmental Management—Life Cycle Assessment—Principles and Framework. International Organization for Standardization, Geneva, Switzerland (2006). ISO 14040:2006
38. Guinée, J.B.: Handbook on life cycle assessment: operational guide to the ISO standards. Springer (2002)
39. Rebitzer, G., Ekvall, T., Frischknecht, R., Hunkeler, D., Norris, G., Rydberg, T., et al.: Life cycle assessment: Part 1: framework, goal and scope definition, inventory analysis, and applications. Environ. Int. **30**, 701–720 (2004)
40. Banar, M., Cokaygil, Z., Ozkan, A.: Life cycle assessment of solid waste management options for Eskisehir, Turkey. Waste Manage. **29**, 54–62 (2009)
41. Bravo, Y., Carvalho, M., Serra, L.M., Monné, C., Alonso, S., Moreno, F., et al.: Environmental evaluation of dish-Stirling technology for power generation. Sol. Energ. **86**, 2811–2825 (2012)
42. Cleary, J.: Life cycle assessments of municipal solid waste management systems: a comparative analysis of selected peer-reviewed literature. Environ. Int. **35**, 1256–1266 (2009)
43. Aranda Usón, A., Ferreira, G., Zambrana Vásquez, D., Zabalza Bribián, I., Llera Sastresa, E.: Environmental-benefit analysis of two urban waste collection systems. Sci. Total Environ. pp. 72–77 (2013)
44. Taghizadeh, F., Safavian, T.: Network analysis based designing for a municipal solid waste collection and transportation system. J. Food Agric. Environ. **9**, 758–762 (2011)
45. Fthenakis, V.M., Kim, H.C.: Photovoltaics: life-cycle analyses. Sol. Energy **85**, 1609–1628 (2011)
46. López, J.M., Gómez, Á., Aparicio, F.: Javier Sánchez F. Comparison of GHG emissions from diesel, biodiesel and natural gas refuse trucks of the City of Madrid. Appl. Energy **86**, 610–615 (2009)
47. Genon, G., Brizio, E.: Perspectives and limits for cement kilns as a destination for RDF. Waste Manage. (Oxford) **28**, 2375–2385 (2008)
48. Ministerio de Industria Energía y Turismo. La energía en España 2011. Available from: http://www.minetur.gob.es/energia/es-ES/Documents/Energia_Espana_2011_WEB.pdf (2012)

in Hamburg, Germany

MIX
Papier aus verantwortungsvollen Quellen
Paper from responsible sources
FSC® C105338

If you have any concerns about our products,
you can contact us on
ProductSafety@springernature.com

In case Publisher is established outside the EU,
the EU authorized representative is:
Springer Nature Customer Service Center GmbH
Europaplatz 3, 69115 Heidelberg, Germany

Printed by Libri Plureos GmbH
in Hamburg, Germany